GIS and Crime Mapping

Mastering GIS: Technology, Applications and Management series

Location-based Services and Geomatic Engineering, Allan Brimicombe and Chao Li (forthcoming)

Geodemographics, GIS and Neighbourhood Targeting, Richard Harris, Peter Sleight and Richard Webber

Landscape Visualisation: GIS Techniques for Planning and Environmental Management, Andrew Lovett, Katy Appleton and Simon Jude (forthcoming)

Integration of GIS and Remote Sensing, Victor Mesev (forthcoming)

GIS for Public Sector Spatial Planning, Scott Orford, Andrea Frank and Sean White (forthcoming)

GIS Techniques for Habitat Management, Nigel Waters and Shelley Alexander (forthcoming)

GIS and Crime Mapping

Spencer Chainey
Director of Geographic Information Science, Jill Dando Institute
of Crime Science, University College London

Jerry Ratcliffe
Department of Criminal Justice,
Temple University, Philadelphia

John Wiley & Sons, Ltd

Other Wiley Editorial Offices

John Wiley & Sons Inc., 111 River Street, Hoboken, NJ 07030, USA

Jossey-Bass, 989 Market Street, San Francisco, CA 94103-1741, USA

Wiley-VCH Verlag GmbH, Boschstr. 12, D-69469 Weinheim, Germany

John Wiley & Sons Australia Ltd, 33 Park Road, Milton, Queensland 4064, Australia

John Wiley & Sons (Asia) Pte Ltd, 2 Clementi Loop #02-01, Jin Xing Distripark, Singapore 129809

John Wiley & Sons Canada Ltd, 22 Worcester Road, Etobicoke, Ontario, Canada M9W 1L1

Wiley also publishes its books in a variety of electronic formats. Some content that appears in print
may not be available in electronic books.

Library of Congress Cataloging in Publication Data

Chainey, Spencer.
  GIS and crime mapping / Spencer Chainey, Jerry Ratcliffe.
    p. cm. — (Mastering GIS)
  Includes bibliographical references and index.
  ISBN 978-0-470-86098-4 (cloth : alk. paper) — ISBN 978-0-470-86099-1 (pbk. : alk. paper)
  1. Crime analysis—Data processing. 2. Geographic information systems. 3. Digital mapping.
  I. Ratcliffe, Jerry. II. Title. III. Series.
  HV7936.C88C48 2005
  363.25—dc22

          2004028500

British Library Cataloguing in Publication Data

A catalogue record for this book is available from the British Library

ISBN 978-0-470-86098-4 (HB)   978-0-470-86099-1 (PB)

Typeset in 11/13pt Times by Integra Software Services Pvt. Ltd, Pondicherry, India
Printed and bound in Great Britain by TJ International Ltd, Padstow, Cornwall

Spencer Chainey: To Victoria and Oscar
Jerry Ratcliffe: To Philippa

Contents

Acknowledgements   xiii

1 Introduction   **1**
  1.1 The geography of crime   1
  1.2 A brief history of GIS and crime mapping   2
  1.3 Using GIS in policing and to prevent crime   3
  1.4 The audience for this book   4
  1.5 The content and structure of the book   5
  1.6 Putting it all in perspective   7
  Case study: Crime mapping in Lincoln, Nebraska   7
  Further reading   10
  References   11

2 Mapping and the Criminal Justice Environment   **13**
  Learning Objectives   13
  2.1 Introduction   13
  2.2 The terminology of services in the criminal justice
       environment   14
  2.3 The spatial hierarchy of the criminal justice system
       and crime reduction services   19
  Case study: Policing across the spatial hierarchy in the
       UK – The National Intelligence Model   21
  2.4 The geographical jurisdiction of law enforcement
       and crime reduction services   22
  2.5 The use of crime mapping in law enforcement and
       crime reduction   24
  Case study: Using GIS to monitor the effect of
       alley-gating schemes   28
  2.6 Summary   31
  Further reading   33
  References   33

Contents

3   The Basics of Crime Mapping   **37**
Learning Objectives   37
3.1   What is a GIS?   38
3.2   How does a GIS work?   40
3.3   GIS files   41
3.4   Coordinate systems and projections   43
3.5   Getting crime data into a GIS   46
Case study: Using GPS technology to capture
        environmental crime incidents in
        North London, England   50
3.6   Geocoding in the real world   52
3.7   Address data cleaning   56
3.8   Address reference files   56
Case study: Geocoding crime data at the point of
        record entry in Dumfries and Galloway, Scotland   58
3.9   Geocoding functions   59
3.10   Geocoding and fitness for purpose   60
3.11   Measuring geocoding accuracy   61
Case study: Handling uncertainty and incompleteness in
        crime records   64
3.12   Mapping and unreported crime data   65
3.13   Editing data in a GIS   68
3.14   Performing queries on data in a GIS   69
3.15   Performing spatial functions and integrating data
        in a GIS   69
3.16   Asking spatial questions before mapping or
        analysing data   70
3.17   Summary   75
Further reading   75
References   76

4   Spatial Theories of Crime   **79**
Learning Objectives   79
4.1   Introduction   79
4.2   Early environmental criminology   81
4.3   The space and time of offences   86
4.4   Offender–offence interaction   96
4.5   Spatial crime theory in practice   107
4.6   Summary   109
Further reading   109
References   110

5 Spatial Statistics for Crime Analysis **115**
Learning Objectives 115
5.1 Introduction 116
5.2 Spatial processes 116
5.3 Centrographic statistics 119
5.4 Estimates of spatial dependence 126
Case study: The application of Moran's *I* on burglary at the
state level in the United States of America 132
5.5 Spatial regression models 134
Case study: A spatial lag model of anonymous narcotics tips
in Philadelphia, USA 137
Case study: Local spatial processes with Geographically
Weighted Regression 138
5.6 Summary 140
Further reading 141
References 142

6 Identifying Crime Hotspots **145**
Learning Objectives 145
6.1 Introduction 145
6.2 When is a hotspot 'hot'? 147
6.3 Point maps 148
6.4 Geographic boundary thematic mapping 150
6.5 Grid thematic mapping 153
6.6 Continuous surface smoothing methods 155
Case study: Mapping hotspots of thefts of vehicles
in Camden, London 162
6.7 Local Indicators of Spatial Association (LISA)
statistics 163
6.8 Considering the underlying population 172
Case study: Identifying street crime risk hotspots
in the West End of London using pedestrian counts 174
6.9 Predictive crime mapping 177
6.10 Summary 179
Further reading 180
References 180

7 Mapping Crime with Local Community Data **183**
Learning Objectives 183
7.1 Introduction 184
7.2 What are crime reduction partnerships? 184
7.3 Mapping and the benefits of partnership working 186

Contents

Case study: Comparing the perception of where crime
    happens with where crime actually happens   187
7.4   Partnership data   189
Case study: Crime And Disorder Information Exchange
    (CADDIE), Sussex, England   195
7.5   Information sharing   199
Case study: The Amethyst Crime and Disorder Information
    Hub, Cornwall, England   211
7.6   Combining data from different geographic units   213
7.7   Summary   219
Further reading   220
References   221

8   Mapping and Analysing Change Over Time   223
Learning Objectives   223
8.1   Introduction   223
8.2   The timeline   225
8.3   Temporal resolution and querying a temporal
    database   228
8.4   Comparing two distributions   231
8.5   Mapping temporal change with graphs   235
8.6   Using animation   240
8.7   Quantifying change over time   245
8.8   Aoristic analysis   251
Case study: Aoristic analysis of vehicle crime in Sydney's
    Eastern Suburbs   253
8.9   Summary   254
Further reading   255
References   255

9   Mapping for Operational Police Activities   257
Learning Objectives   257
9.1   Introduction   258
9.2   CompStat   259
Case study: CompStat mapping in the Philadelphia Police
    Department   264
Case study: CompStat from a management perspective   268
9.3   Intelligence products in the UK   271
9.4   Repeat victimisation   274
9.5   The hotspot matrix   277
Case study: A street crime hotspot matrix   281
9.6   Summary   283

Further reading 283
References 284

10 Tactical and Investigative Crime Mapping Applications 287
Learning Objectives 287
10.1 Introduction 288
10.2 Understanding offenders 290
10.3 The journey to crime 296
Case study: The journey to crime and the
'self-containment index' 302
10.4 Geographic profiling 302
Case study: Geographic profile for Operation Lynx 305
10.5 Using maps as evidence 309
Case study: Using maps as evidence in a murder trial
in Florida 310
10.6 Detecting offenders through their self-selection 312
Case study: Self-selection of offenders through illegal
parking in disabled parking bays 315
10.7 Summary 317
Further reading 318
References 318

11 Policing the Causes of Crime 323
Learning Objectives 323
11.1 Introduction – the level of strategic crime control 324
11.2 Policing for crime reduction 325
Case study: Supporting strategic crime analysis in London,
England 332
11.3 Analysing the underlying drivers of crime 333
11.4 The geography of neighbourhood studies 340
Case study: Street corner geography for street corner
problems? 345
11.5 Summary 348
Further reading 348
References 349

12 Crime Map Cartography 353
Learning Objectives 353
12.1 Introduction – the purpose of the map 353
12.2 Design considerations 355
12.3 Visual variables and colour 366
12.4 Thematic maps of areal data 373

Contents

12.5 Thematic maps of point data 380
12.6 Getting away from paper: The digital age 381
12.7 Summary 385
Further reading 386
References 386

13 The Management and Organisation of Crime Mapping Services **389**
Learning Objectives 389
13.1 Introduction 390
13.2 Implementing crime mapping 391
13.3 Understanding the role of crime analysis 399
Case study: Crime mapping and analysis in the Glendale
    Police Department, Arizona 406
13.4 Organising the production of crime mapping
    products 409
Case study: Project Spectrom – a new operational policing
    model for West Midlands Police, England 412
Case study: The importance of management to support
    crime analysis 415
13.5 Summary 418
Further reading 418
References 419

Index 423

Acknowledgements

We would like to thank the following who have provided case study material, comment on the book's content and supported us in its production: George Rengert, Philippa Ratcliffe, Ron Wilson, Rachel Weeden, Suzanne Siegel, Patricia Giorgio-Fox, Charles Brennan, Tom Casady, Ned Levine, Michael McCullagh, Brian Flood, Ralph Taylor, Jim LeBeau, Stewart Fotheringham, Gloria Laycock, Kate Bowers, Shane Johnson, Kim Rossmo, Mark Partick, Lee Kilroy, Stuart Arnott, Tim Hemsley, Phil Davies, Sylvia Chenery, Andy Brumwell, Steve Rose, Nancy LaVigne, Julie Wartell, Bryan Hill and Dave Flitcroft.

1
Introduction

1.1 The geography of crime

Crime has an inherent geographical quality. When a crime occurs, it happens at a place with a geographical location. For someone to have committed a crime they must have also come from a place (such as their home, work or school). This place could be the same location where the crime was committed or is often close to where the crime was perpetrated (Frisbie *et al.*, 1977; Brantingham and Brantingham, 1981; Rossmo, 2000; Wilcs and Costello, 2000). 'Place' therefore plays a vital role in understanding crime and how crime can be tackled.

The study of crime has traditionally been the preserve of other disciplines such as sociology and psychology (Georges, 1978) and it was not until the late 1970s that the 'place' and the spatial dimension to crime began to be more fully explored. The police have long recognised the inherent geographical component of crime by sticking pins into maps displayed on walls, where each pin represented a crime event, but it was studies such as those from the 'Chicago School' of the 1930s (Shaw and McKay, 1931) that first demonstrated the importance of geography in understanding crime.

What has taken time, and only seriously surfaced in the 1970s, was the realisation that crime could be explained and understood in more depth by exploring its geographical components. New techniques emerged – techniques that included identifying patterns and concentrations of crime; the exploration of the relationships between crime and environmental or socio-economic characteristics; and techniques to assess the effectiveness

GIS and Crime Mapping   Spencer Chainey and Jerry Ratcliffe
© 2005 John Wiley & Sons, Ltd

of policing and crime reduction programmes that are targeted to geographical areas. What has materialised from this emergence of academic and practitioner activity is the field of crime mapping – a progressive blend of practical criminal justice issues with the research field of Geographical Information Systems (GIS).

1.2 A brief history of GIS and crime mapping

Since the 1960s GIS has emerged as a discipline in its own right. From its origins in land use applications in Canada to an all-pervasive technology used today in applications as diverse as in-car navigation, retail store site location, customer targeting, risk management, construction, weather forecasting, utilities management and military planning, GIS has become ubiquitous in modern life.

The infancy of GIS grew through applications such as planning for the US Census of Population in 1970 (and other national censuses in many other countries since) and from the national mapping agencies that began using this technology to help automate their cartographic draughting. Imagery of the earth from satellites has also played a significant role in the development of GIS, particularly through the military where GIS was the platform into which imagery could be displayed and analysed for the purpose of intelligence gathering. Indeed it was the military that were responsible for the first uniform system of measuring location, driven by the need for accurately targeting missiles. The military were also initially responsible for the development of the Global Positioning System (GPS). However, it was not until the 1980s that reductions in the price of computer technology created a conducive environment for the development of the GIS software industry and the subsequent growth in cost-effective GIS applications (Longley *et al.*, 2001).

These reductions in the cost of computer hardware were complemented by improved operating systems, electronic storage media and developments in computer software, and have had a wide and significant impact in introducing GIS technologies to new areas, such as policing and crime reduction. The computerisation of police records has come with a realisation that this material can be used for crime and intelligence analysis (Ratcliffe, 2004), and in turn used to better recognise patterns of crime that can be targeted for action, patterns that evidence suggests police officers are not necessarily aware of (Ratcliffe and McCullagh, 2001).

The early use of GIS for mapping crime was often held back by organisational and management problems (Openshaw *et al.*, 1990), issues with sharing information (Chainey, 2001), technical problems (Hirschfield *et al.*, 1995) and geocoding problems (Craglia *et al.*, 2000). These problems were shared with many of the other industries and disciplines trying to implement GIS, and it took several innovators to resolve these issues and show how they could be overcome. In reality many of these problems have not simply gone away, and several new ones have emerged, so in this book we try to help the reader by pointing to ways in which many of these technical and organisational issues can be overcome.

Much of the innovation in crime mapping was driven in the United States by the National Institute of Justice's Crime Mapping Research Center (CMRC). Renamed in 2002 as the Mapping and Analysis for Public Safety (MAPS) programme, the impact of this US government initiative was not isolated to the USA, but has also been the foundation for the development of crime mapping in many other countries, including the United Kingdom, Australia, South Africa and across South America. The MAPS programme has raised awareness of crime mapping through arranging seminars, conferences and producing publications, and by developing crime mapping software tools and funding new fields of crime mapping research. The MAPS programme, joined now by many other institutions and organisations, continues to be an active player in support-ing the development of crime mapping with the result that crime mapping is now more widely recognised by government and law enforcement ser-vices as a tool to aid policing and crime reduction.

Many argue that the 'systems' development of GIS has in recent years been overtaken by the 'scientific' development of the discipline (Goodchild, 1992, 1997; Longley *et al.*, 2001). This geographical information science has seen the development of analytical methodologies, techniques and processes for the advancement of spatial understanding, and as a result has contributed to many disciplines where understanding space and place is important, such as with crime (Ratcliffe, 2004). The geographical analysis of crime has also shown strong parallels with the field of spatial epidemi-ology and continues today to learn from this related field, using many of the analysis techniques that were originally designed for the study of disease patterns. We cover a number of these techniques in this book.

1.3 Using GIS in policing and to prevent crime

Crime mapping can play an important role in the policing and crime reduction process, from the first stage of data collection through to the monitoring and evaluation of any targeted response. It can also act as an important mechanism in a more pivotal preliminary stage, that of preventing crime by helping in the design of initiatives that are successful in tackling a crime problem. In subsequent sections of this book we present and discuss a wide range of applications for crime mapping, but as a resumé to demonstrate how it can support many central processes to policing and crime reduction we include the following application areas:

- Recording and mapping police activity, crime reduction projects, calls for service and crime incidents;
- Supporting the briefing of operational police officers by identifying crimes that have recently occurred and predicting where crime may occur in the future;
- Identifying crime hotspots for targeting, deploying and allocating suitable crime reduction responses;
- Helping to effectively understand crime distribution, and to explore the mechanisms, dynamics and generators to criminal activity, through pattern analysis with other local data;
- Monitoring the impact of crime reduction initiatives; and
- Using maps as a medium to communicate to the public crime statistics for their area and the initiatives that are being implemented to tackle crime problems.

Crime mapping is becoming central to policing and crime reduction in the 21st century, and this book aims to make a contribution to its continued growth. As Clarke (2004, p. 60) notes, 'Quite soon, crime mapping will become as much an essential tool of criminological research as statistical analysis is at present.'

1.4 The audience for this book

This book has been written to appeal to a wide range of professionals, academics and students interested in crime mapping. The book's style and content is one that should interest analysts working in policing and law enforcement, and analysts working in other services that support crime reduction, such as those who work in a crime reduction partnership, or those that are involved in analysing or researching crime problems within national or regional government bodies. The book should also appeal to academics and lecturers in GIS, crime science, crime prevention, criminal justice, law enforcement, policing, community safety or criminology, and to researchers and students studying in these disciplines. The book also aims to be of international appeal by focusing on generic themes in GIS and crime mapping, and by drawing from worldwide experiences and developments in the active areas of crime mapping.

1.5 The content and structure of the book

This book has been written to support readers with the essential theory, scientific methodologies, analysis techniques and design processes that they require for applying crime mapping as a tool to help understand crime. The book's content is also illustrated with examples and case studies designed to bring depth to the explanation of certain principles and to demonstrate crime mapping at work. The book can also be used as a reference text which can be dipped into on relevant occasions, and to complement existing texts on the subject.

The book is divided into four main parts: Part One covers the basics required for mapping crime; Part Two covers crime mapping and GIS techniques; Part Three presents and discusses a comprehensive range of methodologies and applications of crime mapping; and Part Four aims to bring together much of what has been described and discussed by explaining how to make maps work to their best effect, and the organisational and management arrangements that can help to ensure that these outputs are used.

'The basics' included in Part One begins in the next chapter which describes the different professional areas where crime mapping is being applied, clarifies the definitions of terms and professions and explains the geographic jurisdictional and hierarchical boundaries of policing and crime reduction services. Chapter 3 further explores the basics of crime mapping by describing the starting point for converting raw crime data into a computer map. This chapter also explains some of the key concepts of GIS and geographic data, and in particular the chapter discusses and addresses the issues with geocoding crime data. Chapter 4 covers the essential theoretical concepts for interpreting and understanding the geographical aspects of crime. It includes an explanation of how thinking about crime and space has developed into the field of environmental criminology and demonstrates many of the key theoretical concepts in a practical sense. This chapter provides the essential backcloth to why there is value in crime mapping.

Part Two emphasises geographic information techniques and processes that can be applied to crime mapping. This part begins by providing a grounding in spatial statistics, from descriptive spatial statistics through to more advanced techniques such as spatial autocorrelation and spatial regression. The theme of spatial statistics continues in Chapter 6 where a range of different methods are discussed for identifying crime hotspots. The concept of using crime data with non-police data is then explored in Chapter 7, which discusses the use of non-crime data sources to complement the picture of criminal behaviour and the possible causes that account for that behaviour. The chapter also explores the data that these other sources may have available to share, and reviews the information-sharing challenges and the technical processes for combining geographic information. The analytical techniques for understanding crime patterns in space are comple-mented in Chapter 8 where we explore ways to visualise the temporal and spatio-temporal patterns of crime. This includes helping the reader to recognise the value that the temporal dimension contributes to understanding crime as well as equipping the reader with a range of tools that can be used to identify and interpret temporal and spatio-temporal patterns.

Part Three focuses on crime mapping applications and approaches in its use, grouping these applications into three separate chapters. Chapter 9 describes the ways in which maps are incorporated into the operational thinking of police organisations, and are used to inform police officers wishing to reduce crime. This is illustrated with examples that describe the principles and functions of the CompStat process; describe the intelligence products of Britain's National Intelligence Model; discuss the importance of repeat victimisation identification in crime prevention research and outline the components of the 'Hotspot Matrix' as a way to understand and establish suitable responses to crime problems based on their spatial and temporal characteristics. Chapter 10 describes how crime mapping can support the tactical and investigative requirements of law enforcement and crime reduction, particularly in terms of catching offenders or gathering facts that can be useful in targeting diversion schemes, focusing the control of behaviour and directing crime prevention strategies. The chapter includes analytical methods for understanding offenders; discusses the importance of understanding the journey to crime; presents the main concepts of geographic profiling and how this important tool has been used to help serial crime investigations; describes the key requirements to consider when producing maps for prosecution evidence and explains the concept of offender self-selection. Part Three's final chapter begins by re-examining the main spatial crime theoretical structures to examine the different types of geographical scale and strategies that provide the link between theoretical understanding of crime and its spatial extent. This chapter includes an overview of the current paradigms in policing and crime reduction and whether these can be effectively employed in a spatial sense to prevent crime. These discussions then extend to the analysis of neighbourhoods and understanding the underlying drivers of crime.

Part Four focuses on pulling together the outputs of crime mapping into a form that makes them work, be effective and reach the right audience. Chapter 12 examines the end of the crime mapping process – getting the message across to the audience. The main focus of this chapter is gearing the output to the needs of the client or audience, and in this chapter we offer the reader tips, techniques and theories to ensure that their effort is not wasted. More specifically, the chapter looks at key principles of cartographic design, the effective use of colour and the incorporation of maps into presentations. The book's final chapter helps to identify the ways in which crime mapping can be managed and organised in policing and crime reduction services. It discusses the GIS and crime mapping implementing process, the role of crime mapping (and analysis) and its effective use and the integration of crime mapping products into information-driven processes.

1.6 Putting it all in perspective

To put this book into a functional perspective, the following case study illustrates the developmental process of getting a crime mapping system up and running. It also shows the myriad of ways in which mapping technology can aid the policing and crime reduction effort. Chief Casady's experiences are probably similar to that of many innovative police departments and show how effective use of mapping does not instantly occur, but rather happens as a result of a combination of technological advancement and organisational change, both of which have been achieved through Chief Casady's leadership.

Case study: Crime mapping in Lincoln, Nebraska

Material supplied by Tom Casady – Chief of Police, Lincoln Police Department

Geography is basic to policing, and all good police officers are intimately familiar with the lay of the land in their area. They know their beat like the back of their hand. Perhaps because 'place' is such an important component of policing, maps are a common tool. Police officers cut them

7

into manageable pieces, they punch them for a ring binder, they clip them out of phone books, they laminate them, and they fold them into their pockets.

Back at police headquarters in Lincoln, someone was sticking coloured pins in a map on the wall when Teddy Roosevelt was President. In those days the pins represented saloons, or horse thefts, stick-ups or burglaries, maybe accidents, houses of ill repute, homes of officers, or any of hundreds of other events, facts or people. The map was a picture in time of a phenomenon, and told its story with a glance.

Today, the pin map has often been replaced by a much more powerful analytical tool – a GIS. Almost all printed maps are produced by high tech computer software today, but the real power of a GIS lies not in the printed output, but in the ability to interact with the data. Law enforcement is a relative latecomer to the use of GIS, but the use of GIS in policing has grown dramatically in the past several years. As this field grows, more and more police departments are discovering GIS as an incredibly valuable resource that in many cases is already at their disposal in their own community (if not hidden elsewhere in their own police department, it can often be found in the city hall or county government office).

This is exactly what happened to the Lincoln Police Department in the 1990s. Well before we began our crime mapping and analysis programmes, the Lancaster County Engineer and the Lincoln Public Works and Planning Departments had developed an extensive GIS, and were anxious to share their mapped data with other government agencies. In 1997 we began our first GIS applications in policing, making extensive use of the accurate basemap components, such as streets, land parcels, aerial photographs, and many other layers of geographic information that had already been developed and were being maintained by several city and county agencies. What we added to this cartographic mix was our police data about places.

Virtually everything we do as a police department revolves around an address or location. All of our dispatch records, incident reports, citations, intelligence reports have a place, and all of these are records collected in the ordinary course of business. GIS software allows mappers to use these computerised records of such things as crimes, by automatically placing the 'pins' on the map.

Each day, the preceding 24 hours of dispatch records, incident reports and field interviews are electronically mapped, as well as the addresses of gang members, registered sex offenders and parolees. The power of the GIS software allows our analysts to query this data rapidly to illuminate trends and patterns that would be lost in the sheer volume of events. With 400 dispatches on a typical day, no one has the full picture of everything that has occurred. Even in a much smaller agency, the differences in shifts and days off make it quite possible that two officers investigating similar thefts from the laundry rooms of two side-by-side apartment buildings are each unaware of the other's case. A GIS pulls these connections together out of stacks of reports by enlisting the power of the computer to extract data through time, date, modus operandi, crime type queries and queries that are geographically based.

We primarily use our GIS for operational and tactical purposes: locating crime series and intervening in these quickly. But we also use these data in many other ways. We often print large maps of crime scenes and the surrounding area as a visual aid in major crime investigations. GIS also helps identify other crimes that may be the work of the same suspect. Adding geography to the modus operandi can reveal information about the offender's target selection. Following an arrest, GIS analysis can help identify other cases that are prospects for a multiple clearance.

GIS also supports strategic decision-making. The power of GIS, for example, dramatically simplifies the time-consuming task of redistricting or adjusting boundaries in patrol areas. In 1998 we were studying the need to add a fifth patrol team area, and considering the need to move one of our major patrol boundaries. Working on one single scenario, our planners spent a full week examining the workload implications of redrawing one boundary between the Center and Northeast Team Areas. When we applied GIS to this analysis, we were able to examine dozens of alternatives and their impacts during a 4-hour management staff meeting. We have used GIS analysis to support decisions on locating substations, targeting crime in fragile neighborhoods, and developing problem-oriented policing projects.

Crime mapping has changed dramatically during the years since we produced our first GIS crime map in December, 1997. Among the major developments for us have been automating our geocoding process, simplifying our interface, and deploying our mapping applications to a much wider range of personnel. With our new applications, anyone in the department who knows how to book their airline tickets online is perfectly capable of doing their own geographic analysis. Officers and employees with basic Internet skills can produce high-level GIS analysis with nothing more than an Internet connection and a browser. Our easy-to-use Intranet-based mapping application is available to everyone on the department at anytime of the day or night. Officers can query crime type, date range, certain MO patterns, and proximity to addresses or landmarks and quickly obtain both maps and tables of matching incidents. In a matter of moments, our employees can produce analysis and maps that a few years ago would have required expensive software, high-powered computers, and extensive GIS expertise.

We also make mapping available to the general public through the Internet. Our public mapping application allows citizens to select crime types and date ranges, and to produce both tables and maps of incidents near a specific address, a landmark such as a school, or within the boundaries of a specific neighborhood association. Although the query functions of this public Internet application are intentionally throttled back and the data reduced considerably to protect the confidentiality of crime victims, it remains a valuable asset to landlords, neighborhood associations, and other interested members of the public. At a recent meeting of the Lincoln City Council, a neighborhood association was testifying during a public hearing on a zoning issue. Their representative was using data and maps about crimes to bolster the association's case.

I recognized the layouts and the data as being from our public web mapping application.

Further reading

LaVigne, N.G. and Groff, E.R. (2001). The evolution of crime mapping in the United States. In A. Hirschfield and K. Bowers (eds) *Mapping and Analysing Crime Data*, pp. 203–221. London: Taylor & Francis.

Weisburd, D. and McEwen, T. (eds) (1997). Introduction: Crime mapping and crime prevention. In *Crime Mapping and Crime Prevention*, pp. 8, 1–23. New York: Criminal Justice Press.

These two references offer a comprehensive history of how crime mapping has developed from its infancy to its use in modern policing and crime reduction.

The United States National Institute of Justice's Mapping and Analysis for Public Safety programme http://www.ojp.usdoj.gov/nij/maps/briefingbook.html.

The National Institute of Justice's Mapping and Analysis for Public Safety (MAPS) programme's website offers a 'briefing book' that outlines many of the key ways in which crime mapping is used. The MAPS website also offers a number of other resources on crime mapping including links to software, online publications and details of crime mapping conferences.

Leipnik, M.R. and Albert, D.P. (2003) *GIS in Law Enforcement: Implementation Issues and Case Studies*. London: Taylor & Francis.

Leipnik and Albert offer a useful crime mapping resource list of internet sites, key publications and agencies that support crime mapping.

References

Brantingham, P.J. and Brantingham, P.L. (eds) (1981). *Environmental Criminology*. London: Sage.

Chainey, S.P. (2001). Combating crime through partnership: Examples of crime and disorder mapping solutions in London, UK. In A. Hirschfield and K. Bowers (eds) *Mapping and Analysing Crime Data*. London: Taylor & Francis.

Clarke, R.V. (2004). Technology, criminology and crime science. *European Journal on Criminal Policy and Research*, 10(1), 55–63.

Craglia, M., Haining, R. and Wiles, P. (2000). A comparative evaluation of approaches to urban crime pattern analysis. *Urban Studies*, 37(4), 711–729.

Frisbie, D.W., Fishbine, G., Hintz, R., Joelson, M. and Nutter, J.B. (1977). Crime in Minneapolis: Proposals for prevention. St Paul, MN: Community Crime Prevention Project, Governor's Commission on Crime Prevention and Control.

Georges, D.E. (1978). The geography of crime and violence: A spatial and ecological perspective. Association of American Geographers: Resource papers for college geography, 78(1).

Goodchild, M.F. (1992). Geographical information science. *International Journal of Geographical Information Systems*, 6(1), 31–45.

Goodchild, M.F. (1997). What is Geographic Information Science?, NCGIA Core Curriculum in GIScience. http://www.ncgia.ucsb.edu/giscc/units/u002/u002.html, posted 7 October 1997.

Hirschfield, A., Brown, P. and Todd, P. (1995). GIS and the analysis of spatially-referenced crime data: Experiences in Merseyside, UK. *International Journal of Geographical Information Systems*, 9(2), 191–210.

Longley, P., Goodchild, M., Maguire, D. and Rhind, D. (2001). *Geographic Information Systems and Science*. Chichester: John Wiley & Sons.

Openshaw, S., Cross, A., Charlton, M. and Brunsdon, C. (1990). Lessons learnt from a post mortem of a failed GIS. The 2nd National Conference and Exhibition of the Association for Geographic Information, Brighton.

Ratcliffe, J.H. (2004). Crime mapping and the training needs of law enforcement. *European Journal on Criminal Policy and Research*, 10(1), 65–83.

Ratcliffe, J.H. and McCullagh, M.J. (2001). Chasing ghosts? Police perception of high crime areas. *British Journal of Criminology*, 41(2), 330–341.

Rossmo, K. (2000). *Geographic Profiling*. Boca Raton, Florida: CRC Press.

Shaw, C.R. and McKay, H.D. (1931). *Social Factors in Juvenile Delinquency*. Washington: US Government Printing Office.

Wiles, P. and Costello, A. (2000). The road to nowhere: The evidence for traveling criminals. Home Office Research Study 207, Research, Development and Statistics Directorate, Home Office. http://www.homeoffice.gov.uk/rds/pdfs/hors207.pdf.

2
Mapping and the Criminal Justice Environment

Learning Objectives

In this chapter we identify how crime mapping fits into the professions and organisation levels of the criminal justice environment. We begin by removing some of the ambiguity regarding the definitions of certain terms, so that the reader has a clearer sense of the terminology used in criminal justice. This includes defining crime prevention, community safety, law enforcement, crime control and crime reduction. We also examine the structuring of criminal justice in terms of its geographic jurisdictional domains and hierarchical boundaries of policing and crime reduction services. The chapter offers a foundation that helps the crime mapper locate how their work can be applied and the contributions that they can offer. We conclude by introducing several situations in which crime mapping is applicable.

2.1 Introduction

The criminal justice environment uses a wide-ranging set of terminology and is structured across many organisations and agencies that act across various levels of hierarchy. To the reader who is new to crime mapping it is useful to begin by reviewing how the application of mapping crime can be applied in different professions and at the various hierarchical and organisational levels in policing and government. For the experienced crime mapper, a review of this type may also help them orientate their skills more effectively into this working environment and identify others whose skills they can complement.

The broad arena of criminal justice runs the gamut of services from early intervention activity to the work of the courts, prison and probation. While we recognise that mapping applications have been applied in other areas, this book concerns itself with crime mapping, and so in the following discussion we will limit the focus to those areas of the criminal justice system that are concerned with the mapping of crime for crime prevention, reduction and detection.

The criminal justice, policing and crime reduction services environment is awash with a range of terms that are used and interchanged so liberally that it can often leave the crime mapper confused as to what it is they actually do and how their craft can be applied in an appropriate context. The terms 'law enforcement', 'crime prevention', 'community safety', 'crime control' and 'crime reduction' often suffer from definitional ambiguity, are commonly misused and misunderstood by those working in or supporting these professions, can suffer from weak distinction, are frequently interchanged without reason and are even used to compete against each other – leaving an environment of clouded contradiction (Brantingham and Faust, 1976). Indeed Wiles and Pease (2000) refer to crime prevention and community safety anecdotally as 'Tweedledum and Tweedledee', referring to the identity crisis between the two terms, which are viewed by some as identical, and by others as two individual positions. These definitional ambiguities can easily lead to a situation where the professional who simply wants to apply their trade can easily become a victim, confused over how their skills can be applied and the context in which they should be working. Here we seek to provide some clarity and explain these terms in a manner that is explanatorily concise and clear in meaning. These definitions can also provide useful food for thought that help point the crime mapper in the direction in which they can apply their mapping skills. In this chapter we start by defining the differences between crime prevention and community safety, and other closely related terms that are used in criminal justice.

2.2 The terminology of services in the criminal justice environment

The criminal justice, policing and crime reduction environments use a number of terms that describe the services they provide or what they aspire to achieve. These include:

- Crime prevention
- Community safety
- Law enforcement and policing
- Crime control
- Crime reduction

2.2.1 Crime prevention

Crime prevention involves any activity by an individual or group, public or private, which attempts to either eliminate crime prior to it occurring or before any additional activity results (Brantingham and Faust, 1976; Lab, 1988). Lab goes on to include fear of crime in his definition: 'Crime prevention entails any action designed to reduce the actual level of crime and/or perceived fear of crime' (Lab, 1988, p. 9). These definitions can be expanded by considering crime prevention between the levels or stages in criminal behaviour at which intervening activity can be implemented, and considering the different types of intervention. Lab (1997) identified five approaches to modern crime prevention:

1. *Saving the less fortunate* – separating those that are at risk and intervening before further activity results.
2. *Altering the social fabric* – building thriving communities that are cohesive and that could control the behaviour of people in the neighbourhood (Shaw and McKay, 1931, 1942; Schlossman and Sedlak, 1983).
3. *Changing the physical environment* – removal of those physical features that are conducive to crime (through physical design) to show that the area is well cared for and protected, and the removal of any signs of incivility ('broken windows') to make the area less promising as a site for crime (Jacobs, 1961; Newman, 1972; Wilson and Kelling, 1982; Lewis and Salem, 1986; Taylor and Gottfredson, 1986).
4. *Organising the community* – using citizen surveillance and action as a means to protect the neighbourhood, including interaction and engagement between the police and active local residents (Palumbo *et al.*, 1997).
5. *Situational crime prevention* – focusing on the settings for crime, rather than upon those committing criminal acts. By directing measures at specific forms of crime that involve making or managing environmental changes for crime to be more difficult and risky, or less rewarding and excusable (Clarke, 1992).

Interpreting the definition of crime prevention by considering these approaches is useful, but Brantingham and Faust's (1976) analogy with public health (as explained below) offers a means of considering crime prevention between the levels or stages in criminal behaviour – defined through the Crime Prevention Model. The Crime Prevention Model describes three levels of activity:

1. Primary prevention
2. Secondary prevention
3. Tertiary prevention.

2.2.1.1 Primary prevention

Primary prevention identifies conditions of the physical and social environment that provide opportunities for, or precipitate, criminal behaviour. This is analogous to a public health intervention that provides community-wide health advice and is not targeted to an individual. The objective of crime prevention interventions in this case is to alter those conditions so that crimes cannot occur. This can be through consideration of the design or modification of the physical environment, modification of the social environment to reduce temptations or impulsions towards criminal behaviour, providing security provisions, social and physical well-being programmes and crime prevention education. Primary prevention is very much about preventing the criminal behaviour at the point of criminal opportunity (Brantingham and Faust, 1976).

2.2.1.2 Secondary prevention

Secondary prevention engages in the early identification of potential offenders, be they groups or individuals. Analogous to public health, this involves identifying those who have a high risk of developing a disease or those that show the early signs of having a disease. Intervention is then through specialist treatment that is designed to reduce the risk, prevent it from developing or prevent it from further developing. In crime prevention terms the intervention seeks to ensure that identified high-risk offenders never commit a crime or to reduce the risk of their future involvement in more serious criminal activity. These interventions could include police intelligence operations, stop-and-search tactics, diversion programmes, mentoring and supervision programmes, screening of individuals, neighbourhood education programmes in high-risk areas or crisis intervention through welfare programmes. Secondary prevention is about preventing crime by reducing the risks of those vulnerable to involvement in crime, or preventing additional criminal behaviour from occurring or graduating to activity that is more serious (Brantingham and Faust, 1976).

2.2.1.3 Tertiary prevention

Tertiary prevention focuses on dealing with actual offenders. It is analogous to identifying individuals with advanced cases of disease where intervention aims to prevent death or permanent disability. Tertiary crime prevention involves intervening with the lives of these offenders in a manner that prevents them from committing other crimes and includes arrest and prosecution, reform and rehabilitation and institutional education programmes. In essence, tertiary crime prevention is the primary goal of the correctional system (Brantingham and Faust, 1976).

These definitions of crime prevention demonstrate that dealing with criminal behaviour is not something that the police can do on their own. It points to the need for crime prevention activity that is delivered in partnership between police and many other public bodies. We take forward these concepts of partnerships in Chapters 7 and 11.

2.2.2 Community safety

If 'safety refers to the absence, or the sensed absence, of likely or serious harms [...whether caused by a person or otherwise...] the practice of community safety involves the management of risk so as to maximise public safety' (Wiles and Pease, 2000, pp. 22–23). Community safety is not just about managing the risks of crime, but is a term that is broader and refers to all actions that could cause a community some harm, and the consequences that could result from these harms (Ekblom, 2000). Thus community safety also includes transport accidents and other related activities that impact on the well-being of the public, and the requirement for strategies that remove harms and deal with consequences. There is often confusion between community safety and crime prevention, particularly in the UK where the terms have in recent years been regularly substituted. Naturally, crime issues should be considered as a component of community safety, but confusion is promoted in the UK when community safety is presented in government legislation on crime and disorder, such as the Crime and Disorder Act 1998. Indeed, Wiles and Pease (2000, p. 25) note, 'community safety, as legislated, is simply presented as re-badged crime prevention, not as a broader way of ranking and dealing with the dangers facing people' and that 'community safety is now employed as a synonym of crime prevention, with fluffy overtones added'. However, in its broader and more accurate sense, community safety is realised through an integrated consideration of diverse harms to the public, and 'refers to the likely absence of harms from all sources, not just from human acts classifiable as crimes' (Wiles and Pease, 2000, p. 21). Community safety also provides a strategic viewpoint on community harms by focusing attention towards the development of programmes that set targets to manage risks and aims to maximise public safety.

2.2.3 Law enforcement and policing

Law enforcement is the act of ensuring that laws and regulations are followed. In a policing context its aims are to prevent, investigate and solve crimes while protecting public citizens. When a law or regulation becomes broken it involves the use of a police (or other enforcement activity such as security or administrative) investigation that leads or could lead to a penalty or sanction being imposed. The role of policing usually extends further than simply law enforcement. Bittner (1980) identified three primary domains of policing: law enforcement, regulatory control (e.g. traffic management and control of specific licencing activities such as permits for firearms and taxi licences) and peacekeeping (e.g. crowd control, control of civil disorders and noise complaints). In the UK the police domain of law enforcement is relatively clear and transparent. The 43 police services of England and Wales, the eight police services in Scotland and the Police Service of Northern Ireland are supplemented by national bodies such as the British Transport Police and the National Criminal Intelligence Service. The picture is more complicated in America, where there is no agreed number of police services. Estimates range somewhere from 18 000 to just over 21 000 (Maguire *et al.*, 1998) across three general tiers: local police, state police and federal police. It is estimated that about half of all local municipal police services in the US have 10 sworn officers or less (Walker and Katz, 2001).

There is a distinction between police and policing. Policing is an activity that the police conduct, but it can also be conducted by people and agencies that are not sworn officers of the law. Considerable crime prevention of private property is now conducted by private security and non-police agencies, and financial agencies such as major banks and insurance companies are developing their investigative departments. These agencies conduct policing, but are not the police.

2.2.4 Crime control

Crime control considers that crime has already happened and that some management of these criminal activities is required to ensure it does not spiral out of control. It points to the need for maintenance of a problem, one where crime is kept to a tolerable level, and not to a situation where crime can be prevented. Therefore, it is not synonymous with crime prevention. Crime control 'involves halting rapid or accelerating growth in crime risk' (Ekblom, 2000, p. 65) which may be specific to a certain crime or a wider problem. In essence, crime control is about the management of serious and emerging crime problems. Lab (1988) claims that crime control fails to address the fear of crime, although others suggest that by controlling for the harmful consequences of particular criminal events, the consequences can include fear (Ekblom, 2000). Crime control has some overlap with community safety because of the common focus towards dealing with the consequences of crime and the need for strategic management to be applied to crime problems (Ekblom, 2000).

2.2.5 Crime reduction

Crime reduction is concerned with diminishing the number of criminal events and the consequences of crime (Ekblom, 2000). Crime reduction is applied within the bandwidth of an available resource input (e.g. financial input) and needs to be considered as an action that brings net benefits after considering the impact of displacement and diffusion of benefits, fear of crime and the impact from other programmes that may have contributed to any specific crime reduction activity. Crime reduction promotes a spirit of optimism that actions towards a problem will reduce crime or reduce the seriousness of criminal events and makes the assumption that crime is already high. It aims to intervene directly in the events and their causes. Thus as an action, crime reduction shares a philosophy with crime prevention in that crime prevention is a means to achieving crime reduction. There can 'be no act of crime prevention which is not also crime reduction' and that there are 'very few crime reduction actions that do not have a preventive aspect' (Ekblom, 2000, p. 61). Crime reduction is more crime-focused than community safety and more widely applied (i.e. to all crime problems) than crime control.

2.3 The spatial hierarchy of the criminal justice system and crime reduction services

Generally there are three levels of spatial hierarchy that operate in the criminal justice system and across crime reduction services:

1. The federal or national level
2. The state or regional level
3. The municipal or local level.

This spatial hierarchy follows the typical arrangement of government at different geographic scales and has remained fairly stable in modern times, considering the different paradigms and evolution of law enforcement and crime reduction services (US Department of Justice, 2000). From a crime mapping point of view it is useful to review the analytical functions and responsibilities at these three levels to help provide context to the ways in which crime mapping can be used.

The federal level maintains a responsibility for analysing serious and organised crime, and incidents or events that may impact the safety of the nation (e.g. terrorism). The federal level is also where certain new research developments are led, and which may explore crime problems across the entire country, or crimes that are a national problem but where analysis has focused on certain local sites to generate results that will have national interpretation. This activity can include identifying good practice on how to tackle a particular problem, or providing guidance that has been drawn from testing certain approaches at pilot sites. A good example of this is the work conducted in the UK by the Home Office Research, Development and Statistics Directorate. This national government body performs research, as well as many other activities, into crime issues that are of interest to the nation but that support and offer direction for helping those reduce crime across all three spatial hierarchical levels (visit www.homeoffice.gov.uk/rds/index.htm). In the United States, the National Institute of Justice (NIJ) (the research, development and evaluation arm of the Department of Justice) fulfils a similar function (www.ojp.usdoj.gov/nij).

The state or regional level typically operates with a responsibility for analysing crimes that are across the borders of the jurisdictions within their state/region area, but that are contained within the national arena. The cross-border problems that are analysed tend to reflect national priorities and crimes that are of particular concern in these regions.

At the local jurisdictional level, the functions that are performed are relevant to the operations of local law enforcement and any partnerships that are in place to address the local crime issues. These could include functions that are performed on a day-to-day basis, or functions that look into specific problems and try to design a strategic approach to how these problems can be tackled.

All levels usually carry an analytical function of performance review that monitors crime statistics and their performance against targets on crime reduction, detection and clear-up levels, public satisfaction in the service being provided, as well as performance in relation to the use of resources for policing and crime prevention.

Case study: Policing across the spatial hierarchy in the UK – The National Intelligence Model

In the UK three levels of spatial operation have been formally defined in the National Intelligence Model (National Criminal Intelligence Service, 2000). This definition of functions and responsibilities helps to provide clarity over operations at the different levels of spatial hierarchy and also helps to coordinate analysis activities. It is therefore a model that works as a stand-alone system for each level of activity, but also as an integrated model that interacts between levels for the maximum identification of any problems. It is also designed to optimise identification of potential solutions to these problems.

Level 1 describes the local sphere of activity, including the analysis and assessment of local crime problems. Each local policing division (also known as the Basic Command Unit) follows the model for analysis and assessment. This means that when it comes to considering issues that extend beyond their local area (such as an offender who operates beyond the jurisdictional boundaries of the local police division), neighbouring local police divisions can more easily integrate their information into the analyses and assessments of their colleagues.

If the full extent of the crime problem can only be recognised when a higher overview of data and intelligence is provided, this calls for Level 2 activity: cross-border and regional (e.g. a police force function). The Level 2 body has access to Level 1 analysis and assessments and, by integrating these with any additional state or regional-based information, can provide an assessment that identifies the problem, the potential solutions and recommended responses. These responses may be applied at Level 1, or could be a solution that is implemented across the region by those policing at Level 2.

In a similar way to how problems can be identified and integrated across the local level (Level 1), problem identification, analysis, assessment and response can also occur between the regional/force levels (Level 2). This level of activity may not necessarily be region-wide and constrained to the boundaries of the regional hierarchy, but instead could focus on the nature of the problem in a manner that allows geographic boundaries to appear invisible.

Level 3 provides activity at the national level. In National Intelligence Model terms, a product from Level 3 is the National Threat Assessment. In the UK, it is the National Criminal Intelligence Service that produces the Threat Assessment, drawing on data provided by the UK's police forces (such as their analytical and strategic assessments generated at Levels 1 and 2), Customs and Excise, the intelligence and security agencies (e.g. MI5) and other law enforcement-related bodies (e.g. the Serious Crime Analysis Section and the Home Office).

The introduction of the National Intelligence Model is providing some clarity to the analytical functions performed at the different hierarchical levels, and consistency in approach, though there have been some organisational problems associated with the implementation of the Model (for example, see Christopher, 2004). Where the Model functions well, crime mapping has the opportunity to thrive because of the information gathering, evidence-based and spatial intelligence-led requirements for investigating and understanding crime problems.

2.4 The geographical jurisdiction of law enforcement and crime reduction services

With the emphasis of policing and crime reduction services being organised on area-based jurisdictions, comes with it the imposition of geographic boundaries that define each agency's coverage of service provision. The geography of the policing service is, as referred to above, usually defined through the three tiers of national, regional and local levels. For example, Figure 2.1(a) shows the geographical boundaries of the 43 police forces in England and Wales. Each police force has within it several local Basic Command Units (BCU). Figure 2.1(b) shows the 34 BCUs in London. London actually has two police forces within it: the Metropolitan Police Service and the City of London Police – a relatively unusual arrangement for the European continent (Figure 2.1c). However, the geographical boundaries of their operation are distinct even though in practical terms the two police forces do often support each other on many crime and public safety issues in England's capital city.

At the federal level England, Wales and Scotland are administratively grouped together in the governance of many of Great Britain's national

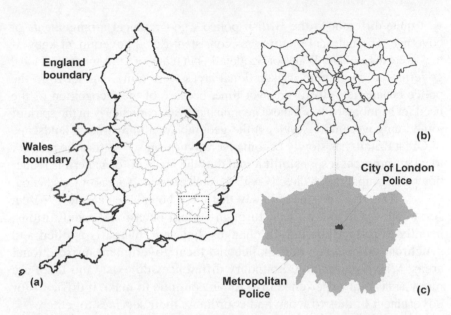

Figure 2.1 The geography of policing in England and Wales. (a) England and Wales' 43 police forces, (b) London's Basic Command Units and (c) London's Police Forces. Reproduced by permission of Ordnance Survey on behalf of HMSO. © Crown copyright 2005. All rights reserved. Ordnance Survey Licence number 100044021

policing services. The devolution of government from Parliament in London to Wales and Scotland can often create some confusion over which national bodies cover which countries. In other words, when using the word 'national' is it with reference to the countries individually, collectively (as Great Britain) or (and to complicate things even further) as just England and Wales? Examples of this confusion include certain national policing services which are provided to England and Wales, while Scotland operates something different.

What is not confusing though is that each country is clearly defined by the extent of their geographic boundary areas. These geographic boundary distinctions are also very precise at the regional and local levels, right down to which side of the road is one BCU and on which side is another. In Great Britain, as in many countries, geographic boundaries are often drawn using the centre of the road as the separating line.

This precise geographic boundary distinction is relatively straightforward to distinguish when only one agency is involved, but once new agencies also want to join up with the police and support them in the effort to reduce crime, confusion can begin to creep into the organisational system. Until approximately eight years ago, Britain's local government landscape was quite different to the British police's geographical arrangements of coverage. The national picture was consistent (local government knew if they were in England, Wales or Scotland!), but many of the regional and local government geographic jurisdictional areas were quite distinct from the police boundaries. However, over time, because of the recognition of the need for harmony between these geographic areas (particularly in the spirit of joined-up government services) the geographic administrative landscape of Great Britain has slowly become more synchronised so that the areas of local government responsibility (and thus the areas of formal crime reduction partnerships) match directly with the police service's areas of jurisdiction.

This does not mean to say that these boundaries are now frozen. Jurisdictional boundaries go through certain occasional redefinitions, usually to reflect population changes, but any change is applied and synchronised between both police and local government jurisdictional areas. Where geographic boundary differences do exist, and there are many areas that experience this, it can continue to make it difficult for government bodies to work and coordinate their services together.

The system in the UK is infinitely easier to map and understand than in the US, where organisational boundaries and jurisdictions overlap and interweave in such a confusing way that rarely is it possible to draw a definitive map of police service provision. As an example, consider the situation in the state of Pennsylvania where a population of about 12 million is served by in excess of 1100 police services. In the largest city, Philadelphia, there exist state police services, the Philadelphia Police Department, a school district police, a couple of local transport police bodies (one for Amtrak, the national rail carrier, and one for the local transit network), a housing police, a number of the city's universities maintain their own armed police services and running around in between all of these are the federal police bodies. Handling the differences between different geographic boundary areas for the purposes of data exchange and statistical analysis is a subject we further pursue in Chapter 7.

2.5 The use of crime mapping in law enforcement and crime reduction

The hierarchical structure of law enforcement and crime reduction in both geographic and organisational terms is useful to understand in order to appreciate the diverse range of applications that crime mapping can support. Chapter 1 provided a resumé of several applications. We examine many of these in more detail in subsequent chapters, but start by introducing several situations in which crime mapping can be applied.

2.5.1 Command and control

Responding to an incident is a key role for any police service. Quickly knowing where the incident has happened is therefore essential to ensure a rapid response to the scene of the crime. Many police forces use computer mapping systems within their response centres to identify where the caller is calling from and the location where an incident has happened. Pin pointing this location on a map and ensuring that the location where an incident has happened is identified with minimal ambiguity is vital to ensure a quick and direct response. Many police services use address lists, known as 'gazetteers', to complement the computer mapping available in their command and control systems. Command and control systems, when linked to vehicle-integrated or handheld GPS, can also provide a real-time link to police officers in the field. A GPS-enabled display can show field officers' location and their movement relative to the location of the incident (although in practice, interference, obstacles and the reflection of signals off buildings, particularly common in urban areas, persist as a problem for the accurate calculation of positions). Centrally located staff are also able to see the location of the officer, alongside any supporting attribute information that can help direct a suitable response to an incident. Attribute information may include who the officer is, their specialist skills and what equipment they may have which would aid a particular type of response.

2.5.2 Crime mapping and administrative or management analysis

Managers like information and reports. If the report includes a map they can suddenly become your best friend or start to besiege you with further requests for similar reports. Reports that describe if crime has gone up or gone down, where the problems are, and information that helps to diagnose the problem can be provided in a number of different reporting styles. Displaying some of this information in map format or by using a computer-based mapping system to extract area-based information, such as a graph showing how crime levels have changed in a particular area, can be a powerful means of communicating information. In Chapter 12 we explain the best ways to maximise the impact of maps and graphics.

2.5.3 Operational analysis

Operational analysis, also referred to by others as tactical analysis (Osborne and Wernicke, 2003), works within the time frame of what is happening now, or has recently happened, and what can be done about a problem for it to have an immediate effect. Understanding what can be a large volume of information and attempting to quickly establish if any patterns exist mean making use of geographical analysis to explore the patterns and provide information to those working on the front line to effectively deal with the problem. For this to be timely requires an ability to turn information into intelligence as quickly as possible. Many police forces utilise crime maps in their daily or regular meetings to assist in the briefing of police patrols, as well as draw on maps to help in organising the deployment of police officers for particular operations. The CompStat process draws heavily on crime mapping and is an example of operational analysis, where the need is to access and visualise up-to-date information in a timely fashion, and to monitor if responses that have been prioritised and targeted are having the positive effect of reducing crime (Bratton, 1998). Operational analysis is closely linked to the concept of crime control (section 2.2.4).

2.5.4 Crime auditing

Auditing is a measure and analysis of crime problems – offering descriptive statistics, details of the crime issues and their interpretation. Crime audits can either be measures that are carried out on an annual (or longer) period basis or more continually (e.g. monthly or quarterly). In either case the aim is to provide an assessment of the issues that complement the short-term localised view provided from an operational analysis. Crime audits tend to focus on describing the existing crime levels, how these crime levels have changed over time, where the hotspots are, the patterns that exist across different temporal spectrums (i.e. seasons, day of the week and time of the day) and profiles of offenders. In this case, offender profiles include describing the typical age, gender and ethnicity profiles of offenders active in certain crimes. The profiles of offenders can be organised in relation to different crime types, and can include an appreciation of offender residential addresses that might reveal offender density hotspots. Victims can be profiled in a similar way. Audits can also use data from non-police sources to help support the measure and analysis of local crime problems. For example, police can use data from the fire service on vehicle arsons to complement the fire-related crime data that the police agency may collect. The role of the crime mapper in the auditing process is to make use of geographical analysis and mapping as a display media to aid the measure and (spatial) analysis of crime problems.

2.5.5 Problem-solving crime analysis

Problem-solving crime analysis focuses on specific, recurring crime problems, aiming to remove the causes of crime by identifying the problems that require attention (Clarke and Eck, 2003). Its rationale is centred on the belief that prevention is better at reducing crime than enforcement, employing a focused action-research approach that scans for crime problems, conducts an in-depth analysis of the crime issue, finds a practical response to the problem and assesses the impact of the chosen actions designed to solve the problem (Clarke and Eck, 2003). We cover problem-solving crime analysis in more detail in Chapters 3 and 11. In the context of crime mapping, problem-solving analysis draws heavily on an exploration of the spatial aspects of crime, and so crime mapping is often central to problem-oriented crime strategies.

2.5.6 Geographic profiling

Geographic profiling is an investigative methodology that uses the locations of a connected series of crimes to determine the most probable area of offender residence (Rossmo, 2000). Originally developed to help police locate serial killers, rapists and arsonists, geographic profiling can be applied to any circumstance where an unidentified person is known to have carried out criminal activities at a series of known geographic points. The technique is now being used to explore high volume crimes such as robbery and distraction burglary when a number of offences are known to be a linked series where the same offender has committed all offences in the series.

2.5.7 Monitoring, assessment, evaluation and performance review

Many crime prevention initiatives and police operations are area based. Resources are targeted to a particular area in an attempt to deal with the crime problems in that region. Crime mapping offers a useful tool to help monitor if the initiative or operation has been a success by analysing the *before* and *after* picture of the crime levels in the area. With the use of the right analytical methods, an investigation can be conducted to discover if there has been any displacement or diffusion of benefit as a result of the crime prevention scheme.

Similarly, the performance of policing or a crime prevention service in an area can be analysed and visualised geographically to reveal if their performance targets are being met, and to identify those area-based units that may need some encouragement and support to improve.

Case study: Using GIS to monitor the effect of alley-gating schemes

The British Crime Survey indicates that in over a half of all house burglaries, the burglar entered through the rear of the property. In some areas, particularly those with a high proportion of terraced housing, up to 75% of all burgled homes were entered from the rear. The backs of some houses, particularly terraced housing, lead onto quiet footpaths and alleyways. These are often unlit and allow burglars to gain access to houses without being observed. Networks of alleys can [also] provide convenient escape routes for burglars. (Neighbourhood Renewal Unit, 2002)

Alley-gating involves the installation of security gates across alleyways or footpaths that run behind properties (Figure 2.2). The central aim is to prevent potential burglars from accessing the rear of properties, particularly in areas that are vulnerable to this type of crime. In addition, alley-gating aims to offer potential benefits for reducing the fear of crime and reducing opportunities for other illicit or delinquent behaviour (e.g. drug taking and arson). They create a safer environment for children to play in, improve the sense of ownership of the alleys, keeping them cleaner and tidier, and help to improve the alleys' appearance.

Targeting the best locations for installing alley-gating has required the use of GIS. The mapping system has been used to identify areas that are suffering from burglary problems and where the housing stock and physical layout includes alleys and footpaths. As a result of the GIS analysis, a number of priority areas were identified in Merseyside as opportune locations to install alley gates. After the completion of several site visits and consultation with the local residents, 20 areas were selected, where 208 alley gates were installed, protecting 3442 properties. Similar GIS-based studies helped to identify suitable areas in Nottingham, Merton (London) and Kingston-upon-Hull. Although there are good reasons for

Figure 2.2 An alley-gate restricting access to the rear of the houses in the photograph

thinking that alley-gating should reduce burglary (Johnson and Loxley, 2001), it was not until the evaluation of these studies that any hard evidence was available that demonstrated their impact.

Evaluating the impact of alley-gating schemes also called on the use of GIS and spatial analysis techniques. With the help of GIS, each of the schemes were able to report the following:

- *Merseyside*: A 50% reduction in burglary rates in the year following the gate installations (Young, 1999).
- *Nottingham (Forest Fields project)*: A 41% reduction in the target area (Armstrong, 1999).
- *Merton (London)*: An evaluation of alley-gating in the Abbey ward, conducted halfway through the project, indicated a 50% reduction in rear-entry burglaries (Neighbourhood Renewal Unit, 2002).
- *Kingston-upon-Hull (Dukeries area)*: Over 60% burglary reduction in the nine months following the installation of gates (Holden McAllister Partnership, 2002).

GIS has recently been employed in a further Merseyside (UK) evaluation to test if any crime has been displaced. This analysis used concentric buffer rings around the alley-gating areas to explore if and how far any burglary was displaced (Figure 2.3). The results revealed that there was a diffusion

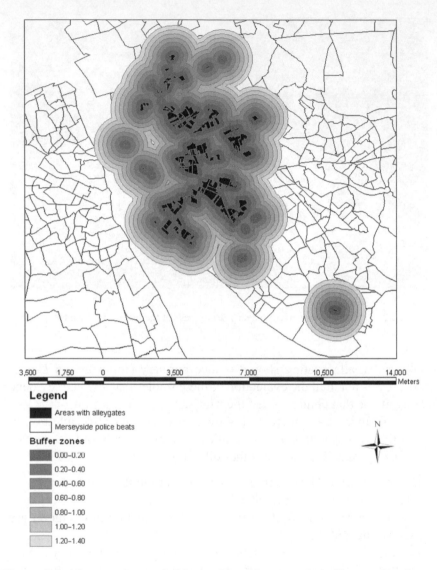

Figure 2.3 Alley-gated areas in Merseyside and concentric buffers used to detect displacement or diffusion of benefit. Source: Bowers *et al.* (2004). Reproduced by permission of Ordnance Survey on behalf of HMSO. © Crown copyright 2005. All rights reserved. Ordnance Survey Licence number 100044021

of benefit to properties in the surrounding areas, this being most dramatic within 200 metres of any scheme. The evaluation also enabled a rudimentary cost–benefit analysis to be calculated, the results of which showed that in the areas where the gates had been installed for some time (12 months or more) the cost–benefit ratio was around 1.86 (i.e. a saving of £1.86

for every pound spent). These results indicate that there was a substantial financial saving from the implementation of the alley-gated schemes (Bowers *et al.*, 2004).

2.5.8 Dissemination of crime information

Without being able to disseminate crime-related analysis and intelligence to decision-makers in the criminal justice system, there is little point in doing the work. Mapping provides a powerful visual means to support policing and crime reduction services. Many of those working in policing or crime prevention produce 'crime awareness' newsletters that are sent to local residents, and include in them maps of recent crime patterns (Figure 2.4). Many policing agencies also now make use of web mapping technology to update the public on crime patterns. The more advanced websites are interactive, allowing the public to select and query crime-related information. Other crime reduction agencies use computer mapping systems as a way to identify those people who live in certain areas when they want to alert this community to a spate of crimes that is affecting their area, or to pass on good news stories that crime in their area is on the decline.

This description of the ways in which crime mapping can be used is not exhaustive but aims to offer a flavour of the range of applications and uses to which the crime mapper can apply themselves. Later in this book we describe crime mapping applications in more detail in relation to their role in supporting operational, investigative and strategic forms of analysis – devoting separate chapters to each form.

2.6 Summary

Understanding the context in which the crime mapper works, both in terms of the activities they support (e.g. law enforcement, crime prevention) and the hierarchy in which they are located, is important in order to help bring clarity to a mapper's purpose. It also helps to identify the ways in which crime mapping can be applied and developed in the environment in which the crime mapper works. The operational, investigative and strategic applications of crime mapping are further expanded in subsequent chapters. In this chapter we have aimed to provide clarity on the definitions of terms used across the professions that the crime mapper works,

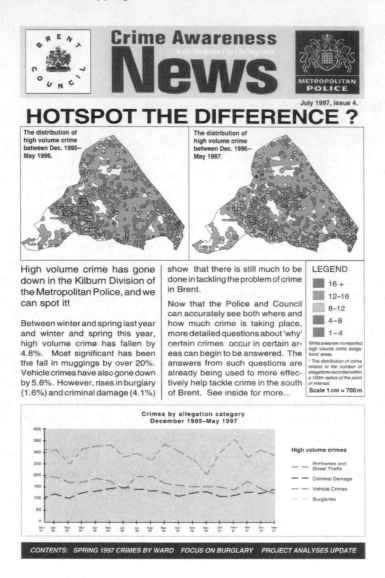

Figure 2.4 Crime Awareness News. The London 'Brent Crime Mapping Project' published a quarterly newsletter that told the community about changes in crime patterns and what was being done about the problems in certain areas

aimed to demonstrate the need for appreciating the different geographical hierarchy and jurisdictional coverage within single agencies and across multiple agencies, as well as describing some of the areas where crime mapping has developed to support services for policing and crime reduction.

Further reading

Osborne, D. and Wernicke, S. (2003). *Introduction to Crime Analysis: Basic Resources for Criminal Justice Practice*. New York: Haworth Press.

This is a useful practical starters' guide for the crime analyst. Deborah Osborne and Susan Wernicke are themselves crime analysts with a number of years of experience working in several US local police departments.

National Criminal Intelligence Service (2000). *The National Intelligence Model*. London: NCIS. www.policereform.gov.uk/implementation/natintellmodel.html.

The National Intelligence Model (NIM) is the UK's 'Model for Policing'. Its aim is to ensure that information is fully researched, developed and analysed to provide intelligence that can be used to provide strategic direction, make tactical resourcing decisions about operational policing, and manage risk. The NIM is being adopted by all police forces in the UK.

LaVigne, N. and Wartell, J. (eds) (1998). *Crime Mapping Case Studies: Successes in the Field*, Volume 1. Police Executive Research Forum, Washington, DC.
LaVigne, N. and Wartell, J. (eds) (2000). *Crime Mapping Case Studies: Successes in the Field*, Volume 2. Police Executive Research Forum, Washington, DC.

Published by the National Institute of Justice Mapping and Analysis for Public Safety programme in partnership with the Police Executive Research Forum. Volume 1 and Volume 2 of *Crime Mapping Case Studies: Successes in the Field* contain real-life uses of crime mapping in law enforcement and community safety. The papers in each volume demonstrate a successful outcome from a crime mapping project. The examples are chosen to represent a variety of crime and disorder problems experienced by agencies of different size and type.

References

Armstrong, Y.A. (1999). Evaluation of the Forest Fields Gating Project. Bristol: University of the West of England (Unpublished manuscript).
Bittner, E. (1980). *The Functions of the Police in Modern Society: A Review of Background Factors, Current Practices and Possible Role Models*. Cambridge, Mass.: Oelschlager, Gunn and Hain.
Bowers, K.J., Johnson, S.D. and Hirschfield, A.F.G. (2004). Closing off opportunities for crime: An evaluation of alley-gating. *European Journal on Criminal Policy and Research*, 10(4), 283–308.
Brantingham, P.J. and Faust, F.L. (1976). A conceptual model of crime prevention. *Crime and Delinquency*, 22, 284–296.
Bratton, W. (1998). *Turnaround*. New York: Random House.

Christopher, S. (2004). A practitioner's perspective of UK strategic intelligence. In J.H. Ratcliffe (ed.) *Strategic Thinking in Criminal Intelligence*, First edition (pp. 177–193). Sydney: Federation Press.

Clarke, R.V. (1992). *Situational Crime Prevention: Successful Case Studies*. New York: Harrow and Heston.

Clarke, R.V. and Eck, J. (2003). *Become a Problem Solving Crime Analyst*. London: Jill Dando Institute of Crime Science. www.jdi.ucl.ac.uk.

Ekblom, P. (2000). The conjunction of criminal opportunity: A tool for clear, joined-up thinking about community safety and crime reduction. In S. Ballintyne, K. Pease and V. McLaren (eds) *Secure Foundations: Key Issues in Crime Prevention, Crime Reduction and Community Safety*. London: Institute for Public Policy Research.

Holden McAllister Partnership (2002). Evaluation of the Dukeries Gating Project. Evaluation report for the Hull Community Safety Partnership (Unpublished).

Jacobs, J. (1961). *Death and Life of Great American Cities*. New York: Random House.

Johnson, S. and Loxley, C. (2001). Installing Alley-gates: Practical lessons from burglary prevention projects. Policing and Reducing Crime Unit Briefing Note 2/01. London: Home Office.

Lab, S.P. (1988). *Crime Prevention: Approaches, Practices and Evaluations*. Cincinnati: Anderson.

Lab, S.P. (1997). Crime prevention: Where have we been and which way should we go?. In S.P. Lab (ed.) *Crime Prevention at a Crossroads*. Cincinnati: Anderson.

Lewis, D.A. and Salem, G. (1986). *Fear of Crime*. New Brunswick: Transaction Books.

Maguire, E.R., Snipes, J.B., Uchida, C.D. and Townsend, M. (1998). Counting cops: Estimating the number of police departments and police officers in the USA. *Policing: An International Journal of Police Strategies & Management*, 21(1), 97–120.

National Criminal Intelligence Service (2000). *The National Intelligence Model*. London: NCIS.

Neighbourhood Renewal Unit (2002). *Alley-gating*. London: Office of the Deputy Prime Minister. http://www.renewal.net.

Newman, O. (1972). *Defensible Space: Crime Prevention Through Urban Design*. New York: Macmillan.

Osborne, D.A. and Wernicke, S.C. (2003). *Introduction to Crime Analysis: Basic Resources for Criminal Justice Practice*. New York: Haworth Press.

Palumbo, D., Ferguson, J. and Stein, J. (1997). The conditions needed for successful community crime prevention. In S.P. Lab (ed.) *Crime Prevention at a Crossroads*. Cincinnati: Anderson.

Rossmo, D.K. (2000). *Geographic Profiling*. Boca Raton: CRC Press.

Schlossman, S. and Sedlak, M. (1983). *The Chicago Area Project Revisited*. Santa Monica: Rand.

Shaw, C.R. and McKay, H.D. (1931). *Social Factors in Juvenile Delinquency.* Washington: US Government Printing Office.

Shaw, C.R. and McKay, H.D. (1942). *Juvenile Delinquency in Urban Areas.* Chicago: University of Chicago Press.

Taylor, R.B. and Gottfredson, S. (1986). Environmental design, crime and prevention: An examination of community dynamics. In A.J. Reiss and M. Tonry (eds) *Communities and Crime.* Chicago: University of Chicago Press.

US Department of Justice (2000). Boundary changes in criminal justice organisations. *Criminal Justice 2000*, Volume 2. Office of Justice Programs, National Institute of Justice.

Walker, S. and Katz, C.M. (2001). *The Police in America: An Introduction*, Fourth edition. Boston: McGraw-Hill.

Wiles, P. and Pease, K. (2000). Crime prevention and community safety: Tweedledum and Tweedledee. In S. Ballintyne, K. Pease and V. McLaren (eds) *Secure Foundations: Key Issues in Crime Prevention, Crime Reduction and Community Safety.* London: Institute for Public Policy Research.

Wilson, J.Q. and Kelling, G. (1982). Broken windows: The police and neighbourhood safety. *Atlantic Monthly*, March, 29–38.

Young, C.A. (1999). The Smithdown Road pilot 'Alleygating' project. Evaluated on behalf of the Safer Merseyside Partnership (Unpublished).

3
The Basics of Crime Mapping

Learning Objectives

This chapter provides the starting point for getting crime data onto a computer map. It begins by explaining some of the key concepts of GIS and geographic data. A number of these concepts may be familiar to the reader who is experienced in the use of GIS. We recommend that the GIS-professional does not skip this chapter altogether because the key concepts are covered concisely and there may be items that even the experts may learn. In many places in this chapter we refer those unfamiliar with GIS to other helpful literature that provides a detailed account of GIS concepts.

The core part of the chapter is focused upon getting crime data onto a map (geocoding) and the processes that may need to be performed in advance so that this geocoding process is as painless as possible (those experienced in geocoding crime data will understand how painful this can sometimes be!). We also describe the components that relate to the data's fitness of purpose for the operations that will be performed on it, discuss the impact of unreported crime on mapping crime patterns and encourage the reader to ask questions about the mapping and analysis they are to perform before they embark on interrogating data. Asking certain spatial questions before striding ahead in eagerness to produce

GIS and Crime Mapping   Spencer Chainey and Jerry Ratcliffe
© 2005 John Wiley & Sons, Ltd

a map encourages the application of a more robust theoretical and scientific approach to GIS and crime applications.

3.1 What is a GIS?

A GIS is 'a computer system for capturing, managing, integrating, manipulating, analysing and displaying data which is spatially referenced to the Earth' (McDonnell and Kemp, 1995, p. 42). A GIS provides the forum within which crime data can be layered with base maps and other geographic data that represent the landscape of the area where the crime data is associated. The base maps could represent the street network, the homes in a housing estate or buildings and open spaces in a town centre. Other geographic information may include population data from a census, the locations of cash point machines (or Automatic Teller Machines) or data describing the local land use patterns. These data can be represented as individual layers that can be manipulated, analysed or displayed as separate entities, or could be combined with other layers to be displayed together, integrated to provide a new perspective of the area that they represent or analysed against each other to reveal a particular relationship. For example,

- *Represented as an individual layer*: The locations of where all crimes have been committed.
- *An individual layer that is filtered*: Rather than mapping all crimes, only burglaries to residential properties are selected as a subset from the 'all crimes' data.
- *An individual layer that is analysed*: The locations of burglaries of residential properties are analysed to identify the burglary crime hotspots.
- *Different layers are combined and displayed against each other*: A map that only shows burglary points and absolutely no other data may not be particularly useful. The data can be layered with base map data that display the street network to show where these burglaries were committed. A third layer could also be displayed that describes the tenure of all the properties across the area (e.g. descriptive information on residential properties in the area and if they are privately owned, rented from a private landlord, or rented from the local government or a housing association). Data of this type could be sourced from a census and are typically represented in a format where the data are aggregated to a geographic unit (e.g. a census geographical boundary area). Visualising multiple layers offers a means of initially exploring relationships between different data variables (Figure 3.1).

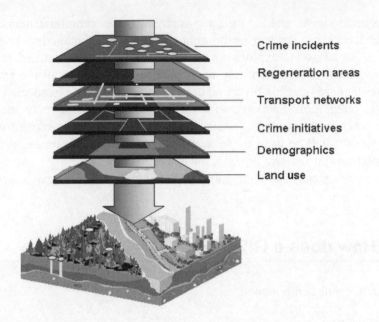

Crime incidents

Regeneration areas

Transport networks

Crime initiatives

Demographics

Land use

Figure 3.1 A GIS is often conceptualised as a computer system that brings together many different layers of data that can be geographically referenced. The above example shows how abstract layers of land use, demography, regeneration areas and the transport network (built from the original reality) can be displayed with crime data and areas targeted for crime reduction, superimposed over each other as georeferenced layers. Source: ESRI (UK)

- *Layers are integrated to provide a different perspective*: Burglary dwelling data could be combined with data that describe the total number of residential properties for an area, sourced from the housing tenure layer, to produce a burglary dwelling rate of the number of burglaries per 1000 households.
- *Layers are analysed against each other*: Burglary dwelling and housing tenure data layers could be analysed to identify if the level of tenure of a certain type in the study area is related to the levels of burglary. For example, the analysis may identify that burglary incidents are highest in the areas where there is a high percentage of properties that are rented from a private landlord.

A GIS is a powerful tool to assist in the management, integration and analysis of a disparate range of data (Laurini and Thompson, 1992; Burrough, 1998; Longley *et al.*, 2001). GIS are particularly suited to crime data because of the inherent geography that exists in a crime incident (i.e. a crime occurs at a location). Many other forms of data also have a spatial component to them, making their geography form the common denominator that enables these data to be brought together and managed, combined and cross-analysed to explore possible relationships between the data.

GIS on their own may not enable us to perform all the spatial queries that are possible on crime data so in some cases it is necessary to call on other tools to assist with these processes. These may include using other programs to help make crime data fit for the purpose for the operations it will be used for, drawing on more advanced database tools and performing certain types of analysis. GIS is typically open in its design which means that linking to these other tools can be a straightforward process.

3.2 How does a GIS work?

A GIS has four components:

1. hardware
2. software
3. data
4. people.

The hardware component's bare minimum is a computer. This would usually be a desktop computer with sufficient processing power and memory. Higher end workstation-based systems (e.g. UNIX or Sun workstations) could also be used. These computers (desktop or workstation) may be connected to a local network or server, or could be stand-alone. GIS hardware requirements would also usually include computing peripherals such as a printer or a plotter. Hardware may also include mobile computing and GPS technology that synchronise data captured in the field to a central database. A case study of this type of application is provided in this chapter.

Computers used for GIS have the standard requirement of an operating system (e.g. a Microsoft Windows-based operating system) on which software applications can be run. Popular GIS software used for crime mapping include ArcView, ArcGIS, MapInfo, GeoMedia and Northgate blue8. These desktop software products possess the majority of core GIS functionality, but also integrate with software extensions that expand functions for analysis and map production. Software extensions that are relevant to crime mapping include CrimeStat, Spatial Analyst (for ArcView and ArcGIS), HotSpot Detective (for MapInfo),

Vertical Mapper (for MapInfo), CrimeView (for ArcView and ArcGIS) and SpaceStat (for ArcView and ArcGIS). Specialist crime mapping software is also available such as Rigel which is used for geographic profiling (geographic profiling is discussed in more detail in other parts of this book).

Data are essential to crime mapping. Without crime data and base maps, a GIS has no crime prevention or reduction application. It is simply a sterile software package awaiting data. Many organisations underestimate the cost of base maps, and the time and effort required to geocode and map crime data. We will discuss the geocoding process in detail later.

The fourth and probably most essential component of a GIS is the user. People who operate GIS usually require some form of training. Many GIS users possess a certified academic education in GIS through university undergraduate or graduate courses; however, most users in the crime mapping field would need to at least have benefited from a GIS software short course that would have trained them in the core functionality of their GIS. These short courses are offered by many GIS service providers and some universities, but are also offered by several police and law enforcement agencies, either as internally run courses for their own staff or by a national body for those agencies that they work with (for example, the US Police Foundation's Crime Mapping Laboratory offer a range of GIS and crime mapping courses for US law enforcement agencies).

For a GIS to work it involves bringing together the computer files that contain the data that are to be used, and possibly editing them, querying against them, linking them, adding new data and generating a display for output. It also requires understanding the main concepts of coordinate systems and projections of the Earth.

3.3 GIS files

A GIS will typically represent data in three ways. These are discussed below.

3.3.1 Attribute file

An attribute file is a table listing the contents of data it relates to. For example, a crime attribute file could be a table that includes the crime reference number, the code that describes the crime type, the date of the crime, the time of the crime and the address of where the crime happened.

3.3.2 Spatial object file

Spatial objects can be represented in three main ways. This may be as a point, a line or a polygon (Figure 3.2).

1. *Points* – points could either be represented as a dot or by using a suitable symbol (most GIS hold an extensive library of symbol fonts). Crime records are usually represented as points in a GIS.
2. *Lines* (also known as polylines) – lines are represented as a straight line object connecting two points. Polylines describing a curved line are represented as a series of straight segments connecting vertices (Longley *et al.*, 2001). Lines and polylines could be displayed in a range of fonts in a GIS, either as a continuous or dashed line, in different colours and thickness. Streets are usually represented in a GIS as lines.
3. *Polygons* – a polygon is a closed area or region represented around its perimeter with a polyline. It exists as a solid object for the region it covers and can be shaded using a range of different fonts. Police beats are usually represented in a GIS as polygons.

The attribute file and the object can be linked together so that when a point, line or polygon is queried in a map view in a GIS the attribute data for this object is selected.

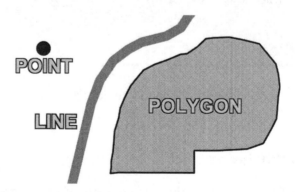

Figure 3.2 GIS spatial objects – point, line and polygon

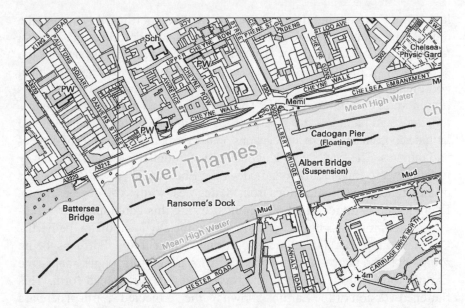

Figure 3.3 A GIS raster file with its greyscale value as the only attribute. Reproduced by permission of Ordnance Survey on behalf of HMSO. © Crown copyright 2005. All rights reserved. Ordnance Survey Licence number 100044021

3.3.3 Raster file

A raster file represents spatial features using cells or pixels that have attribute information attached to them. This attribute information can often be displayed visually as a colour. Many high-end GIS use raster files to store satellite imagery or other remote sensing pictures, and code each cell with a value representing a characteristic of the Earth. This has limited application in crime mapping, and most crime mappers are more likely to come across a raster file as an image file that has a street layout comprised of pixelated cells. The only attribute information that these cells contain is the colour code for the cell. Street atlas-style paper maps (e.g. the street atlas maps produced by Rand McNally, A to Z Geographers and the Automobile Association) are usually represented in a GIS in raster format (Figure 3.3).

3.4 Coordinate systems and projections

Coordinate systems are at the heart of a GIS. These systems are the basis on which crime data and other data that are spatially represented can be layered, integrated and displayed in a GIS. The coordinate system is the digital representation of the landscape and is usually described in spherical terms or in Cartesian form. A crime may happen at an address but it is the conversion of this address (expressed as a line of text) to a geographical coordinate (expressed as a numerical coorindate pair) that enables its presentation on a map in a GIS. The most universal coordinate system is longitude and latitude – this is a spherical coordinate system. Longitude and latitude coordinates reference locations on the Earth using the units of degrees and minutes. The line of zero longitude runs through the Royal Observatory in Greenwich, England. Zero latitude is the Equator. There are 360 degrees of longitude, and latitude varies from 90 degrees North to 90 degrees South. Each degree is divided into 60 minutes. Using this geographical referencing system it is possible to uniquely determine any location on the Earth.

Longitude and latitude coordinates reflect the Earth as being round but for many requirements such as generating paper maps, the Earth's round surface needs to be represented in a flattened form. This flattened version of the Earth is known as the 'projected' Earth. There are many hundreds of projections used for different parts of the Earth (for a comprehensive list, see Bugayevsky and Snyder, 1995), mainly differing in the approach that is used to flatten the Earth's surface. Table 3.1. summarises the main approaches.

All projections distort the Earth's features in some way, but for the tasks that crime mappers would use a GIS, the impact of these distortions has very little impact on the typical functions they would perform. The most important aspects of coordinate systems and projections for crime mappers to know are which system and projection their geographic data is in. This knowledge enables mapping and integration of this data, and the combination of these systems can determine the geographical position of any location to which their address or other locational descriptive data can be referenced.

Table 3.1 Common projection processes. Source: Longley *et al.* (2001)

Type of projection	Description of flattening process
Cylindrical	This is the equivalent of wrapping a cylinder of paper around the Earth and projecting the Earth's features onto this cylinder
Planar	This is the equivalent of touching the Earth with a sheet of flat paper at some determined point (e.g. the equator at zero degrees lóngitude) and projecting the Earth's features onto the flat surface
Conic	This is the equivalent of wrapping a sheet of paper in the shape of a cone around the Earth and projecting the Earth's features onto this cone

Let us consider an example of a particular location and how its position can be geographically represented using coordinates. The Queen of the United Kingdom lives at Buckingham Palace, The Mall, London, England, SW1 1AA. If the Queen was burgled and the crime record was to be presented on a map in a GIS, the crime record would need to have geographical coordinates assigned to it. The longitude and latitude of a point that represents Buckingham Palace is 0 degrees, 8 minutes and 30 seconds (0:08:30) West and 51 degrees, 30 minutes and 4 seconds (51:30:04) North. Buckingham Palace is very close to Greenwich, hence the very low longitude (West) value. However, in England longitude and latitude is not the coordinate system that is most commonly used. Instead, the British National Grid projection is most commonly used.

The British National Grid is a metric Cartesian grid that starts at the value zero metres Eastings and zero metres Northings. This location, known as the Point of Origin, is off the farthest South West land point in Great Britain, somewhere in the middle of the sea. From this point, all locations in Britain can be measured so that they can be expressed in metres East and North of this point. Using this grid a point located at the centre of the back building to the Queen's London residence can be determined as being positioned at 528 988 m East and 179 622 m North (Figure 3.4).

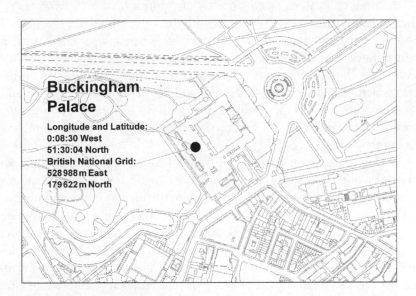

Figure 3.4 Buckingham Palace, labelled with its longitude and latitude coordinates and its British National Grid Easting and Northing coordinates. Reproduced by permission of Ordnance Survey on behalf of HMSO. © Crown copyright 2005. All rights reserved. Ordnance Survey Licence number 100044021

Many countries prefer this type of coordinate system over latitude and longitude. The ability to express a location as a positive number in simple feet or metres is preferable to making distance calculations using degrees of latitude or longitude. Australia, New Zealand and individual states in the United States all use a projected system. Some of the US states are so large that they have to employ more than one projected system for different parts of the state. The principle, however, remains the same and users should try and ensure that their data are mapped in a projected coordinate system.

3.5 Getting crime data into a GIS

The example of determining the geographic coordinates of Buckingham Palace and then assigning these coordinates to the crime record is a process that is referred to as geocoding. Geocoding is an operation that is required on all crime data if it is to be spatially displayed in a GIS.

Capturing geographical coordinates for crime records is a process that is usually performed after the crime details have been captured. Some police crime recording systems do use an approach where addresses and other location references are stored with their geographic coordinates in a look-up table in the system. This means that when an operator is entering address information for a crime, by simply typing certain key words of the address the database performs a search and presents address options from which they can select the correct address. This address is then automatically entered into the crime record with its geographic coordinates. Dumfries and Galloway Police Force in Scotland provide an example of this approach in this chapter.

Geocoding can operate at different levels of spatial precision, and thus does not necessarily require all data to be matched to a precise address or location. Geocoding could be performed to street segments or the centroid (the position at the centre of an object) of a geographic area such as a postcode, zip code, a police beat or a census geographical boundary area. In Great Britain, an Ordnance Survey product called ADDRESS-POINT is popularly used to assist precise geocoding of crime data. ADDRESS-POINT stores the details of all property addresses in Great Britain (Figure 3.5a), including their metre-precise geographic coordinates (visit www.ordnancesurvey.co.uk/oswebsite/products/addresspoint for more details). ADDRESS-POINT can also be integrated into a building polygon layer produced by the Ordnance Survey so that crimes at these locations

(a)

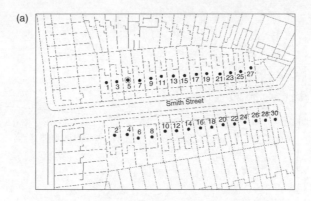

(b)

Figure 3.5 The figures above show the use of reference files (sometimes also referred to as *gazetteers*) that are used to precisely geocode crime data. Each figure demonstrates the example of geocoding to 5 Smith Street. (a) Geocoding in the UK usually makes use of the Ordnance Survey product ADDRESS-POINT or (b) Ordnance Survey MasterMap where ADDRESS-POINT has been integrated into a topographic polygon layer. (c – on page 48) In other countries, such as the United States and Australia, street segment TIGER files are commonly used for geocoding, using an offset to determine the position of the crime. Notice the degree of error between the TIGER approach and the true location of the address. The position along Smith Street was calculated by measuring the length of the street segment (112 m), counting the number of addresses in the odd number range (14) and calculating the proportionate distance of this third address in the odd range along the street segment using the following formula: Distance along the street segment = (The house's numerical position in the range−1)×A, where A = Length of street segment/ (Number of houses along one side of the street−1) (i.e. A = 112/(14−1) = 8.615). Number 5 is the third odd number address. The distance from the start node of the street segment to the proportionate position of 5 Smith Street is (3−1)×8.615, which equals 17.23 m. The line drawn perpendicular to the street segment is 24 m from the start node of the street segment, and the point is positioned at a 10 m offset (meaning it is 10 m from the centreline of the road). Reproduced by permission of Ordnance Survey on behalf of HMSO. © Crown copyright 2005. All rights reserved. Ordnance Survey Licence number 100044021

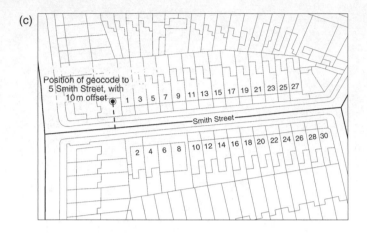

(c)

Position of geocode to
5 Smith Street, with
10 m offset

Figure 3.5 (Continued)

can be represented as a polygon object, rather than a point positioned in
the property (Figure 3.5b). In certain other countries, such as the United
States and Australia, street object files are used for geocoding crime data
to precise spatial levels. The US and Australian TIGER files contain
address ranges along a street segment rather than individual address references (Figure 3.5c). The address range refers to the first and last number
of the addresses along the segment of street, with odd numbers referenced
on one side of the street and even numbers on the other side (visit www.
census.gov/geo/www/tiger for more details on the US TIGER files). The
geocoding process estimates where the crime happened along the street in
relation to the address detail that is captured in the crime record and the range
of addresses. The location that a crime record is geocoded to is a relative
position, proportionate to the address of the crime offence and its ratio
between the length of the street and address range values. Geocoding using
these TIGER files usually also includes an 'offset' attribute that determines
the position and distance where the geocoded crime record is placed, perpendicular to the street. The offset ensures that the geocoded point does not
sit on the centreline of the street, but off to the side, at a location closer to
the likely location of the house. Some approaches for geocoding using TIGER
files also make use of an 'inset' value to adjust the geocoding position along
the street so that crimes geocoded to the first or last address in the address
range are not positioned at the start or end nodes of the street segment.

Even though there are different levels of spatial precision and
approaches that are used in geocoding crime data, what is standard in all
geocoding processes is that data which describe the geographical position of

Table 3.2 Matching geographic information content between data that are stored in a crime record and corresponding reference files

Geographic information content in record that is to be geocoded	File that contains geographic coordinates to assign to record (the reference file)		
	Type of geographic information stored in reference file	Geographic information stored in reference file	Geographic coordinates
Buckingham Palace, The Mall, London	Property addresses	Buckingham Palace, The Mall, London	528988, 179622
Buckingham Palace, The Mall, London	Street segments, with address ranges	The Mall, London	529187, 179762
SW1A 1AA	Postcodes	SW1A 1AA	529090, 179645 (Postcode centroid)
00BKGQ0013	Census output area code	00BKGQ0013	527655, 172488 (Output area centroid)
St James	Ward names	St James	529828, 180284 (Ward centroid)
Westminster	Local government district names	Westminster	527556, 180870 (Local government district centroid)

the event need to be matched to corresponding details in the file that contains these geographic coordinates – this file is often referred to as the reference file (e.g. a TIGER file or Ordnance Survey ADDRESS-POINT) or a gazetteer.

Table 3.2 describes several examples of how address data in a crime record can be matched to a reference. Using the example of a burglary at Buckingham Palace the first record in Table 3.2 shows that the full address is recorded in the crime data file. For this record to be geocoded it must be matched to the corresponding record contained in the reference file. A match is identified and the coordinates for this address that are stored in the reference file are assigned to the burglary crime record. If the only geographic information that was captured in the crime record was the street name in the address, this could be used and matched to a reference file containing these street names and address ranges and the crime could be geographically positioned to an estimated position on the street. Because Buckingham Palace is at one end of 'The Mall', this could be at the starting node of the street segment of 'The Mall' (though automatic processes usually use the centroid – the centre point – of the street segment). We could then continue with progressively less precise geographic references such as the Palace's postcode, the code for the census output area that covers this address, the census ward or the Local Government District boundary. In each case a match would be required between the geographic content that is captured in the crime record and an appropriate reference file that contains like-for-like geographic information and appropriate geographic coordinates for these geographic references. All the geographic coordinates listed in Table 3.2 are relevant to Buckingham Palace, the differences relating to the spatial scale at which the crime is geographically represented.

When a crime record is being recorded from a victim, attempts should be made to ensure that as much address information as possible is captured, rather than capturing just a geographic reference such as a postcode or other generalised address data.

Case study: Using GPS technology to capture environmental crime incidents in North London, England

Material supplied by Lee Kilroy at ESRI (UK) and the North London Strategic Alliance

Environmental crimes, or 'envirocrimes', such as graffiti, fly tipping (i.e. rubbish dumping), car abandonment and anti-social behaviour have a significant impact on citizens' quality of life (DEFRA, 2003). Since 2003, the UK's Anti-Social Behaviour Act has given local authorities more powers to tackle these types of crimes and support the police with their responsibilities for making communities safer.

One of the biggest challenges in tackling envirocrimes is identifying and, more fundamentally, recording where these crimes happen. Crimes of this type quite naturally often occur at non-addressable locations. For example, graffiti could be sprayed on a bridge or a car abandoned on a derelict land site. Geocoding these crimes is therefore a challenge.

In North London, a group of four local authorities (Enfield, Haringey, Barnet and Waltham Forest) have teamed up to take a proactive approach to tackling envirocrimes. Known as the North London Strategic Alliance (NLSA), they have recognised the challenge of geocoding a location and responding to the problems of envirocrimes, and have employed a mobile and real-time solution to capture these crime details. Their approach uses Pocket PCs with GIS software, loaded with Ordnance Survey mapping data (Figure 3.6). These handheld units are

Figure 3.6 A Pocket PC, used by the North London Strategic Alliance, equipped with GIS software and GPS to capture the geographic coordinates of environmental crimes and other incident details. Source: ESRI (UK)

equipped with GPS, which allows the user to capture accurate geographic coordinates for an incident, and General Packet Radio Service (GPRS), which enables the information to be uploaded back to the central database in real time. NLSA estimate that it would usually take three to four days to process and deal with these types of crimes, but the new system means that corrective action can be organised as soon as the incident is reported. In the case of abandoned vehicle details, this enables response teams to recover these vehicles before they may become burnt out by vandals or thieves, reducing the recovery cost from £5000 to just £30.

This mobile GPS solution is bringing benefits that help improve the end-to-end process of reporting an incident to taking corrective action, particularly in terms of using this information to quickly identify crime patterns and initiating a faster response. Although this example demonstrates its use for envirocrimes, the same technology is also being applied by police forces in other parts of the UK to help them capture geographic coordinates and other details of crime incidents.

3.6 Geocoding in the real world

The practical experience of geocoding crime data is rarely as straight-forward as the address match example for Buckingham Palace. Crime data pose particular challenges for geocoding. We demonstrate this with the following example.

Table 3.3. lists 12 crime records that require geocoding against a reference file that contains corresponding address data. These records have been arranged in each crime record so that different parts of the address are written into separate fields. First sight of this data may lead the reader to think that geocoding these crimes will be straightforward. However, there are a number of problems that exist in geocoding these data, problems that are not untypical in crime data.

Table 3.3 A list of 12 crime records that require geocoding. At first sight it may appear that these records will be straightforward to geocode, but in practice these 12 records pose particular geocoding challenges

ID	Field 1	Field 2	Field 3	Field 4	Field 5	Field 6
1		125	Mandrake Rd	Tooting Bec	London	SW17 7PX
2	Ground Floor Flat	36	Brenda Road		London	SW17 7BJ
3		10	St James Road	Upper Tooting	London	SW17 6DV
4		5	High Street		London	
5		14	Beachcroft Road	Upper Tooting	London	
6	Sainsburys Supermarket		Balham High Road	Balham	London	
7			15 Fircroft Road	Upper Tooting	London	
8			Trinity Road	Wandsworth	London	
9			Wandsworth Common	Wandsworth	London	
10	Car park at Library		Balham High Road		London	
11	Pavilion Square	1	Beechcroft Road	Upper Tooting	London	SW17
12		58	Brenda Road	Upper Tooting	London	

3.6.1 Problems with handling abbreviations

The first record in Table 3.3 appears complete, but the abbreviation of the word 'Road' to 'Rd' means that a match to the equivalent address in the reference file will not be possible because the reference file address entry is written as 'Mandrake Road'. Computer-based geocoding systems often require data to follow exact like-for-like address-matching. This problem is also apparent with record 3. The spelling of the word 'Saint' is written in full in the reference file. The reference file entry for this address does not include a full stop after the 'St', meaning that a match between record 3 and its corresponding reference file entry may not be possible.

3.6.2 Names that are local aliases

In record 1 the locality is written as 'Tooting Bec', but Tooting Bec does not exist in the reference file. Tooting Bec is a local alias name for the area. Its official locality name is 'Upper Tooting'. Local aliases such as these cause problems when the geocoding requires strict record matching. This alias problem is also apparent in record 2. 'Ground Floor Flat' is different to the textual content of this apartment in the reference file, which contains the more common entry of 'Flat 1'.

3.6.3 Multiple listings

Record 4 does not include a full postcode, which may not necessarily pose a problem if we discount matching this field between the address data from our crime record and the address content of the reference file. However, 'High Street' is a common street name in the United Kingdom, and in London there are many streets that have the name 'High Street'. The content of record 4 makes it difficult to distinguish at which '5 High Street' the crime occurred. The crime may therefore be geocoded to the wrong one, or may be left ungeocoded because of confusion between which 'High Street' to match to.

3.6.4 Incorrect spellings

Record 5 appears to contain sufficient data to be able to uniquely match it to a corresponding reference file record. However, the word 'Beachcroft' has been incorrectly spelt. The correct spelling of this street is 'Beechcroft'. A similar problem exists with record 6. Organisation or business names are included in the reference file and are listed in the first address field of this file. However, 'Sainsburys Supermarket' is not the entry of this business in the reference file. Instead it is written as 'J Sainsburys PLC'.

3.6.5 Address detail written into different corresponding fields (or written into a single field)

Record 7 shows a common problem with address data that is recorded for crimes. In some cases the components of the address may not be split into a format that corresponds with the arrangement of address fields in the reference file. In this example the street number is written in the same field as the street name. This can make it difficult for some GIS to geocode data that are formatted in this way.

3.6.6 The address detail is incomplete

Record 8 shows a common problem with crime data where the highest resolution of the address is the street name. Address entries such as these can be common with crime data, particularly with crimes that happen to people in the street, crimes such as robbery or thefts. If you were robbed when walking along a street would you stop to see if it happened outside number 5 or number 105 on the street? For crimes such as robbery it may not be possible for the victim to know the exact address. This often results in the crime only being recorded with the street name. When this record is geocoded, it may be possible to match to just the street, but where does one decide to position this crime on the street? The street in this example, 'Trinity Road', is two miles long. A common practice is to either geocode the crime to the beginning or end of the street, or to a point halfway along the street. This may cause subsequent problems with analysis which will be discussed later in this chapter.

3.6.7 The address describes an area of open space

So far in our geocoding descriptions we have only tried to match to street addresses. However, crimes do not just happen on streets or in buildings but also in areas of open space. Records 9 and 10 present examples of these incidents. Wandsworth Common is a park. It has no entry in the reference file that is being used to geocode these crimes as the reference file only lists street addresses. This means that it would not be possible to match to the reference file unless the reference file was edited and this area of open space was included. But Wandsworth Common is a large park covering several square miles, so what geographic coordinate would be used to determine the location of a crime that happened in this park? In these cases it may be best to enter and include a number of listings of Wandsworth Common in the reference file that relate to certain localities or points of reference. These could include features in the park such as the children's playground, the band stand or the park pavilion, or references linked to the streets that surround the park. For example, the 'Bellevue Road entrance' to the park is opposite number 2 Bellevue Road. Record 10 lists a similar open space problem. Car parks are not routinely included in most reference files, so it would be difficult to match to unless they were edited in. Again, this could be done by adding it as a new, separate reference entry or by linking it to an address on the street segment.

3.6.8 The address does not exist or the reference file is not up to date

Record 11 lists an address that appears as though it can be uniquely matched to a record in the reference file. However, this record is not listed in the reference file because it is a new housing development and the reference file has yet to be updated with these new addresses. This record would therefore remain ungeocoded.

3.6.9 Incorrect address entry

And finally, when it appears that our final example, record 12, looks as if it contained all the details that we need to be able to successfully geocode this crime record, we learn that the address has been entered incorrectly because '58 Brenda Road' does not exist – the numbers only go up to '48'.

Of the 12 records listed in Table 3.3, none can be exactly geocoded. The term used to describe the level of success from geocoding is known as the 'hit rate', and is represented in percentage terms. In this example the hit rate would most likely have been a rather depressing 0%: we could not confidently geocode any of the records without some adjustments to the data. The crime records presented in this example are not untypical of crime data that exist in practice, largely because crime records are not necessarily recorded with the crime mapper or any subsequent analysis in mind. This often means that a cleaning process on crime address data is required as a stage prior to geocoding.

3.7 Address data cleaning

Cleaning (also sometimes referred to as scrubbing) of address data contained in crime records is a useful pre-geocoding stage that improves the hit rate of geocoding crime records. These data cleaning processes can be automated into software products. The typical types of corrections that need to be performed on address data in crime records are those that relate to the experiences in geocoding the 12 sample records from Table 3.3. These include:

- *Abbreviations*: Correcting abbreviations into a format that corresponds to how these abbreviations are recorded in the reference file.
- *Spelling errors*: Correcting the spellings of any words recorded in the address field of the crime record.
- *Standardising addresses*: Recognising and converting words into a standardised address word. For example, converting any versions of the business name that includes the word 'Sainsburys' (e.g. 'Sainsburys Supermarket') into a standardised format of 'J Sainsburys PLC'.
- *Formatting*: Correcting the formatting of the address fields in the crime record into a format that corresponds with the arrangement of fields in the reference file (e.g. correcting fields that record an address in a single field such as '15 Fircroft Road', by splitting it into two fields, '15' and 'Fircroft Road').
- *Incompleteness*: A data cleaning tool may also be able to correct address details that are incomplete by determining the content of the missing information, based on the address information that is complete. If only a street name and town name is recorded in the address, then of course it would not be possible to determine the street number. However, if parts of the address in a crime record, such as the locality name, town name or postcode/zip code, are missing, it may be possible to use the reference file to determine what these are and add these to the address content of the crime record. Whilst this may not help improve the success of geocoding these crime records, it does help to standardise the content and completeness of these records, and may be useful for future analysis requirements.

3.8 Address reference files

The success of geocoding is also determined by the content of the reference file that is used to geocode the crime records. In many countries the local government may maintain a local address reference file. In some countries the national government may take this responsibility and maintain a national address reference file (e.g. the TIGER files in the US, Australia and New Zealand), or work with the local municipalities to administer the process of linking these local address reference files to form a national version. These reference files can be used as the source against which crime records can be geocoded. Several countries have developed (or are developing) and maintain local and national point-level address reference files. These countries include England, Wales, Scotland, Northern Ireland, Denmark, Czechoslovakia Republic, Finland, several of the provinces in Canada (e.g. New Brunswick and British Columbia), some states in the USA, parts of Australia and New Zealand. However, the content of these reference files may not be ideally suited for geocoding crime records (e.g. the English National Land and Property Gazetteer does not specifically contain organisation names). So certain edits may need to be made to these to help improve the geocoding success rate. These could include:

- *Adding data from other registries*: Using other address registries to complete missing data. For example, the English National Land and Property Gazetteer could be integrated with a business names dataset to fill the gaps where organisation names are missing.
- *Including aliases*: Place names such as localities may exist as only a local colloquialism rather than an official address that is recorded in a gazetteer or reference file. These types of new entries could also include the aliases that are used to refer to certain addresses, such as the names of shops that may differ to the name that is recorded in the reference file. For example, a reference file lists the entry 'McDonalds Restaurant, Balham High Street, Balham, London'. An alias for 'McDonalds Restaurant' could be 'McDonalds'.
- *Non-addressable locations*: Entries that describe areas of open space or useful points of reference that are not included in the reference file could also be included.
- *Street junctions and major highway junctions*: The reference file may not include the junctions of streets, in particular the junction references for major highways. These could be added from certain available data sources (e.g. several commercial mapping companies produce this data, and although it is mainly used for in-car navigation applications, it could be sourced and integrated into a property gazetteer).
- *New construction developments*: Keeping the reference file up to date is also important to ensure that new building developments are included in the reference file as soon as they are completed.

Case study: Geocoding crime data at the point of record entry in Dumfries and Galloway, Scotland

Material supplied by Stuart Arnott, Dumfries and Galloway Police

Recording crime incidents and completing the associated paperwork can take up a significant amount of time. The address data captured in a crime record can also lack the detail required for subsequent analysis and is a process that is often prone to error. Dumfries and Galloway Police Force recognised these problems when they began mapping crime data for analysis and realised that a large amount of their data was being geocoded into the Irish Sea! In a drive to improve the efficiency and accuracy of capturing information, Dumfries and Galloway Police have developed an intranet-based Incident Management And General Enquiry system (IMAGE).

IMAGE is integrated with a GIS and comprehensive gazetteer of property and non-addressable locations (e.g. car parks and other areas of open space). Before the introduction of IMAGE, to find and capture the location of a crime incident required a combination of gazetteer look-up, the use of paper maps and local knowledge. The gazetteer was poorly maintained and incomplete, the paper maps took time to review and the local knowledge was not always existent. Through the investment in detailed base mapping data for their GIS and a gazetteer that is comprehensive, up to date, centrally administered and well-maintained, the capture of address data or location data of where crimes happen is now significantly more efficient and accurate. The gazetteer is used to automatically populate the address/location fields of the crime record, and if the crime location is not recognised by the gazetteer, the officer can click on the location on a map and the crime is geocoded automatically. The gazetteer administrator subsequently validates this location entry. This method of approach is ensuring that Dumfries and Galloway are achieving geocoding hit rates of 100% and that the quality of recording the location of where crimes happen is extremely precise and accurate.

The IMAGE address capture system is quick to use, browser-based, accessible to all officers and support staff from 320 networked computers in 17 locations and requires very little training. The cost of cleaning data and making amendments is now much reduced. Dumfries and Galloway Police estimate cost savings of over 70% when compared to their old approach of address data capture, and can now produce geocoded crime data of a significant higher quality while being able to meet and more proactively develop their operational needs and crime mapping abilities.

3.9 Geocoding functions

Several of the problems that were described against the geocoding attempts of the 12 example records listed in Table 3.3 can neither be solved by cleaning addresses nor by improving the content of the reference file. In some cases geocoding may not be possible because the address content that is recorded in a crime record may be incorrect or incomplete. However, certain processes can be performed on the address data to improve the geocoding success of records of this type:

- *Multiple listings*: A common problem in geocoding crime records is determining between multiple reference file entries of similar addresses. In the case of the examples used in Table 3.3, the crime with the address '5 High Street, London' is difficult to geocode because there are many addresses of the same name in London. A geocoding function could though help this record to be geocoded. This function could use other data in the crime record's address fields, such as a locality name (e.g. 'Balham') to identify this High Street, or use data such as the postcode or part of the postcode to determine between multiple listings of High Streets (e.g. if the postcode field contained 'SW17' this may be enough to determine between multiple High Streets). If this additional data are not available the reference file could be constrained or cookie cut to just the area for which crime data is to be geocoded.
- *Geocoding to the next level of precision*: In some cases an error in the crime record's address may prevent geocoding. For example, record 12 from the 12 samples in Table 3.3 could not be geocoded because the street number was incorrect. Rather than not geocoding this record at all, it may be best to geocode to the next level of spatial precision, such as the street. A geocoding function could recognise this address content, determine if this street information was reliable and determine a coordinate location for the point at the centre of the street as the geographic coordinates for this crime record.

A useful function to include in any geocoding process is the ability to assign confidence codes that describe the accuracy and precision of geographic coordinates that have been determined for each crime record.

3.10 Geocoding and fitness for purpose

The previous sections have demonstrated that geocoding crime data can be a challenging task. Whilst it is important to ensure that the geocoded crime data are of good quality, it is also important that in the search for data perfection, the level of data quality that is required of crime data is proportionate to the purposes to which it will be applied. For example, it may be pointless to spend significant time and effort in geocoding crime data to metre precision when the analysis required only seeks to reveal general patterns over a large area. This may mean that geocoded data at an area level such as wards or police beats may suffice for this analysis and be easier to geocode.

Geocoding data that are fit for purpose is a balance between achieving levels of:

- *Accuracy* – data are geocoded to the position it is meant to be geocoded.
- *Precision* – precision is often confused with accuracy. Precision describes the spatial scale of a record and refers to the exactness with which a value is expressed, whether the position of the value be right (accurate) or wrong (inaccurate).
- *Consistency* – consistency ensures that the placement of geocoded points follows standardised processes.
- *Completeness* – completeness refers to the fullness of content of a record for the purpose it will serve.
- *Reliability* – it is vital that crime data that are used for analysis are reliable so that there is confidence in the interpretations that are made from the data to support further decision-making.

An aspiration for anyone working with crime data in a GIS is to achieve geocoded crime data that are 100% accurate, precise to the exact location of where the crime occurred, have been processed consistently, are complete and can be confidently relied upon. The practicalities of the real world mean that all these conditions are not always possible. To determine fitness for purpose levels we require a comparison between its five determining variables:

1. *Reliability*: There are very few occasions when we cannot accept data to be anything but reliable. Levels of reliability are determined by the levels of decision-making that are required from the application of data. If the data are only used for a cursory glance then while reliability is important, the level of necessary reliability is less than if these data are to be used to determine the future funding programmes of an entire

police force. Completeness, consistency, accuracy and precision all help to determine levels of reliability.

2. *Completeness*: A record's content could be full, but if it does not meet its purpose then it is not complete. If a record's content is not full, but it fully meets its purpose then it is complete. Completeness can be influenced by consistency and helps to determine the levels of precision and accuracy. Completeness is also subjective and dependent on the analytical task. The same record will be complete for one task, but may be incomplete for another.

3. *Consistency*: Neither accuracy nor precision determine consistency, but consistency informs reliability. Consistency is determined by the processes that are performed on geocoding crime data. For example, if a crime record can only be geocoded to a street, its geographic placement (be it the street centroid, starting node or end node) needs to be consistent for the whole study area.

4. *Precision*: Data can be inaccurate, yet still very precise. Precision helps to determine reliability particularly in terms of the geographical scale at which analysis can be performed. For example, if crime data are only available at the police beat level of precision, analysis of crimes at this precision may not reliably inform patterns of crimes on housing estates or within neighbourhoods – the police beat level may be too coarse to reveal these sub-beat patterns. Precision can be measured in relation to the spatial scale of data.

5. *Accuracy*: Data need to be accurate, but inaccuracies can be handled, managed and still make data reliable. It is useful to flag the level of accuracy in crime data as this will help determine the confidence in any results that are generated from analysis of data, identify any errors that could be corrected and inform the refinement and improvement of geocoding processes. Measuring accuracy does require an assessment of the data to determine if the processes that have been applied to it have been performed correctly.

3.11 Measuring geocoding accuracy

Measuring the spatial accuracy of geocoded crime data can be performed by following a simple analytical procedure on a *representative* sample. Determining how representative a sample should be is based on the levels of confidence required in the results. There are a number of Internet sites that can be used as a guide to determine approximate sample size (type 'sample size calculator' into a search engine and you should easily find one). For example, an online tool estimated the sample size required for a crime dataset that contained 1000 records, where a 95% confidence threshold was required, and where the results were to be accurate to within a margin of error (also known as the confidence interval) of plus or minus 5%. The estimated required sample would need to contain 278 records.

This simple analysis for measuring geocoding accuracy involves performing the following steps:

1. Determine the sample size and select a random sample of geocoded records that make up the sample.
2. Map the sample in a GIS against a background of detailed base mapping data and a mapped version of the reference file that was used for geocoding.
3. Identify a sample point in the GIS.
4. Observe where the point is mapped.
5. Compare this mapped location of the crime record to the address information stored in the crime record, the details on the map and the geographic position of the equivalent reference file entry. The geocoded crime record should be exactly the same as the equivalent reference file entry, or judged to be positioned correctly if a TIGER file is used. Mapping the reference file entries allows a check against the reference file's geographic coordinates to ensure they are correct. Three possible outcomes will be presented.

   (a) The crime point is mapped to the correct place according to the base map and the relevant reference file entry position.
   (b) The crime point is mapped to an incorrect place according to the base map, but is mapped to the matched reference file position. This reveals an error in the reference file or the address matching process, such that:

      (i) Error in the matching process – the geocoding process may have wrongly matched the address content of the crime record to the reference file.
      (ii) Error in the reference file – the match of the address content of the crime record has been correctly matched to the reference file, but the geographic coordinates of the reference file are incorrect.

   (c) The crime point is mapped to an incorrect place according to the base map and is incorrectly matched to a reference file entry or no reference file entry can be found.

6. For records that have been correctly mapped, note that the record is accurate and proceed to the next sample record back at step 3.
7. For records that have been incorrectly mapped, calculate the difference in distance between the mapped location of the crime record and the true address that is identified on the base map. For TIGER files this error distance can be either the distance to a cadastre base map that shows the location of the address the crime record refers to or the likely position on the line segment for the street that contains the relevant address ranges.
8. If the address information is not precise (e.g. only a street name is provided in the crime record), note the coordinates of the spatial object to which the crime data should be referenced. For example, this could be the centroid of a postcode or for street segments this could be the coordinates of the street centroid. Calculate the difference in distance between the mapped incident location and the position that determines the accurate location of crimes for this level of precision. Note the difference and return to step 3 to repeat for the next record.
9. After all sample records have been tested, calculate the number of records and percentage share where the difference is:

    (a) within 50 metres
    (b) within 250 metres
    (c) greater than 250 metres
    (d) no difference in geographic position.

10. The measure of accuracy is determined from these percentage values. If the full crime dataset consisted of 1000 records and a sample of 278 records was used to perform an accuracy analysis, and 265 crime records showed no difference between their geocoded position and the correct position they should have been mapped to, then the data can be described as 95% accurate (within a confidence interval of plus or minus 5%). Analysis of the error distance for the remaining 13 records will help reveal the scale of these errors and their possible source (to aid future geocoding improvements).

The impact of the offset variable used for geocoding crime records against TIGER files can also be measured. In a study on geocoding crime data to Australian TIGER files Ratcliffe described the need for care in deciding the offset that is used. He found that more than 50% of geocoded records may be given coordinates that fall within the land parcel of a different property (Ratcliffe, 2001). The Australian study goes on to suggest a street offset of 25 m and an end offset of 15 m, though these figures may vary by location (Ratcliffe, 2001, p. 484).

Case study: Handling uncertainty and incompleteness in crime records

Achieving a high hit rate for geocoding crime data often requires a preliminary data cleaning process to raise the success level of matching crime records to the geocoding reference file. This task can be very time-consuming if completed manually, requiring each record to be checked and, if necessary, corrected. A number of companies, provide tools that are used by the police in the UK and the United States that automatically clean the address fields of crime records. The cleaning process records an audit trail of the processes performed when correcting each crime record's address content and how it was address matched and geocoded. The audit trail is a sequence of codes that can be interpreted to describe the geocode precision of each record, therefore allowing an analyst to display data in a GIS and flag each record according to this level of precision.

An additional function that the software performs is one that attempts to handle the problem of crime records that only contain street names in their address fields and that are subsequently geocoded to a single point on a street. For short streets this does not tend to cause a problem for most analytical applications, but for long streets, particularly those that pass through high-crime neighbourhoods, many geocoding processes may position a large number of crime records to a single place on the street (e.g. the centre of the street). This has the knock-on effect of creating 'false hotspots'.

The software overcomes this in a manner similar to how a crime mapper would, if they were to reposition these crime records manually. The software function applies a weighted approach to reallocating these points along the street. This routine works by first identifying those incidents that have been successfully and more precisely geocoded on the street in question (i.e. the street number, building name or organisation name has been combined with the street name to locate these incidents to their exact location). The location of these precise records determines the weights for re-processing the affected records. The greater the number of incidents that have been precisely located to a position on the street, the greater the weight and draw to that location for points that require reallocation. The weighting algorithm also includes an element of random distribution along the street so that no particular location can be over-exaggerated from this routine. The algorithm is then applied to the affected records, new grid coordinates are determined and the change is recorded in the record's audit code.

Studies conducted on crime data that has been processed using this software report hit rates close to 100% and geocoding accuracy levels above 94% (Chainey, 2001). The application of an automated cleaning and geocoding process and the capturing of an audit trail helps the crime mapper understand the fitness of purpose of their crime data by being confident that consistent processes have been applied to the data, that they can measure the completeness and precision in their data, knowing their crime data are accurate, and that the output is reliable for the types of applications they will perform on these data.

3.12 Mapping and unreported crime data

This chapter has discussed the technical challenge of geocoding crime data. At the outset it is also important to recognise the differences between crime data that may finally end up on a map and the actual levels of crime.

Crime data recorded in police information systems offer only a partial view of crime in society (Smith, 1986), and not all crime reported to the police ends up being recorded as a crime. The Australian Institute of Criminology has estimated that for every 100 crimes that are committed in Australia, only 40 are reported to the police and only 32 are actually recorded by the police (see Carcach, 1997 for more details of underreporting in Australia).

A similar picture is seen in the United Kingdom. The British Crime Survey (BCS) asks questions about crime to a large representative sample of Britain's general public. Its main aim is to provide a measure of crime and attitudes to crime that compares and contrasts against crime reports recorded by the police. It includes a count of crimes not reported to the police and is not affected by any changes in the way that police record crime. It therefore offers a means to describe crime that is not reported. The BCS estimates that only 42% of crime is reported to the police and that the police only ever record 74% of the crime which is reported to them. This means that only about 31% of all crime is reported and recorded by the police (Dodd *et al.*, 2004).

The reporting rate for crime differs between different crime types. For example, in England and Wales, 95% of thefts of vehicle are reported to the police while only 30% of common assaults are reported (Figure 3.7).

The BCS cites the main reason for not reporting incidents relates to the victims' perception of the incident being too trivial or that the police

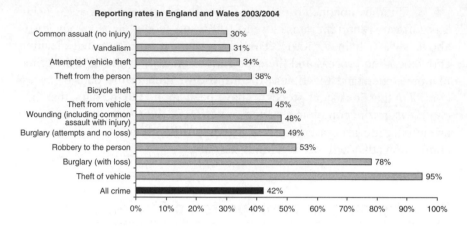

Figure 3.7 Crime reporting rates by crime type in England and Wales. Source: Dodd *et al.* (2004)

could do little about the crime. In almost a quarter of cases the victim felt that the incident was a private matter and could be dealt with by themselves (Figure 3.8) (Dodd *et al.*, 2004).

So if less than half of all crime ends up in a police crime recording system, is there any point in mapping and exploring the geography of what is an incomplete picture? To concede this point would be to suggest that responding to crime patterns had no value, but this is not the case. It is likely that the locations where reported crime concentrates is also where unreported crime concentrates because existing, measurable crime problems are reasonable indicators of chronic crime problems. However, it

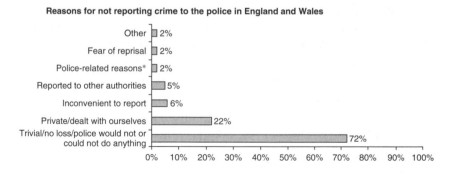

Figure 3.8 Reasons for not reporting crime to the police. * Police-related reasons included fear, dislike or previous bad experience with the police or courts. Source: Dodd *et al.* (2004)

is important that this underreporting is acknowledged in any analysis or interpretation of crime.

To date there are no studies that describe if the underreporting of crime is geographically uniform at the small area (local) level, although studies that have explored perceptions of where crime concentrates have been completed (see Chapter 7 of this book and Rengert and Pelfrey, 1997; Ratcliffe and McCullagh, 1998, 2001; Paulsen, 2004), and studies in areas such as census nonresponse rates and household interview surveys (Groves and Couper, 1998) could offer some useful initial insight by identifying community groups that tend to be reluctant to report information to the authorities.

Placing crime data on a map and pressing a button in a GIS will not provide the answer to why crime happens and how a crime problem can be solved. Tackling any crime requires drawing on professional expertise and experience about crime. It will often require a wider knowledge of the problem to fully understand what is causing it and may force an analyst to look beyond the crime data alone. Detailed analysis of crime data does, however, offer a mechanism to identify patterns in crime and the environmental situations that provoke crime (Clarke and Eck, 2003), and crime mapping offers a means to discern the spatial patterns and limits of criminality. Mapping can also promote questions and initiate discussions on what is not known, how these information gaps can be filled, and if the information that is available is sufficient to be confident enough to make decisions on how resources can be used to deal with the problem.

There may be value in information that can be sourced from other agencies that could offer insight to a particular problem and fill the knowledge gaps that exist with traditional recorded crime data. For example, police records on car thefts may provide a useful leading indicator of a car crime problem, but data from other local sources such as the fire service (vehicle arsons), probation (offender profiles and parolees whose offending history is dominated with car thefts) and the local government (which in many countries are responsible for clearing up abandoned vehicles) may add to the picture and understanding of the car theft problem.

Underreporting of crime does present its problems, and efforts should continue to improve reporting levels across all crimes and help fill the gaps where information is missing, but the existing police and non-police sources of information consistently provide a reliable sample that describes the criminal environment. Why? Because this information has been used time and again for targeting, allocating and deploying policing and crime reduction resources, resources that have brought about a positive impact in reducing and preventing crime, helping address problems with the fear of crime and improving public safety.

3.13 Editing data in a GIS

GIS software offer a number of functions that make it possible to correct, add, delete and format data. Most GIS include the standard software with delete, cut, copy and paste routines that can be applied to all types of data, including spatial objects, that are managed in a GIS. Attribute and spatial objects can be edited, for example, to update the population of a census ward with new data, or to modify the size and shape of a Neighbourhood Watch area due to boundary changes.

New data can also be created in GIS software either as attribute data or as spatial objectives linked to attributes. For example, if the location of cash machines is not available as a geographic dataset then this could be created manually using address details on banks that have these facilities. Another example is the creation of point objects that represent the location of street Closed Circuit Television (CCTV) cameras and the creation of polygons that represent the coverage area that these cameras can view (Figure 3.9).

GIS software also include the standard formats for attribute data that are expected in computer systems – handling character, integer, decimal and date formats. However, very few GIS products have a good concept of time,

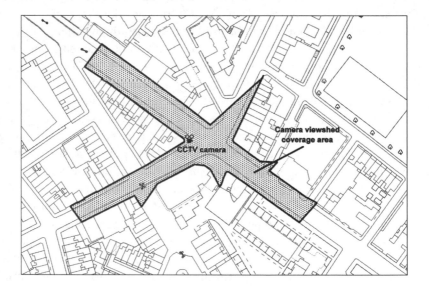

Figure 3.9 GIS functions enable the creation of new spatial objects. This example shows the location of a street Closed Circuit Television camera and its viewshed coverage area. Reproduced by permission of Ordnance Survey on behalf of HMSO. © Crown copyright 2005. All rights reserved. Ordnance Survey Licence number 100044021

and do not handle time attribute fields in a time format that is familiar in database or spreadsheet software packages such as Microsoft Excel. Instead, 'time' is usually handled in 24-hour clock format as a character field (e.g. 18.30, 18:30 or 18–30) or can be handled as an integer (e.g. 1830). This still allows crime records for time periods to be queried, but prevents the software to understand that 00:00 comes after 23:59, that queries within a certain period from a specified time may not be possible (e.g. 20 minutes from 2156) and that calculating time values from each other can be problematic (e.g. 2150 minus 2145 does equal 5 (minutes) but 2200 minus 2155 does not equal 45 (minutes)).

3.14 Performing queries on data in a GIS

GIS provide the ability to query and select specific data of interest. These queries use a standard database language called Structured Query Language (SQL) that builds logical expressions to select data of interest. The types of queries that can be performed include:

- *Selecting by attribute information*: For example, a GIS file lists all crimes but includes a field that stores the code of the crime type and the date when the crime was committed. An SQL expression would be able to select all robberies from the 'all crime' file that occurred between specific dates.
- *Grouping by attribute information*: For example, a GIS file of vehicle crimes lists the makes and models of cars that have been stolen. An SQL expression could group and determine the aggregate number of vehicles by make and model that have been stolen.
- *Selecting by geographic area*: Geographic SQL expressions offer powerful functions to query data in relation to their spatial distribution. For example, a GIS file of crimes mapped as points and displayed with police beat polygons can be queried using an SQL expression to select only those records that are geographically positioned within certain beats of interest. A query of this type can also be extended to determine the number of crimes within each beat.

3.15 Performing spatial functions and integrating data in a GIS

Data that are represented in a GIS with a spatial object can have a number of functions performed on the data to reveal additional information. This may include calculating the distance between two points (e.g. to determine the distance travelled in the journey to crime), the length of a particular street segment (e.g. to determine the volume of street crime per metre along street segments) and the area of a polygon (e.g. to determine the geographical coverage of a crime reduction initiative).

Another powerful function within a GIS is the ability to link data together, data that may have been sourced from disparate databases. Geography is often quoted as being one of the strongest denominators held in all information, in that the majority of data have a spatial reference. Using geography as this denominator and bringing geocoded data into a GIS from many different sources allows a number of possible options that include:

- *Joining datasets using common attributes as the joining link*: For example, a file of crime records lists vehicles that have been stolen and displays on a map the location from where they have been stolen. A separate file shows the recovery sites of several of these vehicles. Using the registration number of the vehicles, the two files could be linked together and displayed on a map to show the spatial relationship between stolen-from sites and recovery sites. A good example of this type of analysis can be seen in Figure 9.2.
- *Joining datasets using their geographical objects as the joining link*: For example, polygon data of police beats in an area record the number of burglary dwellings for each beat. Data derived from the census lists a series of variables, by police beat, describing the population and housing in the area. Using the spatial polygon objects as the common link, these two datasets can be joined together to display crime levels in relation to these census population and housing variables.
- *Populating attribute fields with data from another file's attribute content*: For example, a robbery reduction initiative has been centred on an entertainment complex. The area of this initiative can be defined by a polygon that covers this complex. Using crime data that has been geocoded as points to precise locations for the period before and after the initiative would enable the crime mapper to populate three new fields in the entertainment complexes polygon that describe the number of crimes before, the number after and the difference in crime levels.

3.16 Asking spatial questions before mapping or analysing data

A trap that crime mappers often fall into is that in their eagerness to produce a map they fail to question what exactly it is they are looking to find the answer to and whether the end results they produce are reliable and informative. It is now easier than ever to produce a sophisticated and pretty-looking hotspot map, but the danger in the production of these maps is that the map readers can be falsely lured into the beautiful-looking graphic, failing to question its validity, accuracy and accepting that as it looks nice, it is accurate (Chainey *et al.*, 2002). Crime mappers can also easily fall into the trap of 'paralysis by analysis' – collecting too much data or the wrong type of data, conducting too much analysis or irrelevant analysis and failing to come to any useful conclusions with the outputs they produce (Clarke and Eck, 2003). With these potential pitfalls, it is useful to be guided by methodologies such as the SARA process to help retain focus and ensure the necessary steps are undertaken in the correct sequence: *scan* for crime problems, *analyse* a specific problem in depth, *respond* to the problem by implementing solutions and *assess* the results of the project (Eck and Spelman, 1987). The SARA process helps to manage and organise the tasks required from crime mapping, the subject of which we cover in more detail in Chapter 13.

Before embarking on mapping crime it is important to ask a number of questions that help in the success and reliability of subsequent mapping and spatial analysis outputs. The list below offers questions that the crime mapper should ask and be able to answer before mapping crime or proceeding to the next stage in producing spatial analysis or crime mapping output.

3.16.1 What is it I need to find out?

Start by defining what the problem is and list a number of questions that need to be answered. A not uncommon situation for a crime mapper is one when they are asked to 'produce some maps of residential burglary'. Pause for a moment. Think about this request: What does this really mean? The crime mapper may actually discover that their interpretation of the request is significantly different from the interpretation of the person that made the request. Opening a dialogue with the person making the request may result in a map that more ideally fits their needs and a crime mapping project that is more interesting and productive than the one first requested.

3.16.2 Who is my audience?

Knowing who are going to use the results and how they may use them helps to define the level of spatial analysis that is performed, the data requirements and the map design. Effectively communicating the results from a geographical analysis are essential if the time and effort that the crime mapper has expended are to be worthwhile. In Chapter 12 we provide guidance on map design and the communication of results, but even before the first crime data are mapped it is important that the crime mapper has a clear idea of the output that is required. This should also include an appreciation of the use of graphs, tables and text to complement the map output.

3.16.3 Define the hypothesis that is to be tested

A map or results from a spatial analysis may not be the only part of an analysis that explores the crime problem, however this component still needs to follow the application of strong scientific principles to ensure that the results are robust and reliable. To accomplish this it is useful to define a hypothesis that can be tested against the problem that is to be analysed. A hypothesis is an answer to a question about a problem, and can be true or false. Using a request for residential burglary analysis as an example, the spatial questions that may require answering are 'do certain areas have higher levels of burglary, and if so, why?' The hypothesis to this question could be 'there are hotspots of residential burglary and they are due to the high levels of rented housing that are in this area'. The test for the hypothesis would be to calculate the count of rented housing across the whole study area and compare the residential burglary rates for the hotspot areas in relation to the low crime areas. If the crime levels are higher in areas of rented housing, the hypothesis is true; but if the crime levels are lower, the hypothesis is false.

Setting a hypothesis can help bring structure and definition to a spatial analysis application, and ensure that it remains focused to answering particular issues. This includes:

- helping to suggest the types of data to collect;
- how these data should be analysed; and
- how to interpret the results (Clarke and Eck, 2003).

Hypothesis setting helps the crime mapper to avoid embarking on 'shopping trips' for data. A trap that many crime mappers fall into is to build a long list of the data they would like to collect, spend significant time and effort in collecting and processing it, only to later realise that much of it is of marginal value to their main requirements. Hypothesis setting helps to keep any spatial analysis focused and ensures that the results offer a useful conclusion. Hypothesis setting also helps to appreciate that crime mapping does not start or stop at the production of a hotspot map. This shift in focus ensures that the crime mapper values hotspot mapping as a useful first step and that subsequent analysis can be used to further explore the spatial crime problem.

3.16.4 What do I already know about the spatial patterning of these crimes that will help in exploring the problems?

A crime mapper embarking on a spatial analysis of a particular crime type is not likely to be the first ever crime mapper that has geographically analysed this type of crime. It is important to review what may have been studied before as this will help provide answers to the questions that the crime mapper may be looking to answer in their own analysis. Other studies may also offer comparative work, review methodologies that have been successfully applied to geographically analysing these problems, and reflect the issues or challenges in studying these particular crime issues. This review could help inform how these crime problems or hotspots develop and the data that is useful for revealing these problems. A good starting point is to follow Clarke and Eck's (2003) six defining elements of a problem (CHEERS – Community, Harm, Expectation, Events, Recurring and Similarity) and why items are CRAVED by offenders (Concealable, Removable, Available, Valuable, Enjoyable and Disposable) (Clarke, 1999), a term defined more fully in Chapter 4.

3.16.5 Where is the crime coming from?

Crime mappers often only map the locations of crimes. Other spatial analysis can look to the concentrations of where offenders live, the concentrations of where victims live and the spatial movement or journey to crime that results in victims and offenders coming together at some place. These data can also be combined with other geographic data that describe the geographical characteristics of these areas to help in understanding why certain problems exist. We explore several of these analytical themes in other chapters.

3.16.6 What local initiatives have been targeted in the past to the area of interest?

Knowing what has been tried in the past and whether it has succeeded or failed is vital in helping to understand what the problem was, to learn lessons from the past and to be sure that any problem that was diagnosed was responded to with an appropriate design of response.

3.16.7 What spatial data do I need?

Deciding on the data that are needed should not be asked until the preceding questions have been answered. Data need to be relevant to the spatial questions being asked, and also need to be fit for the purpose in mind. Crime data may not provide the answers to all the questions that need to be asked so it is also important to consider what non-police data will be required for the spatial analysis (the types of data that could be used are reviewed in Chapter 7). At this stage it is also important to consider the base mapping data that will be required and if they are available for the study area.

3.16.8 What are the limits of the data used in the spatial analysis?

It is essential to appreciate that the data may not be perfectly ideal for the analysis that the crime mapper may wish to conduct. The data may be incomplete (including the fact that it may be a crime type that is significantly underreported) and lack the detail required. A scan of other data sources may reveal certain other data products that would help to fill certain gaps, resulting in more than one dataset being needed to answer the spatial question being asked. The limitations in the data also need to be reported concisely to the audience that use the results, to ensure they at least have some appreciation of any data flaws when making decisions on the spatial analysis results that are presented. This does not mean that the crime mapper overwhelms them with data problems, but instead identifies the areas that require future improvement. All data need to be fit for purpose in the first instance, otherwise any results could misinform and be worthless.

3.16.9 Do I need to visit the crime hotspot?

If a crime mapper is seeking to help solve a problem in a particular area, but can only view the problem from a GIS view of the world, the recommendations could be regarded as being too data-centric, not helpful to fully understand the problem in the hotspots, nor offer practical advice on how it can be solved. Leave your desk. Visit the crime hotspot!

3.16.10 What improvements should I start introducing to data and processes that will help in performing a more efficient spatial analysis next time?

It is more than likely that the data that were used to answer the spatial questions could be improved. It is at this final stage that the crime mapper should review the improvements that could be made, assess if they are practical and make recommendations that will benefit the users of the outputs with a better quality product the next time a similar study is required.

3.17 Summary

Getting crime data into a GIS and onto a map is the starting point for many crime mappers, but getting the data onto a map can be a challenging task. This chapter has aimed to raise awareness of these challenges and how they can be overcome to help ensure that crime mappers use geocoded data that are fit for purpose for the subsequent operations. The importance of data quality is often not valued until errors in the data are revealed at the stage of analysis or the final report. Investing in data from the outset can bring long-term returns and improve the quality of all future applications of crime mapping.

In later chapters we will review the range and richness of other spatial datasets that are increasingly becoming available to crime mappers, but although we may enjoy the wealth of these information sets it is still vital to remember to ask clear and concise spatial questions that address the uses of data and the outputs we seek to generate from crime mapping.

Further reading

Geocoding in law enforcement. Report to the Office of Community Oriented Policing Services. The Crime Mapping and Problem Analysis Laboratory, Police Foundation (2000). http://www.cops.usdoj.gov/mime/open.pdf?Item=759.

Introductory guide to crime analysis and mapping. Report to the Office of Community Oriented Policing Services. The Crime Mapping and Problem Analysis Laboratory, Police Foundation (2001). http://www.policefoundation.org/pdf/introguide.pdf.

Bichler, G., Balchak, G.S. and Christie, J. (2004). Address matching bias: Ignorance is not bliss. Presented at the 12th International Symposium of Environmental Criminology and Crime Analysis 1–4 July 2004, New Zealand. http://ccjr.csusb.edu/Reports/ECCA_paper_Bichler.htm.

The United States COPS (Community Oriented Policing Services) Office and the US Police Foundation have produced two excellent texts that add to the content on geocoding described in this chapter. Bichler and colleagues' paper explores the geocoding functionality in popular GIS software and offers a review that helps the user become aware of some pitfalls.

Harries, K. (1999). *Mapping Crime: Principles and Practice*. Washington, DC: United States Department of Justice. http://www.ojp.usdoj.gov/nij/maps/.

Keith Harries' book is a useful complementary text to readers of this book. The book is available for free download from the US Mapping and Analysis for Public Safety programme website. The United States MAPS programme promotes research, evaluation, development and dissemination of GIS technology for criminal justice research and practice. Their services include grant funding, annual conferences, information on training centres, publications and research.

The GIS Dictionary: http://www.agi.org.uk/resources/dicitionary/content.htm.

The United Kingdom's Association for Geographic Information provides an online dictionary resource for GIS. The dictionary covers many of the definitions and terms used in GIS.

The GIS Files: http://www.ordnancesurvey.co.uk/oswebsite/gisfiles/.
GIS.com: http://www.gis.com/.

These two links point to online resources that explain a little more about GIS principles and the range of ways it is applied.

References

Bugayevsky, L.M. and Snyder, J.P. (1995). *Map Projections: A Reference Manual*. London: Taylor & Francis.

Burrough, P.A. (1998). *Principles of Geographical Information Systems for Land Resources Assessment*. Oxford: Oxford University Press.

Carcach, C. (1997). Reporting crime to the police. Australian Institute of Criminology Trends and Issues in Crime and Criminal Justice No. 68.

Chainey, S.P. (2001). Combating crime through partnership: Examples of crime and disorder mapping solutions in London, UK. In A. Hirschfield and K. Bowers (eds) *Mapping and Analysing Crime Data*. London: Taylor & Francis.

Chainey, S.P., Reid, S. and Stuart, N. (eds) (2002). When is a hotspot a hotspot? A procedure for creating statistically robust hotspot maps of crime. In *Innovations in GIS 9*. London: Taylor & Francis.

Clarke, R.V. (1999). Hot products: Understanding, anticipating and reducing demand for stolen goods. *Police Research Series*, Paper 112. London: Home Office.

Clarke, R.V. and Eck, J. (2003). *Become a Problem Solving Crime Analyst*. London: Jill Dando Institute of Crime Science. www.jdi.ucl.ac.uk.

Department for Environment, Food and Rural Affairs (DEFRA) (2003). Neighbour Noise: Public opinion research to assess its nature, extent and significance. London: MORI. http://www.defra.gov.uk/environment/noise/mori/pdf/mori.pdf.

Dodd, T., Nicholas, S., Povey, D. and Walker, A. (2004). Crime in England and Wales 2003/2004. Home Office Statistical bulletin 10/04, Research, Development and Statistics Directorate, Home Office. http://www.homeoffice.gov.uk/rds/pdfs04/hosb1004.pdf.

Eck, J.E. and Spelman, W. (1987). *Problem Solving: Problem-Oriented Policing in Newport News*. Washington, DC: Police Executive Research Forum.

Groves, R.M. and Couper, M.P. (1998). *Nonresponse in Household Interview Surveys*. New York: John Wiley & Sons.

Laurini, R. and Thompson, D. (1992). *Fundamentals of Spatial Information Systems*. London: Academic Press.

Longley, P., Goodchild, M., Maguire, D. and Rhind, D. (2001). *Geographic Information Systems and Science*. Chichester: John Wiley & Sons.

McDonnell, R. and Kemp, K. (1995). *International GIS Dictionary*. Cambridge: GeoInformation International.

Paulsen, D. (2004). Falling on deaf eyes: Assessing the use of crime maps by patrol officers. Paper presented at the 7th International Crime Mapping Research Conference, Boston, USA. http://www.ojp.usdoj.gov/nij/maps/boston2004/papers.htm.

Ratcliffe, J.H. (2001). On the accuracy of TIGER type geocoded address data in relation to cadastral and census areal units, *International Journal of Geographical Information Science*, 15(5), 473–485.

Ratcliffe, J.H. and McCullagh, M.J. (1998). The perception of crime hotspots: A spatial study in Nottingham, UK. In N. LaVigne and J. Wartell (eds) *Crime Mapping Case Studies: Successes in the Field*. Washington, DC: PERF, Crime Mapping Research Center.

Ratcliffe, J.H. and McCullagh, M.J. (2001). Chasing ghosts? Police perception of high crime areas. *British Journal of Criminology*, 41(2), 330–341.

Rengert, G.F. and Pelfrey, W. Jr (1997). Cognitive mapping of the city center: Comparative perceptions of dangerous places. In D. Weisburd and T. McEwen (eds) *Crime Mapping and Crime Prevention* (pp. 193–218). Monsey, New York: Criminal Justice Press.

Smith, S.J. (1986). *Crime, Space and Society*. Cambridge: Cambridge University Press.

4
Spatial Theories of Crime

Learning Objectives

In this chapter, we explain how thinking about crime and space has developed into the field of environmental criminology. The chapter provides the reader with an overview of the key theoretical ideas that have shaped our understanding of criminal behaviour and identifies some of the important players responsible for these theoretical developments. The chapter concludes with a hypothetical example designed to demonstrate many of the theoretical concepts in a practical sense.

4.1 Introduction

Crime mapping is all about the geography of crime and, as Patricia and Paul Brantingham noted in 1981, there are four dimensions to every crime:

1. a legal dimension (a law must be broken);
2. a victim dimension (someone or something has to be targeted);
3. an offender dimension (someone has to do the crime); and
4. a spatial dimension (it has to happen somewhere).

They defined the spatial dimension as a place in *space* and *time* where an offence occurs, and these are components of the crime equation that can be mapped. We now have GIS to help us plot criminal activity, and techniques that you will read about in this book can map the temporal

GIS and Crime Mapping   Spencer Chainey and Jerry Ratcliffe
© 2005 John Wiley & Sons, Ltd

component as well. However, what can we possibly discern from mapping crime? If crimes were a random occurrence that had an equal chance of happening anywhere at anytime then there would be no point in mapping criminal activity. There would be no evidence that future crimes would occur in the same place as past crimes, and there would be no opportunity to predict the potential locations of offenders. However, this is not the case and the following chapter will explain why crime is not randomly distributed across space but is concentrated into hotspots of activity.

Readers with a background in geography may well feel an urge to skip a chapter that proposes to discuss criminological theory. Similarly, crime prevention practitioners may feel some agreement with Marcus Felson and Ron Clarke when they said that 'criminological theory has long seemed irrelevant to those who have to deal with offenders in the real world' (Felson and Clarke, 1998, p. 1). Indeed, many criminological theories provide little practical insight into effective crime reduction. A number of criminological theories concern themselves with what motivates individuals to commit crime. For example, Robert Merton (1938) argued, through the notion of Strain Theory, that deviant behaviour is caused by a social system that holds out the same goals to all its members, yet at the same time does not give members of that society equal means to achieve those goals. Though theories such as these are interesting, they are often difficult to convert into a practical crime reduction strategy from a policing or practitioner perspective. After all, it is difficult to tell just by looking at people walking past on the street how divorced they are from their means to achieve their life goals and who therefore might be a potential offender. However, a solid appreciation of why crimes occur where they do and the spatial behaviour of victims and offenders will help the analyst better understand crime in society; how individuals can protect themselves; how communities can prevent crime; and how the police can detect and solve crime.

The main theoretical area that underpins crime mapping is a practical subset of mainstream criminology called *environmental criminology*. This has nothing to do with oil spills in Alaska, but is the study of criminal activity and victimisation and how factors of space influence offenders and victims (Bottoms and Wiles, 2002).

This chapter begins by exploring how the importance of this spatial influence on people came to be recognised, before examining the spatial dynamics of offenders and the interaction of the offender and victim in time and space. The chapter will conclude with a look at some of the different ideas that have circulated to improve crime prevention and detection, resulting from a greater understanding of criminal behaviour across time and space.

4.2 Early environmental criminology

Researchers have long known that there is variation in the spatial arrangement of crime. Although there have been recorded spatial studies of crime for nearly 200 years, a number of key research periods have punctuated the history. While these periods have overlapped across time, for convenience they can be thought of as distinct schools of thought. Each is examined in the following sections.

4.2.1 The cartographic school

One of the earliest maps of crime originates from France and was published in 1833 when Andre-Michel Guerry published a book of maps showing, amongst other features, the distribution of violent crime and property crime in the various départements of France (Guerry, 1833). These maps indicated that not only was there spatial variation in the crime menace but that the risk of property crime and violent crime was often different in the same areas (Brantingham and Brantingham, 1981a). Guerry also noted that seasonality was a factor. Analysing French data around the same time was Adolphe Quetelet, who supplemented his maps with statistics showing spatial variations across France as well as between different social groups, including beggars, smugglers and strumpets! (Quetelet, 1842). These early crime mapping efforts were possible only due to the reformations that took place within criminal justice in the UK and France at the start of the 19th century. An emphasis on data collection from state agencies allowed researchers to explore the operation of the criminal justice system for the first time, and many current researchers still rely on administrative data from organisations within the justice arena. Early data collections were organised along spatial grounds, and still are. English counties were the spatial unit for early British researchers who used occasional judicial data dating from 1805 and annual data from 1834. French départements formed the basis of the analyses by Guerry and Quetelet. These early pioneers are credited with founding what is termed the Cartographic School.

Throughout the 19th century studies into the spatial arrangement of crime and criminals continued, one of the most renowned being an examination of the infamous London rookeries (Mayhew, 1862). These thieves' quarters were areas of high offender concentration, situated on the boundary of the City of London. In the 19th century (as today), the City of London covered a single square mile at the heart of the much larger metropolitan area of Greater London. Like today, the City of London, made up of mainly prosperous financial buildings and equally affluent residences, was policed by a separate police service to the rest of London. The Greater London area had been policed by the Metropolitan Police since its inception in 1829. The rookeries were poised to exploit any cross-border policing difficulties between the City of London Police and the Metropolitan Police. This demarcation continues to exist today as demonstrated in Chapter 2.

4.2.2 The Chicago School

In the last century, more innovation followed with the research conducted by the Chicago School. The Chicago School was both a group of researchers and a series of sociological theories that originated from the School of Sociology at the University of Chicago in the early to mid part of the last century. This group of researchers included Clifford Shaw and Henry McKay at the Institute for Juvenile Research who drew on the spatial and temporal ideas of social ecology, forged by their predecessors, notably Ernest Burgess (Burgess, 1916). Shaw and McKay mapped, by hand, the residences of juvenile delinquents across Chicago over many years (Shaw and McKay, 1942). This pioneering mapping was used to explore the socio-cultural triggers of crime as Chicago expanded during a period of great economic growth. They drew on Burgess's work by comparing socio-economic factors and physical factors in different zones across the city.

Burgess introduced the zonal (or concentric) model in 1925. It suggested that a city was encompassed within five concentric circles, each one representing a different stage of the city's development. As can be seen in Figure 4.1, Burgess's idealised model had concentric zones radiating outwards in bands from the city centre. The innermost zone (Zone I), termed 'the loop', contained the central business district and had little residential development. Adjoining this was the 'zone in transition' (Zone II), an area being invaded by business and light manufacturing industries, which also included the factory zone. The third zone ('Zone of working-men's homes') was occupied by factory workers who had managed to escape the zone in transition, but were still tied to the city due to the need to work in the factories. Travel cost and time was a factor for these workers, so they resided in Zone III. The 'residential zone' (Zone IV) comprised of high-class apartments or single family suburban dwellings where the occupants accepted the travel costs as a price for quality of life. Finally, beyond the city limits was the 'commuters zone' (Zone V) where people lived in suburban areas or even satellite towns with a commute of up to an

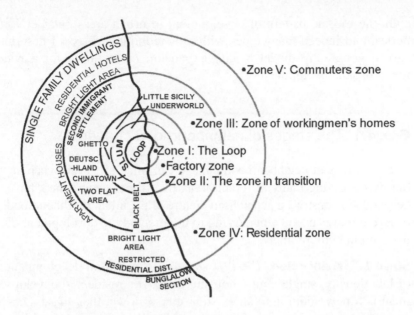

Figure 4.1 The concentric (zonal) circle model. Source: Adapted from Burgess (1925, pp. 51, 55). Reproduced by permission of the University of Chicago Press

hour (Burgess, 1925). To demonstrate the model, Burgess charted 1920's Chicago and overlaid his model onto an expanding and vibrant cityscape (Burgess, 1925, p. 55).

From a criminological perspective, the zone in transition was the area of most interest. Here mobility was greatest, the availability of stimulus peaked and there was a concentration of "juvenile delinquency, boys' gangs, crime, poverty, wife desertion, divorce, abandoned infants, [and] vice" (Burgess, 1925, p. 59). Shaw and McKay (1942) went on to map the concentric zones using different bandwidths for different cities (e.g. in Chicago the bandwidths were two miles, while in Philadelphia they were one-and-a-half miles). Their work spanned decades and formed the basis for much of American criminological thought coming into the latter half of the 20th century, especially in establishing the longevity of delinquency areas discovered through longitudinal studies (Brantingham and Brantingham, 1981b).

The Burgess model worked well for American cities in the 20th century, but was less applicable outside North America. The development of American cities took place over a fairly short temporal period, whereas European urban development occurred over a considerably longer time. Urban geographers have developed models which may help to better explain the mosaic pattern of development in urban areas outside North America, and indeed many cities within the continent. Box 4.1 describes an urban geography model of city expansion more fitting for a wider range of urban developments.

Box 4.1 The stages of neighbourhood change

The processes of neighbourhood change can often seem idiosyncratic, but by observing many different areas over time, geographers have been able to construct a generalised picture of the urban neighbourhood lifecycle that is useful when trying to understand the development of the criminal environment.

Stage 1: Urbanization. The first stage sees the initial development of low density, single family housing units. If a residential development is a new addition to an existing city or town, then these areas are often on the outskirts of the town. Usually less expensive, they attract young families of middle, or better, social status. These families can afford transport to the city or town, but chose to live in the new development for the housing.

Stage 2: In-filling. Vacant land blocks are populated by multi-family units and rental properties, drawn by the land prices and the attractiveness of the neighbourhood. The new developments increase the population density of the area, and decrease the homogeneity of the population demographics.

Stage 3: Downgrading. As time progresses, the housing stock ages and there is a slow decline of the area caused by degradation of the properties and general deterioration. This is emphasised by a lack of maintenance on the rental properties that appeared in stage 2. This is usually the longest stage of the lifecycle.

Stage 4: Thinning out. The population turnover that increased in stage 3 has now reached a peak, with rapid turnover and significant social and demographic change. This stage also sees the conversion or abandonment of some of the residential units.

Stage 5: Rehabilitation or Gentrification. Although not a guaranteed last stage, gentrification can breathe new life into an area with new swathes of housing which extends the lifecycle of the neighbourhood.

Continued on page 85

Continued from page 84

This stage sees conversion of larger properties into a more high-density format. There is also reinvestment in this stage, reflecting the relatively central location of the neighbourhood (assuming that development has continued further outwards). This urban renewal stage can bring in a more homogenous population, drawn to a central location.

Source: Knox (1994).

From a crime perspective, the initial development in stage 1 brings in many young families. While peace may reign for a few years, a suburb populated by a homogenous group, all arriving at the same time, will see many of the children reach their teenage (i.e. highest crime-risk) years at the same time. 10 to 15 years after stage 1 (often during stages 2 or 3) it is possible that crime will therefore increase. Stage 2 also has the capacity to increase crime by reducing the social cohesion of an area, due to the introduction of different socio-economic groups. This process continues to increase in stages 3 and 4 due to social and structural neglect accelerating opportunities for property crime. Stage 5 may see a reduction in crime, caused by the reintroduction of a more affluent population that can afford upgraded security features on cars and homes.

4.2.3 The GIS school

The last couple of decades have seen an explosion of interest in environmental criminology, spatial crime analysis and the investigation of offender patterns using geographic tools. The catalyst for this rebirth in enthusiasm has been attributed to the development of two complimentary ideas: Crime Prevention Through Environmental Design (CPTED) and Defensible Space (Brantingham and Brantingham, 1981a: p. 18). CPTED, in particular, has grown into a significant discipline that addresses space management and architectural design, and urban planning (see Crowe, 2000, for an overview).

Within the framework of CPTED and Defensible Space a number of important advances have been made, most noticeably with:

- the theoretical developments from Marcus Felson and his colleagues (Cohen and Felson, 1979);
- the work of George Rengert in his examination of residential burglary behaviour (Rengert and Wasilchick, 1985) and the spatial arrangement of drug markets (Rengert, 1996);

- the numerous advances in theory made by Patricia and Paul Brantingham at Simon Fraser University in Vancouver (Brantingham and Brantingham, 1981c, 1984);
- the spatio-temporal analyses of rape patterns by Jim LeBeau (LeBeau, 1987, 1992); and
- the examination of crime across different spatial scales by Keith Harries (Harries, 1980).

These significant stages help to explain the growth of certain ideas, but do not explain the status of crime mapping in both academia and the field of the crime reduction practitioner. Nor does it explain the third age of environmental criminology. For that, we have to thank the development of affordable GIS and the increasing technological developments within policing. The new environmental criminologists (such as those referred to in the above list) have been able to exploit the increasing digitalisation of law enforcement, and map crime and calls for service with GIS. They have been able to explore the spatial dimensions of criminality in new and exciting ways. This next section will examine the theoretical developments of the new environmental criminologists, a group we have called the GIS School, and see how they have shaped our current ideas regarding the spatial dimension of offending.

4.3 The space and time of offences

Any police officer knows that the distribution of crime is not uniform across either time or space. There are hotspots of activity. For example, if you wanted to pick a hotspot for a drunken brawl then some time around midnight in an area with a high density of bars would be a pretty good bet. While it may be simple to guess the likelihood of some offence patterns, explaining why they exist is a different matter. The drunken brawl example might be easy, but how do we explain why some areas are plagued by burglary while others see little burglary but are awash with vehicle theft? Certain theories can help us understand these patterns. These theories do not just consider the risk of crime, but also cover offender behaviour. After all, if the crime patterns are not random then the offenders cannot be acting in a random way. It is of course true that some criminals will break the mould and do things differently, but the majority tend to act in a predetermined manner. These tendencies to react in a similar way to the same opportunities across space are termed *aggregate criminal spatial behaviour* (Brantingham and Brantingham, 1984, p. 355).

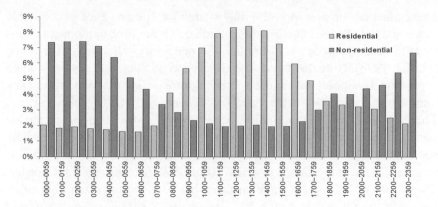

Figure 4.2 Hourly probability of burglary in the Australian Capital Territory, 1999–2000. Source: Ratcliffe (2001). Reproduced by permission of the Australian Institute of Criminology

Some offences have clear patterns. For example, while Hollywood would have us believe that residential burglary is committed by offenders creeping round the house in the middle of the night, residential burglary actually tends to occur in the middle of the day. This is demonstrated in Figure 4.2. The patterns of residential burglary over a typical day in the Australian Capital Territory peak in the early afternoon to mid-afternoon hours. This contrasts to non-residential burglary which Figure 4.2 shows is the reverse time pattern. (Chapter 8 explains how to calculate these values.) These patterns are confirmed for other countries (for example Sorensen, 2004, and the interviews with offenders conducted by Rengert and Wasilchick, 1985 and Cromwell *et al.*, 1999). Why might these patterns occur? The following sections examine theoretical explanations for offence patterns and the offender and victim behaviour that cause these types of patterns.

4.3.1 Routine activities

Routine activity theory originally started as a macro-level explanation of predatory crime (Cohen and Felson, 1979), but has progressed over the years to provide a worthwhile mechanism to examine criminal opportunity and crime prevention in a variety of settings. The original work examined changing patterns of employment and the new criminal opportunities that are created when there are fewer people staying at home during the day. The simple idea is that behaviour of victims explains the occurrence of crime, and that for a crime to occur, three components are necessary.

There must be the presence of a likely offender, the presence of a suitable target and the absence of a capable guardian. These three components must meet in time and space to formulate the necessary 'chemistry' for crime (Felson, 1998). Note that there is no discussion of a target necessarily being a person. The practical nature of this approach is that buildings, cars, mailboxes, people or a wide variety of objects and things can be targets. There is also no clear definition of what is construed to be a capable guardian. This could mean a person, as in the usual meaning of guardian, such as a police officer, security guard or even shopkeeper, or could also include CCTV surveillance systems. Routine activity theory does not just discuss offenders, targets and guardians, but adds the important, but often forgotten, qualifiers. Not all offenders are *likely* offenders, as some will lack the technical knowledge and skill to attack certain types of premises. In a similar vein, not all targets are *suitable* targets, as they may be inaccessible (such as rooftop apartments) or too well-defended. As said, many objects and people can be guardians; however, at different times they may not be *capable* guardians. Researchers in Scotland noted that CCTV cameras worked to prevent criminality most of the time, unless the offenders were under the influence of alcohol when they did not care about the cameras (Short and Ditton, 1998). The thinking behind routine activity theory is that the risk of crime changes over time with the movement of people throughout the daily routine activities of their lives. We will discuss this in more detail in a later section of this chapter. For the moment, routine activity theory can be summarised by the following simple equation:

Likely offender + suitable target − capable guardian = crime opportunity

Most researchers take the existence of likely offenders as a constant in our society and attempt to explain and prevent crime by examining the two remaining components of the equation. Crime prevention practitioners may therefore examine the possibilities for introducing suitable guardianship into an area as a way to prevent crime or to make existing guardians more suitable.

4.3.1.1 Enhancements to routine activity theory

Most of the early work in this area concentrated on the nature of targets and guardians; however some developmental ideas regarding offenders have taken place. Whether an offender will be a likely offender can depend on the presence or influence of 'handlers' (Felson, 1995). A handler is a person, a third party, who can influence the behaviour of the offender.

A parent may therefore be a handler, as could a teacher or any other person who knows or who could determine the name of the person. This makes sense. Would a person commit a crime in the presence of someone who might know them and who could tell the police? A handler may also be a person whose respect the offender might not wish to lose. While the handler may not tell the police, they may not approve of criminal behaviour. In this way, a school friend or work colleague can become a handler for a potential offender.

While a handler has some influence over an offender, a guardian has some influence over the likelihood of crime occurring. Remember that guardians can be formal, such as police officers, or informal, such as the presence of a friend as company on the walk home at night. These two types of controllers provide control: handlers control offenders and guardians protect targets. John Eck has since introduced a third type of controller – the *place manager* (Eck, 1995). A place manager is someone who is able to control a place even if they are not formally in charge of the area. Certain individuals have the capacity to discourage crime in particular areas. These people include landlords, street stall owners, store owners and ticket clerks. With place managers we now have three types of controllers, and to summarise in the words of Marcus Felson, 'Crime opportunity is the least when targets are directly supervised by guardians; offenders, by handlers; and places, by managers' (1995, p. 55). This can be depicted visually with the *crime triangle* (Figure 4.3). The crime triangle helps to focus analysis and problem-solving towards the causes of crime, from a routine activity perspective, and the mechanisms that can influence those causes and so prevent crime.

Figure 4.3 The crime triangle. Source: Adapted from Clarke and Eck (2003). Reproduced by permission of Elsevier

4.3.2 The spatial arrangement of attractive targets

Target suitability can change over space and time. While routine activity theory provides a general model with which to consider the likelihood of crime occurrence, even available targets can differ in their attractiveness to a criminal. During his time studying at university one of the authors was the proud owner of a 15-year-old, clapped-out Citroen 2CV car. It was usually left unlocked, often with the window open, in one of Nottingham's highest vehicle crime areas, yet it remained untouched. Apparently Nottingham car thieves have no appreciation for classic French engineering, and it shows the reality that some commodities that are available to steal are not attractive to offenders. Ron Clarke examined records of stolen property and summarised the basic characteristics of a suitable, or *hot*, target. Working from Cohen and Felson's VIVA (Value, Inertia, Visibility and Accessibility) acronym (Cohen and Felson, 1979), he expanded the idea of hot products to cover most property offences (Clarke, 1999), suggesting that a hot product is 'CRAVED' in that it is:

- *Concealable* – things that are difficult to conceal are harder to steal. Offenders may be stopped by the police if seen carrying goods down the street, and shopkeepers will notice if large items are being carried out of the store.
- *Removable* – the easier a thing is to remove, the greater the chance that it will be a hot product. Given that theft is the appropriation of property belonging to someone else, the necessity to take it to another place is fairly important.
- *Available* – burglars do not spend much time in a house as this is when they are most at risk of capture, so objects that are visible and not secured are most at risk.
- *Valuable* – this appears to go without saying; however, valuable is a relative term. Young offenders will target goods that they define as valuable, and this can include clothing and sports footwear. These goods are not necessarily for resale, so the value of items is dependent on whether the offenders will use the items themselves or sell them on.
- *Enjoyable* – it may seem strange, but burglars tend to steal televisions, videos and CD players rather than kitchen items, even though these have a similar value (Clarke, 1999, p. 24). The degree to which a product can be enjoyed marks its potential theft risk – an indication of both its value to the offender and possibly its resale value on the stolen goods market.

- *Disposable* – many items are stolen so that they can be sold or traded to others, therefore disposability is an important characteristic for stolen goods. Some, but only some, stolen goods are traded for drugs, and these items are usually sold on again, creating a lucrative stolen goods market.

So how can we explain crime patterns with these ideas of hot products and routine activity theory? Ratcliffe (2002) examined crime patterns for vehicle crime in the eastern suburbs of Sydney, Australia, and found that vehicle crime in the affluent suburbs near the beaches was focused during the overnight periods. The residents in this area are generally wealthy with expensive cars, and expensive cars are most definitely 'craved', having a high value for both joyriding (enjoyable) and resale or 'ringing'. Ringing is the process of creating a new apparently legitimate vehicle by altering the documentation of a stolen car (also termed 'rebirthing' in Australia). Although the Eastern suburbs are an affluent part of the city, the density of housing means that few residents have space for private garages. Vehicle crime is therefore concentrated into the overnight period because the items are CRAVED, they are suitable targets at night (the cars are elsewhere during the day) and the night time hours provide for few capable guardians.

Attractive products are not distributed evenly throughout urban space. Given that some offenders are known to work with 'fences' who can provide a ready market for stolen goods, some items are stolen to order (Rengert and Wasilchick, 2000). Professional car thieves will tend to target affluent suburbs in search of specialised vehicles, while a joyrider or car thief in the early stages of his career may search a less-wealthy suburb in the hope of finding an older car that is easier to steal and less conspicuous. Beyond car theft, most white-collar crime is concentrated in the central business district as this is where the major financial institutions can be found.

4.3.3 Rational choice

All of this might appear to suggest that offenders will take any opportunity to steal anything and everything that is not chained down! Well, this may be the case for a few offenders, but most make some sort of a decision to commit a crime by weighing up some of the pros and cons. What are the rewards, against the chance of being caught? This suggests that to commit a crime is a (fairly) rational decision, and that an offender will commit an offence while trying to achieve some sort of desire or goal. The goal may be to derive personal gain, as in burglary or theft, or personal pleasure, as in the crime of joyriding. If the legitimate means of obtaining that goal are not available, then a decision may be made when a criminal opportunity becomes available. As Rengert and Wasilchick noted after their interviews with 31 burglars, 'the decision to commit burglaries was a purposeful, rational decision in almost every case' (2000, p. 60). The decision may not be one that is fully calculated, as offenders may not weigh up all of the consequences. Interviews with younger offenders generally find that they do not consider the implications of capture (Fleming, 1999; Shover, 1999), though some criminals will consider the risks in their choice of criminal behaviour. For example, vehicle crime is often seen as attracting lower penalties from the criminal justice system than burglary.

It is argued that criminal decision-making is in two parts. There is a long term, multistage decision to become generally involved in criminal activity (criminal involvement decision) and a shorter-term, more immediate decision (the criminal event decision) to grasp an opportunity that is presented (Cornish and Clarke, 1986). Of course, factors such as drink, drugs, peer-pressure or limited education do mean that not all offenders' decisions are purely rational, resulting in what has been termed 'limited rationality' also known as 'bounded rationality' (Newman, 1997). These terms are an acceptance from researchers that some offences are committed with less-than-military planning behind them. Although the effects of drugs and alcohol can limit the rationality of offenders, the immediate decision-making of a burglar (for example) is primarily based on the environmental cues from the prospective target, cues that can change from place to place: can the offender be seen breaking in, is there anyone home and is there an easy way into the house? (Cromwell *et al.*, 1999). This characteristic, termed *rational choice perspective* or *rational choice theory* (Clarke and Felson, 1993), provides a framework to consider offender decision-making when a crime opportunity is presented. It can also be used to consider likely strategies that will influence the decision-making of the offender.

4.3.3.1 Situational crime prevention

Crime prevention is a huge field and, in modern society, a huge business. In Chapter 2 we offered a definition of crime prevention but also described variants of the theme. One of these is situational crime prevention, possibly the most spatial crime prevention concept and one that is a natural geographic corollary of rational choice theory. Situational crime prevention does not worry about why people commit crime, but concerns itself with preventing the opportunities for crime.

It suggests a number of opportunity-reducing tactics that are crime-specific, that involve manipulating the immediate environment in a systematic and permanent way, and that are intended to increase the effort and risk while reducing the rewards that are perceived by a range of offenders (Clarke, 1992).

Geographical Information Systems lends itself to situational crime prevention studies due to the place-specific nature of the technique and the ability of GIS to deliver a spatial analysis. For example, Holzman and colleagues examined assault rates against women in two public housing areas in the US. The study used GIS to map assaults within the housing estates, determining that the architectural design afforded some offenders more privacy and accessibility than in other places and that this influenced the rate of assault against women (Holzman *et al.*, 2001). George Rengert and his co-workers went a stage further and designed a high definition GIS to map crime within the confines of a university campus (2001). A student survey of unreported crime was combined with crime recorded by the campus police to better understand the real pattern of crime across the university. They even went as far as to propose a method for mapping crime within buildings.

In providing a way to influence someone *not* to commit a crime, three main strategies are suggested by situational crime prevention. First, it may be possible to *increase the risks*. By making the chance of capture much higher, offenders may make a rational choice to seek a less well-defended target. Tactics such as videorecording customers at bank cash machines are an attempt to increase the risk of capture of cash card robbers. The cameras work by either recording the picture of the offender as they steal the card or cash from a customer at the machine, or by recording the offender when they try and withdraw cash with a stolen card.

A second method of influencing the decision-making of offenders is to *increase the effort* required to commit a crime. A number of new cars have multiple levels of security including alarms, engine immobilisers and tracking systems that tell police where to find the car (sometimes with the thieves inside). These situational crime prevention tactics (called situational because they often only work for that particular car or building) all work to increase the effort required to steal the car.

Finally, it is possible to *reduce the rewards* of crime. Again, situational crime prevention tactics such as tagging expensive clothes with tags that stain the fabric with a permanent dye if removed from the store, or by fitting mobile phones that require a secret personal number before they will operate, are all ways to reduce the rewards of crime.

Clarke initially proposed 12 basic techniques to situational crime prevention, with four techniques for each of the three core areas of

Table 4.1 Clarke's techniques of situational crime prevention. Source: Adapted from Clarke (1997, p. 18) and www.popcenter.org. Reproduced by permission of Criminal Justice Press

Increasing the perceived effort	Increasing the perceived risks	Reducing the anticipated rewards	Remove excuses for crime
Target hardening	**Screen entrances and exits**	**Remove targets**	**Set rules**
Steering column locks	Electronic merchandise tags	Removable car radio	Harassment codes
Anti-robbery screens	Baggage screening	Phone cards	Customs declaration
		Women's refuges	Hotel registration
Control access to targets	**Formal surveillance**	**Identify property**	**Alert conscience**
Entry phones	Red light cameras	Vehicle licensing	Roadside
Parking lot barriers	Burglar alarms	Property marking	speedometers
	Security guards	Cattle branding	Road safety campaigns
Deflecting offenders from targets	**Surveillance by employees**	**Reduce temptation**	**Control disinhibitors**
Bus stop location	Park attendants	Rapid graffiti cleaning	Drink age laws
Street closures	CCTV systems	Off-street parking	Car ignition breathalyser
Control crime facilitators	**Natural surveillance**	**Deny benefits**	**Assist compliance**
Plastic beer glasses	Street lighting	Merchandise tags	Litter bins
Photos on credit cards	Defensible space architecture	Graffiti cleaning	Public lavatories
		PIN for car radios	

increasing the effort, increasing the risks and reducing the rewards. These were later increased to 16 techniques, by including the concept that it is possible to *remove the excuses* for crime. Table 4.1 shows Clarke's techniques with summaries of his suggested tactics. As can be seen from the table, situational crime prevention can be applied across a range of scales from national border protection right down to library tags to prevent book theft. This area of crime prevention is continually developing, and a good way to keep up with the latest ideas is at the website www.popcenter.org.

4.3.4 Territoriality and defensible space

Not all crime control has to be toughened glass, barriers, CCTV and police officers. Crime can also be controlled more informally. Jane Jacobs recognised that increases in the number of people on the street on busy roads increases the number of potential witnesses to a crime and the number of bystanders who might intervene. The shear volume of people acts as a crime inhibitor (Jacobs, 1965). Might it be possible to use this informal social control in a broader manner? To do so, the informal control must work by tapping into a spatial component of human psychology, that of *territoriality*. In residential areas, people seek some degree of privacy from the communal and public areas of our world. This private space is the last refuge, and an area that we seek to protect. 'Territorial functioning' is the mechanism by which we aim to exclude unwanted persons, through the use of boundaries, fences and other signals that indicate to others that a certain area is private and not for everyone to wander through (Taylor, 1988).

Taylor has described human territoriality functioning as a system of attitudes, sentiments and behaviours that are specific to a clearly marked place. This system signifies that a group or individual has some expectation of excludability of use. In turn it also indicates a responsibility and control of activities in the specific location. From a spatial standpoint, Taylor concentrates on the face-to-face perspective and at the block level. And while there would appear to be limits on the extent of territorial functioning up to the suburb level (Ratcliffe, 2003), those who have a stake in a certain area, such as a small park or communal area, will police the area and care for it, looking out for troublemakers.

According to the main originator of defensible space, Oscar Newman, the catalyst that starts this active surveillance of certain spaces is a distinct demarcation between those areas that are deemed to be public and the areas that a group are prepared to defend (Newman, 1972). The private areas that are considered part of the territory of an individual or group become the 'defensible space' and are actively policed. Other areas, such as those that are clearly public or are 'confused space', areas where ownership is not clear to onlookers, tend to be ignored by the residents or occupiers (Coleman, 1985). While Jane Jacobs examined the implications for urban planning, Newman viewed the problem through an architect's eyes, looking more specifically at building design. In the UK, these ideas were further explored by Alice Coleman (Coleman, 1985), who studied different crime types across various housing estates. However, it should be noted that the ideas of Newman, in particular, have been quite strongly criticised for producing evidence that is methodologically flawed (Bottoms, 1974) or theoretically unsound (Hillier, 1988). Indeed, while Newman recanted a number of his early ideas from the 1970s (Brantingham and Brantingham, 1993), there is no doubt that defensible space has certainly sparked debate regarding territoriality and how we view the space around us.

4.3.4.1 Crime prevention through environmental design

The work of Jacobs and Newman brought the spatial focus closer to home, to examine the immediate environment around the house. Around the home, after all, is where territoriality should be working at its maximum rate. But what about the workplace, or at school? C. Ray Jeffery expanded the spatial defensibility notion to promote ways of preventing crime in a wider range of areas. Crime Prevention Through Environmental Design (CPTED) is both the name of a book by Jeffery published in 1977 and a crime prevention philosophy increasingly adopted by researchers, practitioners and urban planners to consider the best designs of public and commercial areas to reduce criminal opportunity. CPTED (pronounced *sep-ted*) appeals to many working in the crime reduction area as it blends architecture, environmental criminology theory and urban planning in an attempt to reduce offending in public or 'confused' space.

4.4 Offender–offence interaction

Armed with an understanding of both offence patterns and criminal behaviour, it is now possible to start to build a model for the interaction between offenders and victims. This interaction must occur at some point in space and time, but can we explain why the victims were unlucky enough to meet the criminals then and there?

4.4.1 Crime pattern theory

While routine activity theory gives us a model to predict if a crime has all of the chemistry to occur, and rational choice perspective enables us to determine some of the thinking behind an offender's ultimate decision to commit a crime, where will the offence happen? We are now exploring the theoretical area that will interest many GIS researchers, because the interaction between the offender and the target has an inherently spatial dimension. The crime must happen somewhere, and that somewhere can be mapped and analysed.

A helpful convergence of routine activity theory and rational choice theory can be found in the area of *crime pattern theory*. This helps to bring the two areas of offender spatial distribution and offence spatial distribution together by examining the 'relationship of the offence to the offender's habitual use of space' (Bottoms and Wiles, 2002, p. 638). Crime pattern theory (sometimes also referred to as offender search theory) suggests that offenders are influenced by the daily activities and routines of their lives, so that even if they are searching for a criminal opportunity, they will tend to steer towards areas that are known to them (Brantingham and Brantingham, 1984). In their day-to-day activities they will be watching for targets that have no guardians or place managers.

4.4.1.1 Awareness spaces

Like offenders, all of us have various routine activities in our lives. Most of us have to go to work, college or school, and we usually go there from home. We may also go to shops and restaurants, bars and movie theatres. These repetitive journeys create within us a 'cognitive map' (Brantingham and Brantingham, 1984, p. 358) of places, routes and associations, and these cognitive maps become a general list of well-known areas, areas in which we feel comfortable. This environment consists of not just the physical things, such as buildings and subway stations, but also the social and economic infrastructure which we pass through. Cities become an urban mosaic to us, places where we have no knowledge, interspersed with well-known places. We also become familiar with the routes between these known areas. These islands of knowledge, and the routes that link them, become our 'awareness space' (Brantingham and Brantingham, 1981c, p. 35; Rengert and Wasilchick, 2000, p. 61).

Like us, offenders also have awareness spaces. They also move between places such as work, school, shops and home, and for some offenders, the search for criminal opportunities takes place around these areas. Opportunities are not spaced evenly throughout a city, and some offenders will only be able to take advantage of some offence opportunities. Also, some of the awareness areas will not be conducive to crime due to the presence of guardians or place managers. Therefore for each offender we can generate a model of awareness space and criminal opportunity space (with the implicit absence of guardianship), and where they intersect we will find the areas of crime occurrence. Crime pattern theory is therefore strongly connected with the interactions of criminals and their physical and social environments. This hypothesis was modelled by the Brantinghams (Brantingham and Brantingham, 1984, p. 362), an adaptation of which can be seen in Figure 4.4. Here we can see that awareness space consists mainly of the places that are routinely frequented, as well as the routes between those places. We may be more familiar with distant places that we frequent regularly than local places just around the corner that we never visit. Proximity does not always mean the same thing as familiarity.

The adaptation to the original Brantinghams' model is in the inclusion of the location of friends as an influential component on offender

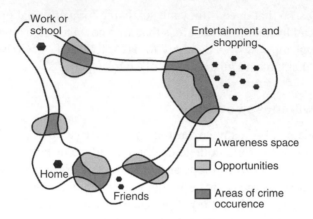

Figure 4.4 Hypothetical model of the creation of criminal occurrence space where offender awareness space and opportunities coincide. Source: Adapted from Brantingham and Brantingham (1993, p. 10)[1]

awareness space. Although for most people, work or school plays a significant part in daily life; this can be less so for offenders. Costello and Wiles (2001) found that many offenders who could have been in the workforce had never had full-time employment, and Rengert and Wasilchick reported interviewing a number of offenders who had quit legitimate employment in order to pursue a professional criminal career, including one individual who made more money as a burglar than as a full-time computer programmer (2000, p. 46). These examples focus on residential burglars only; however, it is likely that the attractiveness in both lifestyle and income of a life of crime is probably more attractive than poorly paid legitimate employment for many burglars, drug dealers and robbers. Having a place of work is not a necessary requisite to forming an awareness space. The Sheffield study found that as many offenders had never been employed, they were often transient and heavily influenced by the location of friends and criminal peers (Costello and Wiles, 2001).

While we can see in Figure 4.4 that areas of criminal occurrence happen when awareness space and opportunities intersect, that is not the whole story. In the diagram we can see that some opportunities are outside the awareness space of the offender and are essentially unavailable, while some within the awareness space have variable attractiveness. For example,

---

[1] Reprinted and adapted from the *Journal of Environmental Psychology*, 13(1), Brantingham, P.L. and Brantingham, P.J. (1993). Nodes, paths and edges: Considerations on the complexity of crime and the physical environment, pp. 3–28, with permission from Elsevier.

for a residential burglar, all homes within their awareness space are theoretically potential targets, but some are more attractive than others. The areas with a greater proportion of targets that are potentially safer to burgle and more profitable will form part of the offender's 'search space' – shown in Figure 4.4 as the opportunities area. Within this search space, certain targets will be more attractive than others, and the micro-search for a particular target occurs in the actual area targeted, termed the 'criminal activity space' (Rengert and Wasilchick, 2000).

4.4.1.2 Templates and cognitive behaviour

So why do some areas remain outside an offender's cognitive space, and why do they commit offences in familiar areas? Offenders' routine activities can be quite limited. In a study in Sheffield (UK), it was found that many offenders were unemployed, and indeed had never worked. This meant that they were often short of money. While this lack of employment and cash was probably a motivating factor in committing crime, it also meant that their routine activities were severely curtailed from a spatial point of view. They had no reason or resources to venture into unfamiliar areas, and their cognitive map was quite small as it did not include a workplace or many recreational opportunities (Wiles and Costello, 2000).

There are several other reasons why offenders might commit offences in familiar areas. It is helpful to know the layout of an area so that if you need to make a quick getaway, you will not run straight up a dead-end street. Secondly, it has been suggested that offenders value feeling 'comfortable' in an area and not feeling as if they stand out. This has been suggested as a reason why offenders who live in poor neighbourhoods do not often commit offences in affluent areas (Rengert, 1989). In a similar vein to the 'poor–rich' distinction is the 'white–black' neighbourhood distinctions found in the burglary patterns of offenders in the US. A number of studies have noted that black offenders avoid white suburbs and white offenders steer clear of black neighbourhoods, each group feeling that the avoided areas were unsafe (Rengert and Wasilchick, 2000; Wright and Decker, 1994). It is also interesting that the desire for spatial exploration to extend criminal opportunities is rare. Offenders are often constrained by time and financial resources and lack the freedom to explore other opportunities or to search further afield for fresh prospects.

While our daily patterns take us to school, work, shops, cinemas and home, we may also go to the coast for our summer holidays. If we do, then perhaps offenders also do. Yet if crime pattern theory suggests that offenders go to places that are known to them, why don't offenders regularly travel from the city to the coast to commit offences at the places where they take their summer holidays? At this point the *least effort principle* becomes a factor. Say, for example, that we need a carton of milk. There are lots of shops in the city that sell milk, but like most people we immediately consider the shops closest to home. To travel across the city and back just to get something that is available half a mile away is pointless. Buying a carton of milk from the nearest shop around the corner requires the least effort. The least effort principle comes into play for offender behaviour. There is no point in travelling 50 miles to steal a television when there is reasonable certainty that a similar television could be found in a house a few streets away. Physical space can be likened to a friction surface in that it requires effort to cross it. The further the distance to travel, the greater the cost in time and possibly money.

Travelling a greater distance to commit crime also incurs an additional possible penalty to buying milk. A person will not be arrested for walking down the street carrying a carton of milk, but a stolen television is a different matter. Increased distance to commit crime increases the effort, increases the risk and increases the possibility that the offender will stray into an unknown area. The least effort principle is a useful mechanism for thinking about offender behaviour, and is a fundamental concept behind *journey to crime* studies and *geographic profiling*, more of which will be discussed later.

It should be noted that crime pattern theory is a general theory to explain the crime patterns of offenders, and there will always be exceptions. Studies have found that a few offenders are influenced by others and do not commit crime within their own awareness space, either as a result of influence by people such as 'criminal fences' for stolen goods who direct the offender to new opportunities (Rengert and Wasilchick, 2000) or by peers who introduce the offender to new areas (Wiles and Costello, 2000). When this happens, the spatial pattern of offences can be very different. However, for the majority of offenders, the tendency is to offend in awareness spaces close to home. If all criminal patterns were purely random and individualistic, then we would not be able to draw conclusions about the behaviour of the majority of offenders. However, we know that broad behavioural tendencies do exist, and can be explained as aggregate criminal spatial behaviour.

4.4.1.3 Establishing a template

Environmental criminologists generally start from the assumption that offenders are motivated to commit crime. Their interest is in the choice of where and when. If offender behaviour was totally random, there would be little point in such work. However, as this chapter has shown, there is a considerable degree of predictability in criminal behaviour in space.

A degree of predictability also extends to the individual target selection. As the offender, either by prior design or opportunistically, comes upon a potential target, a decision is made to offend or not. This decision is based in part on cues that are perceived from the surrounding environment – cues that include physical, cultural and psychological characteristics (Brantingham and Brantingham, 1981c). Prior experience of good and bad situations will influence the offender's interpretation of these cues, and the offender will compare the current situation with previous situations that were amenable to crime. These prior experiences form a *template* against which the existing circumstances are compared.

As the offender commits the offence, he or she will often undertake the crime using the same or similar methodology to a previously successful criminal venture. After all, there is no point taking a risk with a new technique if an offender has an existing one that works. The process of actually committing the offence runs according to a 'cognitive script' (Rengert and Wasilchick, 2000, p. 87) that is based on previous positive criminal experiences. Both the establishment of a template and the creation of a cognitive script will tend to be fairly fixed once the offender has established a working methodology. It is the tendency to stick with what works that makes the study of offender *modus operandi* worthwhile.

4.4.2 From centrography to the journey to crime and geographic profiling

Some of the criticisms of the Chicago School over the years have included the blurring of criminal offender location and criminal offence location. This assumption, that crime and criminal are co-located in areas, confused the situation when data were aggregated and led to a number of subsequent studies that fell foul of the ecological fallacy – the application of research findings from one level of aggregation to another. However, some geographers were beginning to explore criminal record data with an explicit understanding that offender residence and offence location are different places.

LeBeau (1987) tracked the behaviour of rape offence locations in San Diego between 1971 and 1975, examining the spatial change in pattern with *centrography*. This technique plots the change in the location of the mean centre of all points (see Chapter 5) as well as examines the shape of the standard distance and standard deviational ellipse created when analysing the crime events (also explained in Chapter 5) to discern behaviour patterns. While Porteous (1973) had previously used centrographic techniques to determine gang territory in British Columbia, LeBeau enhanced the technique by tracking changes to the mean centre over time, concluding that spatial offence patterns change over time in different manners for rape types. He then went on to complete some of the earliest work using GIS to model individual offence patterns in relation to both time and offender residence (LeBeau, 1992). LeBeau concluded by raising the predictive benefits of spatio-temporal rape patterns, a theme worked on by Canter and Larkin.

Canter and Larkin's (1993) British study of 251 sexual offences committed by 45 offenders in Britain proposed two models for the spatial behaviour of serial rapists. The commuter hypothesis proposed that one group of offenders, termed *commuters*, roamed out of their home range (the area around their residence) to commit offences in another area, the criminal range. In contrast, the *marauder* hypothesis speculated that a marauder would use the home as the base for offences, and would operate predominantly in the home range area. The commuter hypothesis therefore proposed that there would be minimal overlap between the home range and the criminal range, while marauders would demonstrate considerable convergence in their use of space. These hypothetical behaviours are shown in Figure 4.5.

Canter and Larkin went on to propose a circle theory to examine the distribution of their offenders. By drawing a circle using the two most distant crime events in an offender's crime series as the ends of the circle's diameter, they found that 87% of their offenders had a residence

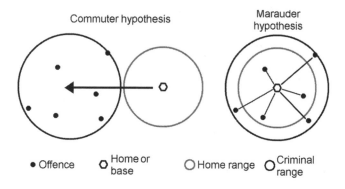

Figure 4.5 The commuter and marauder hypotheses. Source: Adapted from Canter and Larkin (1993). Reproduced by permission of Elsevier

within the circle, supporting the marauder hypothesis for serial rapists in their sample (Canter and Larkin, 1993, p. 67). Although the circle hypothesis has been found to have limited application in crime detection, predominantly because the circle sizes tend to be extremely large and create a huge area for police to cover (Kocsis *et al.*, 2002), the work helped to focus attention onto the predictive power of spatial patterns. At the same time, developmental work on both sides of the Atlantic were about to find an innovative application of crime pattern theory.

Kim Rossmo was a PhD student of Patricia and Paul Brantingham at the same time that he was also a police officer with the Vancouver Police Department. Around the same time, Professor David Canter was working on serial crime patterns in the UK. The centrography of Jim LeBeau had recognised the offender detection potential for a greater understanding of the spatial behaviour of criminals, and both Rossmo and Canter took this a stage further with a piece of reverse engineering. If an understanding of the home base and activity nodes of an offender could give some clues as to the likely offence locations, could it be possible to use the offence locations of an unknown offender to predict where the criminal might live?

To create a working methodology for this type of offender detection system, it was necessary to understand the importance of the *journey to crime*. There is often a theme to police drama television programmes. Over the course of an hour, the detective arrives in a quaint small town to investigate a grisly murder. The horrified locals are all sure the murderer is an outsider, but as the investigation progresses the focus turns inwards until in the last minutes of the programme the detective tracks down and exposes a local person who was the offender. These programmes play on a common theme – that our community is safe and anything nasty is the work of outsiders. However, as the detective often discovers (against the wishes of the townsfolk), the offender is usually local and has not travelled far to commit their crime. Journey-to-crime studies have repeatedly shown that, due to the least effort principle, offender–crime site distances are usually short.

Ratcliffe (2001) found that the average journey from an offender's home to a burglary target was about five kilometres (about three miles) for residential and non-residential burglars. This finding matches broad findings from the United Kingdom and the United States (Rossmo, 1995; Wiles and Costello, 2000). These figures are skewed by a small number of offenders who travel longer distances. In the Australian capital, Canberra, it was determined that one-third of burglaries are committed by offenders who have travelled less than a mile from their home address (Ratcliffe, 2001). Patterns

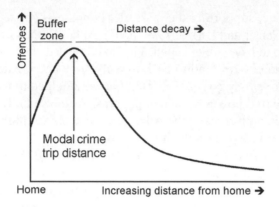

Figure 4.6 Distance decay function with buffer

for individual offenders show that as distance from home or other base increases, offending decreases. This pattern of behaviour is termed the *distance decay effect* (Rengert *et al.*, 1999; Rossmo, 2000).

Of course, if this were universally true then every offender would steal from their next door neighbour. Many actually do, but this does increase the risk of being recognised and arrested. Some offenders therefore will limit the number of offences that they commit close to home, to minimise the risk of recognition by neighbours. This has the effect of creating a buffer around the home address. This can be seen in Figure 4.6 which shows a general model for offender behaviour. Note the distance decay effect as the number of offences drops off rapidly as distance from home increases. Also note that the buffer does not extend far from the home, and that offences still occur close to home within the buffer. The offending does not stop next to the home, it is just tempered by the possibility of identification. The application of studying the journey to crime is further explored in Chapter 10.

Rossmo worked with both the Vancouver Police Department and Simon Fraser University to develop a model for criminal geographic targeting and, with Canter's work, this area of offender behaviour examination has grown into the field of *geographic profiling* – a tool that is increasingly utilised by police investigators to help serial investigations. In Chapter 10 we explore the application of geographic profiling in more detail.

4.4.3 The criminal-spatial landscape

In an earlier section, we showed how offender cognitive behaviour formed a template which was used to judge potential criminal opportunities.

Awareness spaces are not just amoeba-like blobs, as you might have thought looking at Figure 4.4, but are made up of different structures which provide a changing pattern of opportunities to commit offences depending on the *environmental backcloth* – the social, psychological, economic, physical and temporal mosaic of the offender passing through urban space. These different structures change across space and affect the type of criminal opportunities that are available to the motivated offender.

4.4.3.1 Nodes, pathways and edges

In Figure 4.4 it can been seen that awareness space is strongly influenced by the number and location of the *nodes* in a person's daily routine. A node, also known as an *activity node*, is a place that an individual is regularly drawn to, such as home, work or school. For offenders, nodes are also important as these areas tend to be the site of many of their offences. Searching for criminal opportunities in the immediate vicinity of a node is rare, and if it does take place then the search area tends to be quite small. A block or two from the node is more likely. As the young tend to commit more offences than other age groups, the type of node can also be a good predictor of crime. A fancy restaurant with fine cutlery and an expensive wine list may be a node for one select group in the population, but a fast food restaurant is more likely to be a node that attracts an age group in their peak offending years.

At some point, an offender has to travel from one node to another. The routes between nodes are *pathways*, and are also the locations of many offences (Brantingham and Brantingham, 1993). The type of crime and the opportunity may vary; but, from an aggregate perspective, pathways provide an opportunity to go past and assess new criminal opportunities. There can often be an increase in property offences near main street and arterial roads. The movement of offenders through the main roads provides an opportunity to explore a street or two from the main road. This exploration is rarely for more than a block or two, as it can be easy to get lost. Offending on, or near, the main road provides a quick escape route back to a familiar pathway. Of course, an offender may travel into the Central Business District by mass transit and not by private car. This reduces the burglary opportunity, but increases the possibility of other types of offending along a pathway. Robbery, pickpocketing or graffiti are associated with subway travel.

The third component of the urban landscape is the *edges* that exist between different parts of the city. As the Brantinghams (1993) note, these edges may be *physical*, such as the boundaries of commercial developments or the border between a park and a housing estate, or *perceptual*, such as the border between areas of different income or racial mix – a common feature of US cities. Perceptual edges may also be apparent in the territorial functioning (described earlier) of a group of neighbourhood residents (Taylor, 1988). Edges can often provide criminal opportunities, because there is an expectation that outsiders are not usual on the periphery of areas. If they ventured into the centre of the neighbourhood then they are more likely to stand out and appear out of place. The rookeries of 19th-century London provided such an opportunity, by allowing a dense network of housing in which offenders resided to exist directly alongside the wealthy of the city of London. A shabbily dressed offender would stand out in the heart of the city of London, but might be expected to have strayed a street or two at the edge. The rookeries provided a bolthole in the event of police pursuit, as well as an administrative boundary, coming under the jurisdiction of the Metropolitan Police and not the City of London Police.

4.4.3.2 Crime generators and attractors

The spatial arrangement of crime generators and crime attractors is inherently tied to the nodes, pathways and edges of our urban world. A *crime generator* is a particular area or node where large number of people are drawn for reasons that are not related to any particular criminal activity that they might commit (Brantingham and Brantingham, 1995, p. 7). These places are crime generators because they provide times and places where potential targets for criminal activity are concentrated into areas that are conducive to criminal acts. For example, shopping malls draw people who do not intend to commit crime or to be the victims of crime. The combined activity of lots of people drawn to the shopping mall provides an opportunity for the potential car thieves who are interspersed within the law-abiding citizens. These offenders may not be drawn to the shopping mall to commit crime, but find themselves suddenly faced with an unforeseen opportunity when they walk through a poorly lit and unattended shopping mall car park.

By comparison, *crime attractors* are places which create criminal opportunities and, in doing so, attract motivated offenders to the neighbourhood or suburb (or *node*). The lure of a known criminal opportunity draws offenders to the area, enticing them with the knowledge that the area has a reputation for a particular type, or types, of illicit opportunity. For example, red light districts offer the allure of prostitution, as in the area around Kings Cross in both Sydney and London; bar districts can attract individuals or groups looking for hard drinking and a fight; and street drug markets have an obvious attraction to addicts. The area can often draw in offenders from other areas (Brantingham and Brantingham, 1995, p. 8). The attractiveness of certain areas of the city for criminal activity is not a new phenomenon. Offenders are often drawn to the centre of cities, attracted to the features that have long been a part of big city life. In 1925, Ernest Burgess described it thus:

> The great city, with its 'bright lights,' its emporiums of novelties and bargains, its palaces of amusement, its underworld of vice and crime, its risks of life and property from accident, robbery, and homicide, has become the region of the most intense degree of adventure and danger, excitement and thrill. (Burgess, 1925, p. 58)

Probably a fair description of 1920's Chicago! What is useful with crime generators and attractors is the recognition of differential causes for what can often be seen as simply crime hotspots. A certain bar, for example, may be thought of as purely a crime attractor by attracting certain types of offenders who enjoy a drink and a fight. But it is possible that at around midnight a fight breaks out among patrons who came to the bar with no intention of fighting, but who were spurred on by alcohol, close proximity to other bars and belligerent bar staff. In this case, at this time, the location is a crime generator, because the protagonists had no initial intention of fighting when they arrived. At midday the same bar may be a crime attractor, drawing in local pickpockets who take the opportunity to sell stolen goods to local fences. The thieves and the fences are drawn to the bar because it is relatively quiet with no police activity during the day, and it offers seclusion from prying eyes. In the morning and the early evening, with no drunken patrons or pickpockets, the bar may be what is termed as a 'crime-neutral area' (Brantingham and Brantingham, 1995). The crime patterns and the routine activities of the people who interact with the spatio-temporal environment of a location are a considerable determinant factor in deciding the status of the site.

4.5 Spatial crime theory in practice

So how does this all fit together? Here we conclude by considering the behaviour of a single (hypothetical) offender and use this behaviour to summarise many of the key terms used in this chapter.

Smith is a committed offender, motivated and always on the look out for an opportunity. This morning, he decides to go and see a friend. While walking to his friend's apartment he is scouting for opportunities. He knows that Mrs Jones across the road has seen the police come to his door many times and that she is suspicious of him, so he walks for a block so that she will not see if he gets up to something, moving through his *buffer zone*. At the local shops on the way, he sees some CDs on a display by the music shop door. The *routine activities* of the music shop owner provide an opportunity, as the owner is in the back of the shop and therefore not a *capable guardian*, Smith is most certainly a *motivated offender* and the CDs are *suitable targets*. They are *concealable* under his jacket, *removable*, *available*, *valuable*, *enjoyable* and *disposable* (CRAVED) as he may be able to sell them to his friend. However, just as he is about to grab a handful of CDs, the baker from the next shop comes outside to have a cigarette and acts as a *place manager*, removing the opportunity for Smith. Given the risk of capture, Smith makes a *rational choice* to try again another day.

As he approaches his friend's house, he sees the cul-de-sac where a number of old people live. They are always tending their gardens, and have placed large potted plants down their street. He does not like to go down there as the residents tend to display quite a bit of *territoriality*. He once sat on a wall watching the old man at one house, when a neighbour came out and told him to go away or she would call the cops. With so much *defended space*, he decides to seek easier burglary opportunities nearby. He does not want to go far, due to the *least effort principle*, and he often visits his friend so this area is part of his *awareness space*. His friend lives in a fairly poor area, so Smith does not feel as if he stands out when he is in this area. Although he has committed some burglaries in this area, it is getting harder. He used to enter the apartment buildings and knock on all of the doors on a floor. If nobody answered, he would kick a door in. However, now the city council have provided some *crime prevention through environmental design* and made the entrances to the buildings security doors. The only one he can enter belongs to his friend, and he cannot steal from his friend's neighbours because they are all suspicious of him.

Smith and his friend go to the shopping mall. Their *cognitive map* of the mall says that there are lots of electronic and music shops on the upper level, a good opportunity for some shoplifting in the past. In the electronics shop, there unfortunately appears to have been some *situational crime prevention* to prevent shoplifting. The CD players are now behind a locked case, *increasing the effort* of theft. The notebook computers are for demonstration only and have most of their components removed, *reducing the reward*, and there is a security guard at the door, *increasing the risk*. They give up and go elsewhere. In the main foyer of the mall a crowd is gathering. Thoughts of crime gone from his head, Smith wanders over to see what is going on. A local singer is giving a show to sell CDs, but in the crowd Smith identifies a new opportunity. The crowd has acted as a *crime generator* because there are many open handbags. Smith's *crime template* knows that this is a good chance, and his *modus operandi* in the past has always worked. His *cognitive script* tells him to nudge a target from behind, and to steal their purse at the same time. He does this, apologises for the nudge (while pocketing the purse) and moves out of the crowd with his reward.

The stolen purse is full of credit cards, but Smith does not have the skill to make use of them. So he goes to a downtown bar to see another friend. On the way there, he takes a familiar bus route to the bar. He feels comfortable on this *pathway* to one of his *nodes*, and scratches his name into the back of the bus seat for fun. He scans around the bar before entering, taking in the *environmental backcloth*. He can see no sign of trouble, or cops ready to raid the bar, so he goes inside. The bar attracts many criminals, as it provides a safe place to do business. In this *crime attractor*, he converts the stolen credit cards to cash and heads home.

4.6 Summary

The continual development of GIS is the key to understanding the growth in thinking about space as a significant factor in the occurrence of crime. Indeed a positive feedback loop appears to be emerging, where advances in our understanding of criminal behaviour are driving the use of GIS in crime prevention. Concepts such as crime attractors and crime generators are valuable not only in the way that they can be mapped, but also in the way that they help practitioners conceptualise the underlying criminal behaviour. A map of crime is not the end point – it is a starting position on the way to understanding the real motivations and opportunities available to offenders and preventing criminal activity from happening.

Further reading

Felson, M. (1998). *Crime and Everyday Life*, Second edition. Thousand Oaks, California: Pine Forge Press.

A clear and simple explanation of the importance of understanding and reducing opportunity as a way to reduce crime.

Clarke, R.V. and Eck, J. (2003). *Becoming a Problem Solving Crime Analyst*. London: Jill Dando Institute.

A practitioners' guide to crime analysis with a strong emphasis on spatial analysis as a means to preventing and reducing crime. This manual goes some way to explain why some forms of analyses are worthwhile, and does so in bite-size chunks.

Rossmo, D.K. (2000). *Geographic Profiling*. Boca Raton, FL: CRC Press.

Advances understanding of offender behaviour and explores some of the quantitative work currently being undertaken to hunt serial offenders.

Cromwell, P. (ed.) (1999). *In their Own Words: Criminals on Crime*, Second edition. Los Angeles: Roxbury.

Provides a unique insight into the thinking of offenders and how they view criminal opportunities.

References

Bottoms, A.E. (1974). Review of 'Defensible Space' by Oscar Newman. *British Journal of Criminology*, 14, 203–206.

Bottoms, A.E. and Wiles, P. (2002). Environmental criminology. In M. Maguire, R. Morgan and R. Reiner (eds) *The Oxford Handbook of Criminology* (pp. 620–656). London: Oxford University Press.

Brantingham, P. and Brantingham, P. (1995). Criminality of place: Crime generators and crime attractors. *European Journal of Criminal Policy and Research*, 3, 5–26.

Brantingham, P.J. and Brantingham, P.L. (1981a). *Environmental Criminology*. Prospect Heights: Waveland Press.

Brantingham, P.J. and Brantingham, P.L. (1981b). Introduction: The dimensions of crime. In P.J. Brantingham and P.L. Brantingham (eds) *Environmental Criminology* (pp. 7–26). London: Sage.

Brantingham, P.J. and Brantingham, P.L. (1981c). Notes on the geometry of crime. In P.J. Brantingham and P.L. Brantingham (eds) *Environmental Criminology* (pp. 27–54). London: Sage.

Brantingham, P.J. and Brantingham, P.L. (1984). *Patterns in Crime*. New York: Macmillan.

Brantingham, P.L. and Brantingham, P.J. (1993). Nodes, paths and edges: Considerations on the complexity of crime and the physical environment. *Environmental Psychology*, 13, 3–28.

Burgess, E.W. (1916). Juvenile delinquency in a small city. *Journal of the American Institute of Criminal Law and Criminology*, 6, 724–728.

Burgess, E.W. (1925). The growth of the city: An introduction to a research project. In R.E. Park, E.W. Burgess and R.D. McKenzie (eds) *The City* (pp. 47–62). Chicago: University of Chicago Press.

Canter, D. and Larkin, P. (1993). The environmental range of serial rapists. *Environmental Psychology*, 13, 63–69.

Clarke, R.V. (1992). *Situational Crime Prevention: Successful Case Studies* (p. 286). Albany, NY: Harrow and Heston.

Clarke, R.V. (1997). *Situational Crime Prevention: Successful Case Studies*, Second edition (p. 357). Albany, NY: Harrow and Heston.

Clarke, R.V. (1999). Hot products: Understanding, anticipating and reducing demand for stolen goods. *Police Research Group: Police Research Series*, Paper 112, 48. London: Home Office.

Clarke, R.V. and Eck, J. (2003). *Becoming a Problem Solving Crime Analyst*. London: Jill Dando Institute.

Clarke, R.V. and Felson, M. (1993). Introduction: Criminology, routine activity, and rational choice. In R.V. Clarke and M. Felson (eds) *Routine Activity and Rational Choice* (pp. 259–294). New Brunswick: Transaction publishers.

Cohen, L.E. and Felson, M. (1979). Social change and crime rate trends: A routine activity approach. *American Sociological Review*, 44, 588–608.

Coleman, A. (1985) *Utopia on Trial: Vision and Reality in Planned Housing*. London: Hilary Shipman.

Cornish, D. and Clarke, R. (1986). *The Reasoning Criminal: Rational Choice Perspectives on Offending*. New York: Springer-Verlag.

Costello, A. and Wiles, P. (2001). GIS and the journey to crime. In K. Bowers and A. Hirschfield (eds) *Mapping and Analysing Crime Data* (pp. 27–60). London: Taylor & Francis.

Cromwell, P., Olson, J.N. and Avary, D.A.W. (1999). Decision strategies of residential burglars. In P. Cromwell (ed.) *In their Own Words: Criminals on Crime* (pp. 50–56). Los Angeles: Roxbury.

Crowe, T.D. (2000). *Crime Prevention Through Environmental Design: Applications of Architectural Design and Space Management Concepts*, Second edition. Boston: Butterworth-Heinemann.

Eck, J. (1995). A general model of the geography of illicit retail marketplaces. In D. Weisburd and J.E. Eck (eds) *Crime and Place* (pp. 67–93). Monsey, New York: Criminal Justice Press.

Felson, M. (1995). Those who discourage crime. In D. Weisburd and J.E. Eck (eds) *Crime and Place* (pp. 53–66). Monsey, New York: Criminal Justice Press.

Felson, M. (1998). *Crime and Everyday Life: Impact and Implications for Society*. Thousand Oaks, California: Pine Forge Press.

Felson, M. and Clarke, R.V. (1998). Opportunity makes the thief: Practical theory for crime prevention. *Police Research Group: Police Research Series*, Paper 98, 36.

Fleming, Z. (1999). The thrill of it all: Youthful offenders and auto theft. In P. Cromwell (ed.) *In their Own Words: Criminals on Crime* (pp. 71–79). Los Angeles: Roxbury.

Guerry, A.-M. (1833). *Essai sur la statistique morale de la France: Precede d'un rapport a l'Academie de sciences*. Paris: Chez Crochard.

Harries, K. (1980). *Crime and the Environment*. Springfield, IL: Charles C. Thomas.

Hillier, B. (1988). Against enclosure. In N. Teymur, T.A. Markus and T. Woolley (eds) *Rehumanizing Housing* (pp. 63–88). London: Butterworths.

Holzman, H.R., Hyatt, R.A. and Dempster, J.M. (2001). Patterns of aggravated assault in public housing: Mapping the nexus of offense, place, gender, and race. *Violence Against Women*, 7, 662–684.

Jacobs, J. (1965). *The Death and Life of Great Cities*. Harmondsworth: Penguin.

Kocsis, R.N., Cooksey, R.W., Irwin, H.J. and Allen, G. (2002). A further assessment of 'Circle Theory' for geographic psychological profiling. *Australian and New Zealand Journal of Criminology*, 35, 43–63.

Knox, P.L. (1994). *Urbanization: An Introduction to Urban Geography*, Englewood Cliffs NJ: Prentice-Hall.

LeBeau, J.L. (1987). The methods and measures of centrography and the spatial dynamics of rape. *Journal of Quantitative Criminology*, 3, 125–141.

LeBeau, J.L. (1992). Four case studies illustrating the spatial-temporal analysis of serial rapists. *Police Studies*, 15, 124–145.

Mayhew, H. (1862). *London Labour and the London Poor*. London: Griffin Bohn.

Merton, R.K. (1938). Social structure and anomie. *American Sociological Review*, 3, 672–682.

Newman, G. (1997). Introduction: Towards a theory of situational crime prevention. In G. Newman, R. Clarke and S.G. Shoham (eds) *Rational Choice and Situational Crime Prevention: Theoretical Foundations* (pp. 1–23). Dartmouth, UK: Ashgate.

Newman, O. (1972). *Defensible Space: Crime Prevention Through Urban Design*. New York: MacMillan.

Porteous, J.D. (1973). The Burnside teenage gang: Territoriality, social space, and community planning. In C.N. Forward (ed.) *Residential and Neighborhood Studies in Victoria* (pp. 130–148). Victoria, BC: Western Geographical Series.

Quetelet, A. (1842). *A Treatise in Man*. Edinburgh: Chambers.

Ratcliffe, J.H. (2001). Policing Urban Burglary. *Trends and Issues in Crime and Criminal Justice*, No. 213, 6.

Ratcliffe, J.H. (2002). Aoristic signatures and the temporal analysis of high volume crime patterns. *Journal of Quantitative Criminology*, 18, 23–43.

Ratcliffe, J.H. (2003). Suburb boundaries and residential burglars. *Trends and Issues in Crime and Criminal Justice*, 246, 6.

Rengert, G. (1989). Behavioural geography and criminal behaviour. In D.J. Evans and D.T. Herbert (eds) *The Geography of Crime* (pp. 161–175). London: Routledge.

Rengert, G.F. (1996). *The Geography of Illegal Drugs*. Boulder, CO: Westview Press.

Rengert, G.F. and Wasilchick, J. (1985). *Suburban Burglary: A Time and Place for Everything*. Springfield, IL: C.C. Thomas Publishing.

Rengert, G.F. and Wasilchick, J. (2000). *Suburban Burglary: A Tale of Two Suburbs*, Second edition. Springfield, IL: C.C. Thomas Publishing.

Rengert, G.F., Piquero, A.R. and Jones, P.R. (1999). Distance decay reexamined. *Criminology*, 37, 427–445.

Rengert, G.F., Mattson, M.T. and Henderson, K.D. (2001). *Campus Security: Situational Crime Prevention in High-density Environments*. Monsey, NY: Criminal Justice Press.

Rossmo, D.K. (1995). Place, space, and police investigations: Hunting serial violent criminals. In D. Weisburd and J.E. Eck (eds) *Crime and Place*. Monsey, New York: Criminal Justice Press.

Rossmo, D.K. (2000) *Geographic Profiling*. Boca Raton, FL: CRC Press.

Shaw, C.R. and McKay, H.D. (1942). *Juvenile Delinquency and Urban Areas*. Chicago: Chicago University Press.

Short, E. and Ditton, J. (1998). Seen and now heard: Talking to the targets of open street CCTV. *British Journal of Criminology*, 38, 404–428.

Shover, N. (1999). Aging criminals: Changes in the criminal calculus. In P. Cromwell (ed.) *In their Own Words: Criminals on Crime* (pp. 80–86). Los Angeles: Roxbury.

Sorensen, D. (2004). *Temporal Patterns of Danish Residential Burglary*. Copenhagen: Ministry of Justice. www.jm.dk/image.asp?page=image&objno=72081, accessed 5th August 2004, 30pp.

Taylor, R.B. (1988). *Human Territorial Functioning*. New York: Cambridge University Press.

Wiles, P. and Costello, A. (2000). *The 'Road to Nowhere': The Evidence for Travelling Criminals* (p. 60). London: Research, Development and Statistics Directorate (Home Office).

Wright, R. and Decker, S. (1994). *Burglars on the Job*. Boston: Northeastern University Press.

5
Spatial Statistics for Crime Analysis

Learning Objectives

This chapter aims to introduce the crime mapper to the most practical and essential statistical tools required for mapping crime. This chapter is not aimed to replace a whole text on statistics or spatial analytical processes, but will provide an overview of many useful statistical tools and the theoretical constructs that explain why they are used.

One of the key concepts that is contained in this chapter is that of spatial dependency – the notion that crime in one area is related to crime in neighbouring areas. This is an often-overlooked aspect of spatial processes, yet is essential to understand in order to utilise spatial statistics effectively. The chapter then proceeds to explore centrographic techniques and standard deviational ellipses, before getting into nearest neighbour tests that can check for clustering and dispersion in crime series. We provide links to free software that can perform these tests.

The chapter continues with the local Moran's I test, which can test for clustering in areal values from a statistical standpoint, before finishing with an overview of the latest spatial modelling process which are becoming more mainstream in crime mapping. Both spatial regression and geographically weighted regression are likely to become central to the work of crime researchers in the near future, and they conclude this chapter.

GIS and Crime Mapping   Spencer Chainey and Jerry Ratcliffe
© 2005 John Wiley & Sons, Ltd.

5.1 Introduction

Although the mere mention of the word 'statistics' is enough to strike fear into the heart of many crime mappers, there are often times when those working with crime data are required to apply statistical techniques in their work. For example, a police chief might want to know the average number of robberies per week in a local community prior to attending a meeting. When the chief asks for the average, what is actually being asked for is the statistical mean. Crime scientists, be they academic researchers or crime reduction practitioners, similarly may wish to know if burglary has significantly reduced in a housing project after they have spent money reinforcing locks and repairing broken street lights. These types of evaluation require that the analyst does not only explore if the level of crime has reduced, but that this reduction is not just the result of chance. Knowledge of statistics is required to answer these questions.

The crime mapper has the additional entertainment of not just having to know about basic statistics, but also requires some familiarity with spatial statistics. Spatial statistics brings an additional dimension to the study of statistics by introducing the geographic dimension. A trip to a library will confirm that there are dozens of books available on basic statistics, and even a few esoteric academic texts on spatial statistics. This chapter provides an overview of the more applicable spatial statistics, as well as indicating the future trends in spatial analysis that are likely to develop in the forthcoming years.

5.2 Spatial processes

Crime does not occur randomly. As the previous chapter showed, offenders seek out criminal opportunities, and these opportunities occur in different places and at different times. Theories such as Routine Activities Theory and Crime Pattern Theory can explain much of the incidence of crime in terms of victim and offender interaction. At the broader level, regional crime distribution can be explained (to a degree) by factors such as population density and the increased opportunities provided by the anonymity of large urban centres. There are two spatial processes that are fundamental to an appreciation of crime patterns: Spatial dependence and spatial heterogeneity.

5.2.1 Spatial dependency and heterogeneity

Spatial dependence can be described as the degree to which the value of a variable at one location is influenced by neighbouring locations. The basis for the idea stems from Tobler's First Law of Geography which states that 'everything is related to everything else, but near things are more related than distant things' (Tobler, 1970, p. 236). This means that, in general, places closer together are more likely to have a similar value. To use a non-crime example, consider the temperature. Two neighbourhoods in London (UK) are likely to have similar temperatures at a certain time, because there is considerable spatial autocorrelation between the temperature in Kilburn and Maida Vale. However, because they are further apart, there is likely to be less correlation between temperatures in London and Philadelphia in America.

There are many ways that the crime level in an area can be influenced by, or at least related to, the surrounding area. One is in the way that underlying causes of crime can drive the crime rate in small areas. Consider a housing project that is unfortunately designed in such a way that every house is vulnerable to burglary. If the housing project were divided into different census geographical units (such as census output areas or block groups) then it would be an easy task to count the number of burglaries that occur in each geographical unit. These values would not be independent and would most likely be highly positively correlated. If this were the case, then it could be said that the housing project's geographical units display a level of spatial autocorrelation, another term for spatial dependency. This would occur because the underlying driver of burglary (poor housing design) covers all of the areas in the study region and affects them all in the same way.

A second manner in which crime in one measurement area is related to a neighbour is in the influence of edges. As the Brantinghams have hypothesised (see Chapter 4), edges between areas that are significantly different in terms of either socio-economic status or have a noticeable physical boundary will attract offenders who take advantage of the boundaries between these areas. That a neighbouring area is significantly different than a studied location is therefore potentially sufficient to cause an increase in some crime types. Often this spatial dependency can be a measurement issue. Major roads are often used as the boundary between census units. If the presence of a motorway or freeway off-ramp provides a suitable opportunity for a local street drug market, arrests for drug-related offences will occur in the streets around the off-ramp. These arrests will become clear if the arrests are accurately mapped and aggregated to the census units around the off-ramp. Although there is only one off-ramp, arrests will increase in a number of census units. Spatial dependency (autocorrelation) would become clear if the off-ramp were closed for some reason – drug arrests would most likely decrease in all of the census areas around the ramp, and not just in one area. Conversely, if the ramp were re-opened, and no specific enforcement action taken, it is likely that the market would re-establish itself and police activity would increase in a number of census areas. In this way, the level of drug activity in a number of neighbouring areas (all around the off-ramp) can be spatially correlated.

Spatial dependency can not only suggest a clustering of like values, but also the possibility of a dispersion of values, so that high value areas are surrounded by low values areas. One possible example of dispersion, which is actually relative dispersion, are phenomena that tend to occur in the suburbs of metropolitan areas rather than the central city. Some community safety issues (such as fatal motor-vehicle crashes) can be more prevalent in the suburbs, due to different environmental conditions. In this case, mapping urban and suburban areas can produce relatively low rates of a community safety problem surrounded by relatively high rates.

When high levels of a variable in one area are correlated to high levels in neighbouring areas, or when low levels are clustered together spatially, then the series is said to display positive spatial autocorrelation. When high values are associated with low values in neighbours (or low value areas are surrounded by high value neighbours) then the series displays negative spatial autocorrelation. When the pattern of low and high values is randomly distributed, then zero spatial autocorrelation exists. This last case is rarely true of crime patterns.

The converse of spatial dependency is either spatial randomness or spatial heterogeneity. Crime patterns display spatial heterogeneity simply in the variation from one place to another. Spatial heterogeneity is one of the main reasons to study spatial crime patterns, in order to discover why some places are victimised more than others. For example, Figure 5.1 shows vehicle crime rates for counties in Pennsylvania in 2002. Each county is displayed with its average value, which is useful for comparison between counties in this state. However, this fails to convey the complexity of the crime situation within the city of Philadelphia, which in reality is a mosaic of spatial heterogeneity and spatial dependence across different neighbourhoods (Figure 5.2). The challenge for analysts is to understand and map the spatial heterogeneity in crime distributions, so that effective crime reduction decisions can be made and that resources

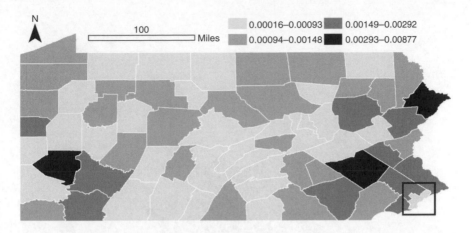

Figure 5.1 Vehicle crime rates for Pennsylvania counties in 2002, per capita. Vehicle crime rates appear to be lower in the centre of the state than in the South-West and South-East. There is a single rate shown for the city of Philadelphia, the area indicated by the black square in the South-East of the state

are targeted to the best areas. This has to be undertaken while also being mindful that the underlying causes of crime that are driving the diverse crime patterns may be spatially dependent.

The remainder of this chapter describes a range of statistical techniques that an analyst may find useful in order to clearly articulate and understand the distribution of crime. These methods do not represent every spatial analysis routine available, but are the more practical and commonly used methods for geographical crime analysis. The further reading section will point the reader towards whole texts dedicated to the subject of spatial statistics in significant detail.

5.3 Centrographic statistics

With the recognition that spatial dependency (autocorrelation) and spatial heterogeneity are processes that should be modelled and understood, there are a number of analytical tools that can be used to describe centrographic patterns in crime data. These centrographic statistics can be useful when a general value of spatial tendency or description of crime data is desired, though it should be recognised that any heterogeneity in the data is rarely reflected in the value of a centrographic statistic.

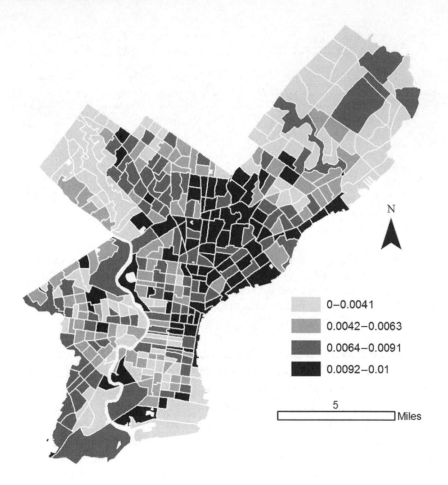

Figure 5.2 Vehicle crime rates for the city of Philadelphia, per captia, 2002. Although Figure 5.1 showed a single rate of vehicle crime for the city of Philadelphia, examination of the crime in the city shows that vehicle crime rates are concentrated in some areas more than others. The main focus for vehicle crime in the city appears to be in the central North of Philadelphia

5.3.1 Mean centre

One global measure that is easy to calculate with Euclidean data is the mean centre. The calculation of the mean for non-spatial data will be familiar to many readers:

$$\bar{x} = \frac{\sum x}{n}$$

where $\bar{x}$ is the mean of the variable x, Σ is the summation sign and n is the total number of records of x. The spatial equivalent, the mean centre, is a simple extension into two dimensions, and is calculated as:

$$\bar{x} = \frac{\sum x}{n} \qquad \bar{y} = \frac{\sum y}{n}$$

where $\bar{x}$ is the mean of the x coordinates of all locations, $\bar{y}$ is the mean of the y coordinates and n remains the number of points in the data set.

The mean centre is easy to calculate either manually, with some minor manipulation of the data in a GIS, or automatically with CrimeStat III (a freeware crime statistics and analysis software package distributed via the US National Institute of Justice MAPS programme). The mean centre can be used as an overall description of the central focal point of the data. For example, LeBeau (1992) used the mean centre as one basic indication of the movement of a data series over time, and Paulsen (2004) has explored its use for profiling the location of serial offender 'nodes'. Some difficulties with the mean centre include the problem that vastly different crime distributions can generate the same mean centre. This is demonstrated in Figure 5.3 where data series A has the same mean centre as series B, although series B is bimodal in distribution with few points near the mean centre.

Interpretation of the mean centre sometimes leaves some problems for analysts. LeBeau's and Paulsen's examples help to demonstrate the application of this statistics, yet it can be argued that the mean centre tells the analyst very little about the data series. The mean centre is vulnerable to the influence of outliers, and with unusual crime series and unusual study

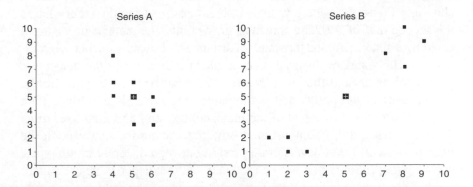

Figure 5.3 Two different series of eight points (units are arbitrary). Although the series have very different distributions, they have the same mean centre

areas the mean centre can even fall outside of the study area. These problems require the mean centre to be used with care, with an assessment as to whether it offers anything of value to a crime series that is being analysed.

5.3.2 Standard distance

A more useful measure is the standard distance. The standard distance can be used in conjunction with the mean centre both to indicate the focal centre of the data series and to provide an indication (albeit a non-directional one) of the distribution of the data around the mean centre. The standard distance is the spatial equivalent of the standard deviation, a statistic usually employed to describe the dispersal of values around the mean. The standard distance is calculated by taking the square root of the squared distance from each point to the mean centre, and then dividing by the number of points. If areal data are used, then the centroid for each polygon can be used in lieu of original point data. This calculation is shown here:

$$ SD = \frac{\sqrt{\Sigma d^2}}{n} $$

where n is the number of points in the dataset and d is the distance from each point to the mean centre, calculated this way:

$$ d = \sqrt{(x - \bar{x})^2 + (y - \bar{y})^2} $$

The standard distance is commonly displayed as a circle, centred on the mean centre with a radius equal to the standard distance. This is shown in Figure 5.4 where vandalism incidents for Lincoln, Nebraska, in the second half of 2002 are shown as grey points, the mean centre is indicated by a black cross and the standard distance is shown by the black circle.

The standard distance gives a visual indication of the dispersion of the points around the mean centre. They can be used to indicate the dispersal of crime points and if the same study area is employed then dispersal and the movement of the mean centre can be plotted over time. Chainey *et al.* (2002) demonstrated how the standard distance could be used to show different levels of dispersion between different crime types, showing that robbery to a person tended to be less dispersed than residential burglary and vehicle crime. This result indicated that robbery was concentrated at a few locations in comparison to the wider spread of residential burglary and vehicle crime. The limitation of the standard distance is the

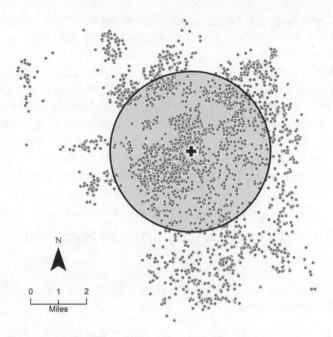

Figure 5.4 Vandalism incidents in Lincoln, Nebraska, for the latter half of 2002. The black cross indicates the mean centre and the black circle shows the standard distance

lack of directional focus. Irrespective of the spread of the points in a particular direction, the standard distance is an equal measure in every direction. A more useful type of global dispersal measure is the standard deviational ellipse.

5.3.3 Standard deviational ellipses

Standard deviational ellipses can be thought of as a directional equivalent of the standard distance. Standard deviational ellipses are elliptical in shape, with a long axis running the longest straight line distance from end to end and through the mean centre, and a short axis which is perpendicular to the long axis and also runs through the mean centre joining the closest edges of the ellipse. The ellipse not only gives an indication of the dispersal of the points (larger ellipses mean a greater dispersion of points) but the direction of the longer axis of the ellipse indicates the predominant alignment of the data dispersal. Long, narrow ellipses are indicative of a linear data pattern, while rounder ellipses suggest that the point pattern is more evenly distributed in all directions around the mean centre. A perfectly circular standard deviational ellipse is the same as a standard distance and, as far as crime mapping is concerned, only occurs in artificial data situations where points are exactly equally distributed in all directions around the mean centre.

The calculation of the standard deviational ellipse is a little more complicated than the standard distance, and the results are not really useful unless plotted graphically. This is most easily achieved using software packages such as CrimeStat. CrimeStat not only performs the calculations but also creates output files that can be imported into GIS programmes such as ArcGIS and MapInfo (CrimeStat can also create standard distance polygons and plot data mean centres) (Box 5.1).

Box 5.1 Calculating centrographic statistics with CrimeStat

CrimeStat is available to download at http://www.icpsr.umich.edu/NACJD/crimestat.html

1. In the 'Data setup' tabbed area, select the data set and select the fields that contain the x and y coordinates. Also in this tab select a coordinate system and the data units.
2. In the 'Spatial Description' tabbed area, choose 'Spatial Distribution'. Select the check boxes for 'Mean center and standard distance (Mcsd)' and 'Standard deviational ellipse (Sde)'.
3. For both options click the 'Save result to...' box and enter a filename for either an Arcview shapefile (which can be opened by ArcGIS) or a MapInfo MIF file. MapInfo cannot read MIF files directly, but can import MIF files into MapInfo tables.
4. The 'Spatial Distribution' area has other options for descriptive statistics, such as median centres, centres of minimum distance, and the directional mean and variance. Details of these measures can be found in the CrimeStat manual, Chapter 4.
5. Click 'Compute'. Once the analysis has concluded, the results are shown as text details in tabbed sections of the output window, though the figures will be of little value to most analysts. More usefully, CrimeStat created output files for GIS depending on the choices the user made in the 'Save result to...' box.
6. CrimeStat automatically adds a prefix to output files, depending on the analysis. The program adds MC to the mean centre and SDE to the standard deviational ellipse. For example, if a shapefile

Continued on page 125

Continued from page 124

output was selected and a filename of burglary.shp chosen, the result of this analysis would produce MCburglary.shp for the mean centre, SDDburglary.shp for the standard distance and SDEburglary.shp for the standard deviation ellipse.

7. Open the GIS files that are the output from CrimeStat. The resulting objects will show the mean centres and standard distances. The visual output in the GIS will be more meaningful than the text output from the program.

Using the same Lincoln vandalism data as in Figure 5.4, Figure 5.5 shows the standard deviational ellipse. Although similar to the standard circle, it can be seen that there is a slight NW–SE tendency in the ellipse. Visual inspection of the data shows that the ellipse is skewed in this direction not only by the pattern of the data but also by the absence of points in the SW section of the map.

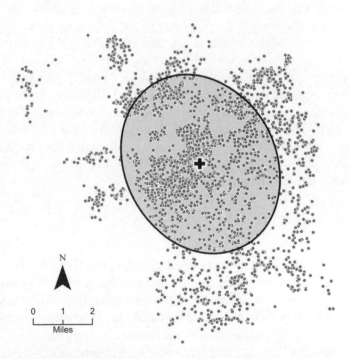

Figure 5.5 Vandalism incidents in Lincoln, Nebraska (July–December 2002) shown with the standard deviational ellipse

There is a common misconception that the standard deviational ellipse will always contain 68.2% of the available points. This misunderstanding stems from the fact that, in statistical terms, one standard deviation will contain 68.2% of the available data for normally distributed data series, and a standard deviational ellipse is a spatially oriented equivalent of a standard deviation. The key point here is that a standard deviation will include this percentage of data points with *normally distributed* data. For a standard deviational ellipse to exactly represent a standardised measure of data series, the points would have to be normally distributed along both the long and short axes of the ellipse. However, even with this limitation, standard deviational ellipses are an improvement over standard distances in terms of indicating point dispersion and direction of that dispersion.

5.4 Estimates of spatial dependence

5.4.1 Nearest neighbour statistic

The Nearest Neighbour Index is a distance statistic for point pattern data sets that gives the analyst an indication of the degree of clustering of the points. A nearest neighbour analysis compares the characteristics of an observed set of distances between pairs of closest points with distances that would be expected if points were randomly placed. As Levine (2004) also showed, it can be used to compare two different crime patterns. During the analysis, the distance from each point to its nearest neighbour is calculated. This value gets added to a running total of all minimum distances, and once every point has been examined, the sum is divided by the number of points. This produces a mean minimum distance, or nearest neighbour distance. The equation looks like this:

$$\bar{d} = \sum_{i=1}^{n} \frac{d_{ij}}{n}$$

where $\bar{d}$ is the mean nearest neighbour distance, d is the distance between the point i and its nearest neighbour j, and n is the number of points in the dataset. On its own, the value of $\bar{d}$ is fairly meaningless. While it tells the analyst the mean nearest neighbour distance, this figure does not give any indication of the type of pattern the data forms.

There are three general types of spatial pattern. A *clustered* pattern is often the most common form of spatial pattern seen with crime data. Given the theoretical outline provided in Chapter 4, a clustered distribution is generally to be expected. Crime events are often a result of opportunities presented during the course of people's routine activities. Neither opportunities nor the routine activity of offenders and victims are randomly distributed, therefore clustered patterns are the most common type of pattern observed.

If an offence pattern is more spread out, it is possible that it exhibits the second type of spatial pattern: a *random* distribution. In this type of pattern, although there may be some local clusters, the overall pattern of the crime series is spread across the study area without any apparent pattern – an event has an equal chance to appear anywhere in the study area. The third type of pattern is a *uniform* one, which is also rarely seen in crime research. This occurs where points are spaced roughly the same distance apart.

In the following discussion of spatial distributions, there are a number of different approaches. While these tests can be used on a single crime distribution to explore the concept of spatial randomness, more usefully these tests can be used to compare the general spatial distribution of one crime type with another or from one time period to another. A number of tests exist for spatial randomness. We will consider the nearest neighbour index in more detail, but point the reader to texts such as Ripley (1981), Diggle (1983) and Bailey and Gatrell (1995) for more examples.

Spatial dependence in a single crime pattern is investigated by examining the observed distribution of nearest neighbour measures and comparing the mean across the data set with an expected, theoretical distribution that would occur if the points were dispersed in a random manner. The random distribution is a function of the size of the study area and the number of points:

$$\bar{\delta} = \tfrac{1}{2}\sqrt{A/n}$$

where $\bar{\delta}$ is the expected mean distance between nearest neighbours and A/n is the point density, expressed as the area of the study region divided by the number of points. The calculation of the nearest neighbour index is a simple ratio of the two calculations:

$$R(\text{NNI}) = (\bar{d}/\bar{\delta})$$

where $R(\text{NNI})$ is the nearest neighbour index expressed as a ratio. CrimeStat is able to calculate all of these values: the mean nearest neighbour distance $\bar{d}$, the mean random distance $\bar{\delta}$ and the nearest neighbour index $R(\text{NNI})$ (Box 5.2). $R(\text{NNI})$ can range from 0.0 for a crime distribution that has all the points at the same location and are separated

---

Box 5.2 Calculating a nearest neighbour index with CrimeStat

1. In the Data setup tabbed area, select the data set and select the fields that contain the x and y coordinates. Also in this tab select a coordinate system and the data units.
2. In the 'Measurement Parameters' area of the 'Data setup' tab, enter the area of the study region (if known) and check that the correct areal units are selected. Square miles, feet or kilometres are the most common.
3. In the 'Spatial Description' tabbed area, choose Distance Analysis and tick the nearest neighbour analysis (Nna) box. Enter 1 in the 'number of nearest neighbours to be computed' box, if the analysis is only to examine the nearest neighbour (it is possible to examine the distance to, say, the fifth nearest neighbour, though there is generally no reason to do this).
4. Click the 'Save result as...' box and choose an output file if desired. Values from the analysis are recorded into a dBase file, though can also be easily read on the CrimeStat output window.
5. There are two options for edge corrections available. Although neither is ideal, they can improve an analysis of study areas that are fairly uniform in shape (see Chapter 5 of the CrimeStat documentation for more details).
6. Click 'compute' to run the analysis. Once complete, the results can be seen in the window. The key value is the 'Nearest Neighbour Index', which can be found about two-thirds of the way down the output. Values less than 1.0 indicate a clustered crime pattern. If the nearest neighbour index is less than 1.0 and the p-values (one and two tailed) are less that 0.05 then the chances that the crime series is clustered to the extent that it is because of random variation is less than 5% – a statistically significant finding (although p-values close to 0.5 should be interpreted with caution, due to some of the problems mentioned earlier).

---

by distances that all equate to zero, through 1.0 for a random distribution of points, up to a maximum value of 2.15. Given that we expect a crime distribution to be clustered, the expected value of R(NNI) is usually less than 1.

CrimeStat will also calculate a statistical significance for the nearest neighbour index. The problem with statistical measures of nearest neighbour significance is the difficulty with correcting for what are termed *edge effects*. Few study areas in crime analysis are perfectly rectangular in shape, and therefore estimates of $\bar{d}$ (the mean nearest neighbour distance) are often larger than in reality. This is because points that lie close to a study boundary are excluded from the (usually real) possibility of having a nearest neighbour just the other side of the boundary, as shown in Figure 5.6.

A second problem exists in the calculation of the study area (A). CrimeStat allows the user to enter the area of the study region in the 'Measurement Parameters' section of the 'Data setup' tab. However, if this is not done, the program assumes a rectangular study area based on the smallest rectangle that could encompass all of the points (what is termed a minimum bounding rectangle). The minimum bounding rectangle rarely reflects the real study area, nor does it consider the possibility that there are areas within the study area that could not have a crime point. As an example, consider examining burglary in a city which has a large lake in the city centre. This all has an impact on the accurate calculation of the mean random distance $\bar{\delta}$.

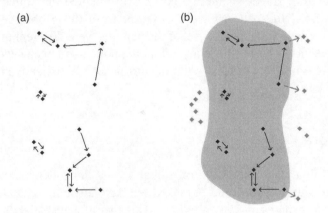

(a)   (b)

Figure 5.6 One of the problems with nearest neighbour measures is shown here. Map (a) shows a group of crime locations, with arrows that indicate the nearest neighbour for each point. Some points are paired with mutual neighbours where they are the nearest neighbour for each other. Edge effects become a factor when the study area is superimposed over the points in map (b). On the right side of the map there are a number of points which would provide closer nearest neighbours to points within the study area, but are excluded because they are beyond the study region boundary. A second consideration is the clear cluster of points on the left of the map, a clear crime hotspot that is excluded from the analysis by also lying just outside the boundary

CrimeStat has a number of built-in options for dealing with the edge effect problem, such as conducting a rectangular or circular adjustment, though these are not ideal. Donnelly (1978) found algorithms to correct $\bar{\delta}$ and its sampling variance, though these equations are complicated and add little value to crime analysis. Nearest neighbour calculations are useful as the basis for a range of other statistical tests, but the nearest neighbour index and associated significance measures on their own have limitations in crime analysis. From Chapter 4 it is clear that given an understanding of spatial crime theory a clustered distribution is expected. Applying a statistical test to confirm that there is clustering across a whole study area does not explain why the cluster exists, nor does it help describe where the clustering occurs.

Nearest neighbour statistics have been applied to estimate changing crime patterns in the Australian capital city, Canberra (Ratcliffe, 2005). Instead of measuring from crime point to crime point, the nearest neighbour distances were measured between crime locations and randomly introduced points. The random points enabled comparisons to be made between the pattern of burglaries before a large police operation and the pattern of offences afterwards. It was necessary to use this variation to cope with the significant variation in the number of crime events in the before and after stages of the operation. This enabled the police not only to determine if the patterns were different from those observed with a random distribution, but also to see if the patterns were different in different areas of the city. The results showed that the police had been successful in reducing burglaries in the city without causing any statistically significant displacement.

5.4.2 Moran's *I*

Moran's *I* is a classic measure of global spatial dependence, and can be applied to both polygons and points which have attribute data attached to them. The test shares many similarities to another global spatial dependence test, Geary's *C*. These two tests are similar, and so because the Moran's *I* test is more commonly applied to crime data (see for example Chakravorty, 1995; Cohen and Tita, 1999; Mencken and Barnett, 1999; Messner *et al.*, 1999; Messner and Anselin, 2004), we will only discuss the Moran's *I* test here.

The advantage of Moran's *I* over nearest neighbour analysis is that while nearest neighbour analysis allows an analyst to measure clustering in points, Moran's *I* can show if there is significant clustering in a variable. This means that the process is able to determine if there is clustering in, for example, burglary patterns or vehicle crime distributions, even if both are aggregated to the same set of polygons. Moran's I can therefore determine if, for example, high vehicle crime areas are also surrounded by high car-crime areas, and if low areas are surrounded by other low crime areas. If so, then the pattern of car crime is said to display positive spatial autocorrelation. If, however, low areas are surrounded by high car-crime areas and high crime areas are surrounded by polygons with little car crime, the series would display negative spatial autocorrelation. If there were no pattern to the distribution of high and low areas, then the series would have zero spatial autocorrelation.

As the test examines the relationship of a variable between neighbouring areas, one of the main decisions to make is how to decide which areas are neighbours. Some programs give the user a choice, while CrimeStat requires that each area be described as an x and y coordinate location, and then gives the greatest influence to points that are located closest to the location being tested (Box 5.3). This does mean that when using CrimeStat, polygons that are not neighbours of the point under examination can have an influence on the overall result, though this may only be a significant problem with highly unusual data sets.

Box 5.3 Calculating a Moran's I in CrimeStat

1. In the 'Data setup' tabbed area, select the data set and select the fields that contain the x and y coordinates. Also, in this tab, select a coordinate system and the data units.
2. Ensure that a Z (Intensity) variable is selected, that corresponds to the crime frequency or rate that is to be tested.
3. Once a Z variable is selected, the Moran's I checkbox in the 'Spatial Distribution' area of the 'Spatial Description' tab becomes available. If examining a number of small areas, such as census tracts or smaller, also check the box 'Adjust for small distances'.
4. Click 'Compute'. A positive Moran's I value indicates positive spatial autocorrelation, with larger values indicating a greater degree of spatial dependence.
5. The two other important values are the Z values. The Z values are an indication of the chances that the Moran's I result could have occurred by chance. The two that are shown in CrimeStat (Normality and Randomisation) are calculated slightly differently

Continued on page 132

Continued from page 131

> but are interpreted in the same way. A simple rule of thumb is as follows: If both the Normality significance (Z) and the Randomisation significance (Z) are greater than 1.96, then the data set displays positive spatial autocorrelation to a significance of $p < 0.05$. In other words, the chance that the positive autocorrelation occurs by random chance is less than 5%.

With CrimeStat it is possible to apply the '*I*' test to a set of distance bands, similar to Ripley's *K*. The distance bands are cumulative, so each band includes the values from the interior areas. The output is called a Moran Correlogram and the latest version of CrimeStat can produce a Moran Correlogram. It is useful for exploring the extent of the spatial autocorrelation for a crime problem, in order to discover how localised or how far the autocorrelation extends. Chapter 6 also demonstrates other methods of achieving this.

Case study: The application of Moran's *I* on burglary at the state level in the United States of America

The Moran's *I* statistic can be used to indicate the presence of clustering. In Figure 5.7 the burglary rate for each of the 48 contiguous US states is shown, calculated as a rate for every 1000 housing units. The crime data are taken from the Uniform Crime Reports for 2002, published by the FBI, and the housing unit data from the 2000 US census. From the map alone, it would appear that there is some clustering of high values in the Southern states and clustering of low values in the Central North and North East of the country. This significance is confirmed by a global Moran's *I* value of 0.65. This is a highly significant value supporting the visual evidence of a North–South divide in burglary patterns.

The Local Moran's *I* statistic can go a stage further and identify which states are in statistically significant clusters, where high values are surrounded by other high values of the variable, and low values are surrounded by other areas that have relatively low values. This is shown in Figure 5.8. Local Moran's *I* is also able to indicate which regions have high values surrounded by low values and low values surrounded by high value areas, although none exist in this burglary example.

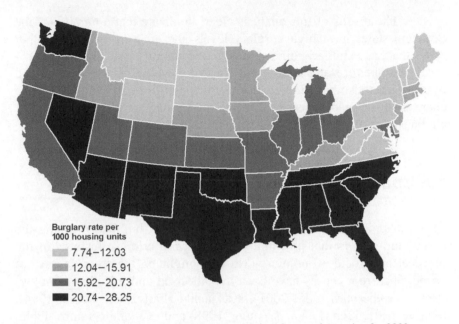

**Burglary rate per
1000 housing units**

7.74–12.03
12.04–15.91
15.92–20.73
20.74–28.25

Figure 5.7 Burglary patterns in the contiguous United States of America for 2002

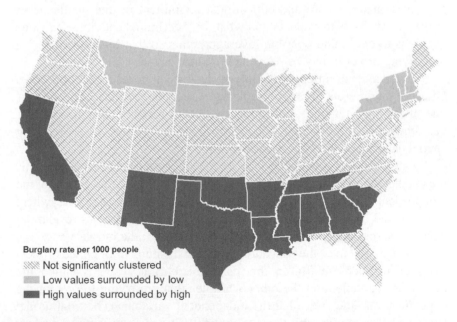

Burglary rate per 1000 people

Not significantly clustered
Low values surrounded by low
High values surrounded by high

Figure 5.8 High and low value clusters of states in the USA of 2002 burglary rate per
1000 housing units, significant to the 0.05 level

133

The results of this analysis clearly indicate that a number of the Southern states have high burglary levels and are surrounded by other states that have similarly high values. In the central North and the North–East, a number of states have significantly low values and are surrounded by states that also have low values. As the significance level is 0.05, the chances of the map appearing like this by chance are less than 5%. The challenge for researchers is to explain these patterns.

5.5 Spatial regression models

Crime mapping is not just about mapping crime, but also mapping the underlying drivers that potentially contribute to crime – features of the physical and socio-economic world that might be the 'root causes' of crime. Such root causes have been hypothesised and include unemployment (Weatherburn *et al.*, 2001), seasonality (Farrell and Pease, 1994), relative deprivation (Lea and Young, 1998) and even temperature (Field, 1992).

Criminologists have expended considerable effort in trying to understand the influence of variables such as unemployment, age, heterogeneity, housing tenure and educational attainment on the incidence of crime in order to better understand what drives criminal activity. The value in this is to determine possible leverage points so that by influencing an underlying driver it may be possible to reduce crime.

As a result, inferential analyses have been popular for some time. Of all of the inferential tests, Ordinary Least Squares (OLS) regression is one of the most popular. Unfortunately, recent work by spatial scientists has drawn into question the value of ordinary regression when exploring spatial data. There are two main concerns.

First, OLS regression is a global linear regression model. This means that when trying to explain a dependent variable such as crime, the independent variables are assumed to have the same influence everywhere in the model. As an example, consider a model that examines residential burglary across a city by measuring in each suburb: the burglary rate, the average age of homes, the education level and the number of police officers assigned to each suburb. In this case, the burglary rate is called the dependent variable, and the other variables that are trying to predict the dependent variable are called the independent variables. The final model from an OLS regression suggests that 80% of the residential burglary level in a city could be predicted by the mean house age in an area, the mean educational level of residents in each area and the number of police officers in each part of the city.

The problem with a global model such as this is that there is an assumption that the influence of house age (for example) is the same in every part of the city. In reality it is certainly possible that house ages in one part of the city are significant in predicting the burglary rate, as it may be that older homes are more easy to burgle. However, in a gentrified part of the city there may be similarly older homes but the new 'yuppie' residents have recognised their properties are vulnerable and have installed alarm systems. In these areas, the influence of property age is less important. OLS regression does not recognise that the influence of variables can change over space and assumes a process of 'spatial stationarity' – a process that is not assured within social science studies, such as crime research.

Secondly, and more worryingly, OLS regression models may be invalid for statistical reasons. While the statistics to explain this are fairly complicated, a relatively simple explanation of the concerns is as follows. There is a growing realisation that spatial data are not spatially independent. We know that crime data are highly clustered in parts of a city (as shown in this chapter and in Chapter 6) and that there are good theoretical reasons for this (Chapter 4). For ordinary regression to be statistically valid, the error terms (that part of the dependent variable that cannot be explained by the model) should be independent of each other. Unfortunately, error terms from OLS regression can be spatially autocorrelated such that high error terms are surrounded by other high error terms, and low error values are in areas of equally low terms.

This positive spatial autocorrelation in error terms means that the terms are not independent, and any statistical inference from OLS regression models is highly suspect (Fotheringham *et al.*, 2002). To combat this problem, there are a number of different approaches that explicitly analyse local spatial processes within regression analysis.

5.5.1 Spatial lag and spatial error models

A classic OLS regression would predict the value at a given location in a manner similar to the following equation:

$$\hat{y} = a + bx$$

where $\hat{y}$ is the predicted value of the dependent variable, a is a regression coefficient (the constant), b is also a regression coefficient and x is the value of an independent variable. The predicted value of y $(\hat{y})$ is not necessarily the one that is actually measured, as there is usually some error between the predicted and the observed value of y. The actual value of y can therefore be expressed as

$$y = a + bx + \varepsilon$$

where ε represents the error between y and $\hat{y}$. A multiple regression simply adds independent variables and their coefficients to the equation, in this manner:

$$y = a + bx + b_1x_1 + b_2x_2 + b_3x_3 + \varepsilon$$

In spatial statistics terms, this can be rewritten (swapping the b and X terms) into a standard linear regression model where the number of independent variables and their parameters is written as a single matrix term (Bivand, 1998):

$$y = XB + \varepsilon$$

A spatial lag model adds to this the impact of the dependent variable in neighbouring areas by including a regression coefficient for the weighted mean value of the local dependent variable (Anselin, 1988):

$$y = pWy + XB + \varepsilon$$

where Wy is the weighted mean of the local values of y in neighbouring areas and p is the parameter. This spatial lag model implies that any geographic clustering of a measure of crime is due to the influence of crime in one area on that crime in another. The output from a spatial lag model is a normal list of independent variables, along with their statistical significance and coefficient (parameter) values, with an additional measure for the significance and parameter value for the crime rate in neighbouring areas.

By contrast, a spatial error model gives the analyst an indication that clustering reflects the influence of unmeasured variables (Messner and Anselin, 2004).

$$y = XB + u$$

where $u = \lambda Wu + \varepsilon$. In this equation, u is a disturbance vector that is spatially autocorrelated and λ is a spatial error parameter. Instead of suggesting that (as in a spatial lag model) the neighbouring crime rate has an influence on the measured crime rate in an area, the spatial error model suggests that the amount of variance in the crime rate that is not predicted by the independent variables is due to spatially autocorrelated missing variables. In reality, it is often possible that both models can partially explain better the distribution of a dependent variable (Anselin, 1988). GeoDa (available at http://sal.agecon.uiuc.edu/) is a software package that includes routines to conduct both spatial error and spatial lag models.

Case study: A spatial lag model of anonymous narcotics tips in Philadelphia, USA

A classic OLS regression ignores the spatial correlations in data, and this can be demonstrated by comparing an OLS regression with a spatial lag regression in Philadelphia. The dependent variable is anonymous narcotics tips recorded by the city authorities, and for illustrative purposes, the independent variables (taken from the 2000 census and mapped at the block group level) are: the unemployment rate; the median household income; the housing occupancy rate; the median year the houses in the area were built; and the heterogeneity of the area (see Chapter 11 for this measure). The tests were conducted using GeoDa.

The OLS regression results in a Moran I value of 0.38, suggesting that there is significant spatial autocorrelation among the residuals from the analysis. This is one of the main problems with an OLS regression – error terms are supposed to be independent and uncorrelated.

Residuals from the spatial lag model have a Moran I value of only 0.02, which indicates the residuals are not spatially autocorrelated. Table 5.1 shows that, on the whole, the same variables that were statistically significant from the OLS regression are significant in the spatial lag model. There are two things to note though. First, the high Moran I value draws into question the accuracy of the significant test for the OLS regression because the residuals are not independent of each other. Secondly, in the spatial lag

Table 5.1 Ordinary and spatial regression results for narcotics tips in Philadelphia

	OLS regression		Spatial lag model	
	Coefficient	**p**	**Coefficient**	**p**
Constant	10.16	0.499	−5.160	0.653
Unemployment	5.006	<0.001	2.462	<0.001
Household income	−0.001	<0.001	−0.001	0.018
Housing occupancy	0.683	<0.001	0.292	0.009
Median year housing built	−0.007	0.478	−0.004	0.576
Heterogeneity	31.738	0.022	20.530	0.051
Spatial weight	–	–	0.6921	<0.001

model, the heterogeneity variable is not significant (to the 0.05 level) but the spatial weight measure is. This means that the number of narcotics tips in neighbouring areas is a good predictor of a local area's number of anonymous tips. The R-squared value of the test rose from 0.22 in the OLS regression to 0.55 for the spatial lag model, indicating the latter model is a better predictor of anonymous tips.

5.5.2 Geographically weighted regression

As an alternative response to the limitations in basic regression, a team at Newcastle University has developed techniques (and software) to specifically analyse spatial regression effects, called Geographically Weighted Regression (GWR). First discussed in 1996 (Brunsdon *et al.*, 1996), GWR recognises that the influence of spatial variables may vary over space, and actively seeks to measure this variation by performing localised regression equations all over the study area. The mathematics of GWR are complicated enough to be beyond the scope of this book, but the following case study shows the value of GWR at a strategic level. The further reading list of this chapter has details of the 2002 book by Stewart Fotheringham and colleagues, a book considered as the seminal work in this area.

Case study: Local spatial processes with Geographically Weighted Regression

Material supplied by Ronald E. Wilson, Mapping & Analysis for Public Safety Programme, National Institute of Justice

Some researchers who have examined social disorganization theory suggest that violence in urban areas results from weak institutional bases. Religion, as one such institutional base, has been studied by such authors as Baier and Wright as a base of social control, showing it to have moderate effects on crime rates (Baier and Wright, 2001). Religious institutions in the US represent powerful organizations that attempt to exert informal social control without directly enforcing laws through the legal system. However, when those institutions are in neighbourhoods that are unstable and have a highly transient population, there may be neither the time nor the participation from the community for the religious location to exert significant influence. There has been little, if any, work that has examined the geographic density of religious institutions in comparison to spatial patterns of violence.

Framed within a structure of neighbourhood boundaries as determined by the City of Philadelphia, a spatial analysis of this premise was undertaken. The unit of spatial analysis selected were census tracts for which the contextual variables of religious institution density were derived. These units were associated with composite variables from social disorganization theory. The primary spatial research question was, does a high density of religious institutions in census tracts limit the frequency of homicide in these tracts?

The difficulty with the classical approach of using a multiple regression (such as negative binomial regression) is the fundamental assumption that the error terms from such a global test are independent. Unfortunately, homicides cluster quite considerably in Philadelphia, therefore this assumption is not valid. Secondly, a global regression test only speaks to the level across the whole study area, and says nothing about local variation across the study area and within neighbourhoods.

Data were drawn from reported homicides collected by the Philadelphia Police Department during 1990. The database of religious institutions was derived from geocoding religious institutions listed in the Yellow Pages, and included premises from all major religious denominations, including churches, mosques and synagogues.

A Poisson model based on a GWR was used to identify local spatial trends and processes between religious institutions, social disorganization variables and their combined effect on the local homicide rates. Unlike classical regression, GWR does not establish which composition variables in a model are significant or not, rather it determines which variables, both contextual and compositional, have significant spatial variation. The result is an analysis of the effects of spatial processes on compositional variables that influence criminal behaviour.

The mapping of independent variable t-values revealed trends across the city that pointed to the presence of non-stationary spatial processes, something the global tests could not detect or depict. Global tests indicated that the density of religious institutions did not have a significant effect on homicide across the entire city. However a spatial analysis that was sensitive to the possibility that some variables are more influential in various parts of the city, revealed that there was a significant influence on neighbourhoods in North and North West Philadelphia (Figure 5.9). In mapping t-values for the covariate number of religious institutions, it was found that there were areas affected by the density of religious institutions, a relationship brought to light only from the application of GWR.

Figure 5.9 The change in *t*-values across the city suggests that the influence of religious institutions on the incidence of homicide is stronger in some parts (the darker shaded areas) than others (see Plate 1)

In this analysis, Geographically Weighted Regression statistics demonstrated that spatial processes differ across space at the local level. Mapping of residuals, as well as local level *t*-values, can illustrate that there are more complex spatial processes at work than can be accounted for with simple global models. Indeed this makes common sense – the influence of some factors will be stronger in some neighbourhoods than others. This study suggests that spatial variation must be controlled for so that the range of other contextual variables can be studied in order to understand, gauge and properly model their influence on crimes such as homicide.

5.6 Summary

Crime analysts are now benefiting from a recent increase in the number of spatial data analysis techniques that are available. The growth has been driven by the availability of cheap information technology that has enabled computer processing to quickly perform tasks that would have taken weeks by hand. There is now no shortage of ways to analyse data, either in an exploratory sense or in an inferential manner in order to uncover relationships in spatial data. The challenge for crime analysts is now to turn this wealth of analytical power into crime reduction. The techniques in this chapter can explain what is going on, and some can help to uncover why crime patterns are happening; however, the role of the analyst is also to use this knowledge to help prevent and reduce crime. Later chapters in this book explore ways in which decision-makers can be influenced into making good crime prevention and reduction decisions from crime maps and geographical crime analysis. One simple way is to show maps of crime hotspots in order to focus attention on areas of significant criminal activity. The next chapter is dedicated to the task of identifying crime hotspots.

Further reading

Ebdon, D. (1996). *Statistics in Geography*. Oxford: Blackwell.

A useful introduction to spatial statistics and descriptive measures. The book is pitched at the introductory level and is a good start for anyone, especially those with less experience in statistics.

Cressie, N.A.C. (1993). *Statistics for Spatial Data*. New York: Wiley.

The other end of the spectrum from David Ebdon's *Statistics in Geography*. The Cressie volume is a classic work that covers a huge array of spatial process, but at quite an advanced level.

Levine, N. (2004). *CrimeStat III: A Spatial Statistics Program for the Analysis of Crime Incident Locations*. Houston, TX: Ned Levine and Associates and Washington, DC: National Institute of Justice. Available at http://www.icpsr.umich.edu/nacjd/crimestat.html.

The manuals for CrimeStat are well-written and are applied to crime data. The examples lead the analyst through the techniques that are provided by the software.

Anselin, L. (1988). *Spatial Econometrics: Methods and Models*. Dordrecht: Kluwer Academic Press.

One of the key texts in spatial regression analysis, though certainly not for the statistically terrified. Not written specifically for crime analysis, it does however contain general terms and methods that are applicable to cutting edge spatial regression models. Be warned that the book contains a lot of equations!

Fotheringham, A.S., Brunsdon, C. and Charlton, M. (2002). *Geographically Weighted Regression*. Chichester (UK): John Wiley.

A recent book that explores the developing area of geographically weighted regression with many examples and applications. The companion software is available for a fairly low price.

References

Anselin, L. (1988). *Spatial Econometrics: Methods and Models*. Dordrecht: Kluwer.

Baier, C.J. and Wright, B.R.E. (2001). If you love me, keep my commandments: A meta-analysis of the effect of religion on crime. *Journal of Research in Crime and Delinquency*, 38(1), 3–21.

Bailey, T.C. and Gatrell, A.C. (1995). *Interactive Spatial Data Analysis*. Harlow: Longman.

Bivand, R. (1998). A review of spatial statistical techniques for location studies, http://www.nhh.no/geo/gib/gib1998/gib98-3/lund.html, accessed 23 March 2004.

Brunsdon, C., Fotheringham, A.S. and Charlton, M.E. (1996). Geographically weighted regression: A method for exploring spatial nonstationarity. *Geographical Analysis*, 28(4), 281–298.

Chainey, S.P., Reid, S. and Stuart, N. (2002). When is a hotspot a hotspot? A procedure for creating statistically robust hotspot maps of crime. In *Innovations in GIS 9*. London: Taylor & Francis.

Chakravorty, S. (1995). Identifying crime clusters: The spatial principles. *Middle States Geographer*, 28, 53–58.

Cohen, J. and Tita, G. (1999). Diffusion in homicide: Exploring a general method for detecting spatial diffusion processes. *Journal of Quantitative Criminology*, 15(4), 451–493.

Diggle, P.J. (1983). *Statistical Analysis of Spatial Point Patterns*. London: Academic Press.

Donnelly, K. (1978). Simulations to determine the variance and edge effect of total nearest neighbor distance. In I. Hodder (ed.) *Simulation Methods in Archaeology* (pp. 91–95). Cambridge: Cambridge University Press.

Farrell, G. and Pease, K. (1994). Crime seasonality: Domestic disputes and residential burglary in Merseyside 1988–90. *British Journal of Criminology*, 34(4), 487–498.

Field, S. (1992). The effect of temperature on crime. *British Journal of Criminology*, 32(3), 340–351.

Fotheringham, A.S., Brunsdon, C. and Charlton, M. (2002). *Geographically Weighted Regression*. Chichester (UK): John Wiley & Sons.

Lea, J. and Young, J. (1998). Relative deprivation. In J. Muncie, E. McLaughlin and M. Langan (eds) *Criminological Perspectives: A Reader* (pp. 136–144). London: Sage.

LeBeau, J.L. (1992). Four case studies illustrating the spatial-temporal analysis of serial rapists. *Police Studies*, 15(3), 124–145.

Levine, N. (2004). CrimeStat III: A spatial statistics program for the analysis of crime incident locations. Houston, TX: Ned Levine and Associates and Washington, DC: National Institute of Justice. Available at http://www.icpsr.umich.edu/nacjd/crimestat.html.

Mencken, F.C. and Barnett, C. (1999). Murder, nonnegligent manslaughter and spatial autocorrelation in mid-South counties. *Journal of Quantitative Criminology*, 15(4), 407–422.

Messner, S.F. and Anselin, L. (2004). Spatial analyses of homicide with areal data. In M.F. Goodchild and D.G. Janelle (eds) *Spatially Integrated Social Science* (pp. 127–144). New York, NY: Oxford University Press.

Messner, S.F., Anselin, L., Baller, R.D., Hawkins, D.F., Deane, G. and Tolnay, S.E. (1999). The spatial patterning of county homicide rates: An application of exploratory spatial data analysis. *Journal of Quantitative Criminology*, 15(4), 423–450.

Paulsen, D.J. (2004). Geographic profiling: Hype or Hope? Preliminary results into the accuracy of geographic profiling software. Paper presented at the 2nd UK Crime Mapping Conference, London. http://www.jdi.ucl.ac.uk/downloads/pdf/second_mapping_conference_papers/D_Paulsen.pdf.

Tobler, W. (1970). A computer movie simulating urban growth in the detroit region. *Economic Geography*, 46(Supplement: Proceedings International Geographical Union. Commission on Quantitative Methods (June, 1970)), 234–240.

Ratcliffe, J.H. (2005). Detecting spatial movement of intra-region crime patterns over time. *Journal of Quantitative Criminology*, 21(1), 103–123.

Ripley, B.D. (1981). *Spatial statistics*. New York: John Wiley & Sons.

Weatherburn, D., Lind, B. and Ku, S. (2001). The short-run effects of economic adversity on property crime: An Australian case study. *Australian and New Zealand Journal of Criminology*, 34(2), 134–148.

6
Identifying Crime Hotspots

Learning Objectives

In this chapter we explore a range of techniques used to identify the geographic location of crime hotspots. This involves visualising spatial patterns of crime in terms of the location, size, shape, extent and relative scale of hotspots. The chapter does not necessarily seek to find the optimal technique, recognising that these mapping techniques can be complementary and have different applicability across a range of crime mapping applications. During this journey you will also learn about several of the more sophisticated techniques that require functionality that is not offered in many standard GIS packages. For these we point the reader to several software extensions that have the functionality to apply these techniques, particularly highlighting software that is free, or refer the reader to other texts that further describe these techniques. Several of these advanced techniques are based on reasonably complex mathematics. In this chapter we do not go into the detail of these algorithms, our focus instead being towards understanding the suitability of these techniques, how they can be applied and how to interpret the hotspot results they produce.

6.1 Introduction

One of the first geographical questions that is asked of crime data is 'where are the hotspots?' A hotspot is a geographical area of higher than

GIS and Crime Mapping   Spencer Chainey and Jerry Ratcliffe
© 2005 John Wiley & Sons, Ltd

average crime. It is an area of crime concentration, relative to the distribution of crime across the whole region of interest (e.g. a city centre, census ward or tract, municipal district, county or state). Hotspots are often clusters of crime that can exist at different scales of interest. Knowing where these hotspots of crime are located offers a first step when exploring why these areas may suffer from persistent problems. However, what may initially appear as a fairly straightforward process of mapping crime data and identifying its hotspots can actually be quite challenging and prone to interpretation error. GIS software products offer a variety of techniques for representing spatial patterns, and the functionality between GIS software products may differ, such that what may be possible in one application may be different in another. Many GIS are also supplied with application extensions that offer a range of additional functions for visualising crime patterns.

Additional considerations include the impact of the Modifiable Areal Unit Problem (MAUP), the range of classes and the numerical thresholds to set in the map legend, the map design and the colours used to represent the spatial patterns, and whether the crime data are best represented as a simple count or concentration value, or whether the data need to be normalised against an underlying population. We also introduce the emerging field of predictive crime mapping, reviewing several techniques that are being developed to help reveal future crime patterns.

In the preceding chapter we described a number of spatial statistics that offer some description of geographic patterns in crime data series. Statistics such as the nearest neighbour index can be used to help describe if clustering exists, while others such as the standard distance can offer an indication of dispersion between different crime series. These types of statistic offer a preliminary insight into the existence of hotspots. We also saw in Chapter 5 that Moran's I can offer insight into whether crime counts that are near to each other are correlated, but since this statistic simply tests for deviations from uniformity (e.g. identifying those areas with high counts that are close to similar areas) its results could just imply areal perturbations in crime, rather than identifying areas where there is a significant high concentration of crime. In addition, many spatial autocorrelation techniques are classed as global statistics, which can be useful for helping to inform the nature of the general distribution of crime, but may only suggest a variety of local spatial patterns that exist in crime data.

In this chapter we begin by exploring the use of point maps for identifying hotspots of crime before discussing how these point data can be aggregated to meaningful geographic boundary areas such as census output areas, census tracts or police beats, and subsequently represented thematically to identify areas of high crime concentration. We also look at quadrat (or grid) mapping and explore the advantages and disadvantages of this technique. The use of continuous surface smoothing techniques such as kernel density estimation are next considered before exploring more advanced techniques (such as Local Indicators of Spatial Association (LISA) statistics) that can provide added robustness, and methods that take into account the underlying population (e.g. dual kernel density estimation and the Geographical Analysis Machine). Our review of these techniques follows an approach similar to that used by Eck *et al.* (2005) and offers useful comparisons. We finish the chapter by reviewing several emerging techniques for predicting crime patterns in space and time.

It is worth appreciating that hotspot mapping is not a beauty contest over which map looks the best – a visually stunning hotspot map does not explain why crime occurs at a location. However, good carto-graphic design is important in hotspot mapping as it clearly identifies areas that persistently suffer from crime. A visually appealing map can help enable a more focused approach to understanding areas that require crime reduction resources, and can offer direction for initiating the next analytical stages that explain the problem and how it can be tackled. Hotspot maps are therefore a blend of good cartographic design and robust methodology, and are a first step towards exploring crime patterns in more detail. The theories that are discussed in Chapter 4 help in understanding why crime incidents concentrate at certain locations, and provide the important basis for thinking about the actions that are appropriate when applying and designing reduction initiatives and policing strategies to hotspots. When mapping crime and analysing spatial patterns it is important to not lose perspective of these theoretical principles. Certain important geographical principles of spatial analysis are also presented in this chapter to ensure that the identification of crime hotspots is as accurate and effective as possible.

6.2 When is a hotspot 'hot'?

A question that often arises in hotspot analysis is how to define a map area as 'hot'. There is no universal or standard numerical threshold that can be used to define the number of crimes that need to have occurred in an area for the area to be defined as 'hot'. Hotspots are relative to the area under study. In other words, a hotspot represents an area of high crime concentration, relative to the distribution of crime across the whole region of interest. This means that regardless of whether crime patterns are being

relative def

studied across a rural, urban or suburban area, the area of high crime concentration relative to the general pattern of crime across the whole area will stand out as the problem crime area. Considered another way, what would count as a hotspot of crime in a rural area might count as a low crime region in an inner city. This does not mean that the rural area does not have a crime problem – a local hotspot is still a hotspot.

The challenge that is explored in this chapter is to illustrate and discuss the application of methods that can identify hotspots of crime. For most crime mappers, these methods need to be easy to use and at the disposal of the majority who map hotspots; however, as new techniques emerge and demonstrate improvements over more established techniques, it will be necessary for these to be built into new crime mapping programmes. We also assume that to identify hotspots in crime data these data need to be fit for the purpose. For example, if crime hotspots at the street level need to be identified, crime data that are accurate, precise, complete, consistent and reliable need to be available. In this chapter we make use of a sample dataset of crime to apply many of these methods for identifying hotspots. This sample data is for no particular geographic area, but has been sourced to help demonstrate and explore the application of hotspots mapping techniques.

6.3 Point maps

The most common method for displaying geographic patterns of crime is by point mapping (Jefferis, 1999). This method is popular because it is a simple digital way to carry out a familiar and traditional method of crime mapping: placing pins representing crime events onto a wall map.

Figure 6.1 shows a sample of robbery and residential burglary data as point maps. As you view the images, ask yourself which areas you would identify as the three main hotspot areas in Figure 6.1a and 6.1b. If you were to ask a colleague to identify the three main hotspot areas, it is likely that they would identify different areas or that the size and shape of the coverage of the area that they identify as a hotspot would be different. So who is right? At this stage answering this question is actually quite difficult.

These maps demonstrate that it is difficult to clearly identify the location, relative scale, size and shape of hotspots when crime data are presented as points. The large volume of data that are mapped makes it difficult to visualise and interpret accurate patterns in the data's spatial

(a)   (b)

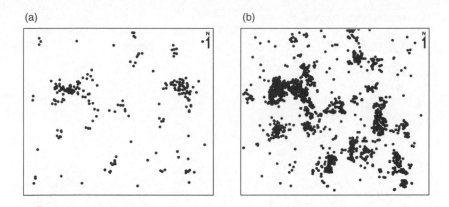

Figure 6.1 (a) Robbery point map and (b) residential burglary point map. Consider which areas you would identify as hotspots. Also consider the degree of confidence that you would attach to your decision: 100% confident or only 10%? A colleague may identify a different area, or disagree with the size and shape of the areas that others may identify as hotspots

distribution. An additional problem is that certain locations on the map appear to be single crime points but may in fact be multiple events mapped on top of each other. This occurs when crime events at the same location have been geocoded to exactly the same coordinates.

While point maps do have crime mapping value when only small numbers of crimes are to be displayed, they can be misleading when identifying hotspot areas, are not the most visually descriptive examples of hotspot maps and may not be the best map design to enthuse discussion and gather interpretation from others (Eck *et al.*, 2005). When it comes to handling large data volumes, a point map can become easily cluttered, especially if these points are also labelled with attribute information. Points mapped at coincident locations can be repositioned by scattering the points around their common location, but this introduces error in the location of crime events. It is usually better to use a symbol of variable size to represent the differences in the number of crimes at each location. For example, a graduated size symbol map could be used to identify properties that are repeatedly burgled (Figure 6.2). Care needs to be taken when interpreting maps of this type because the size of a symbol may be large enough to obscure patterns in surrounding areas. It is also possible that the size of the symbol is so large that it does not exclusively cover the precise location where the crime occurred, and so may lead map readers to falsely interpret where a crime happened. A large symbol can suggest that neighbouring locations were the target of repeated crime events.

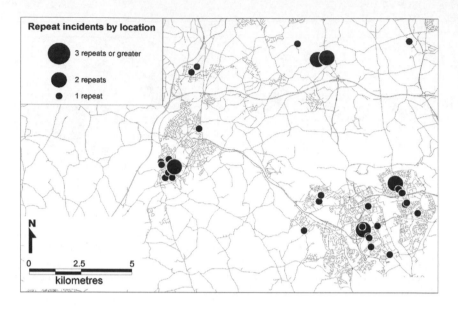

Figure 6.2 A map that varies symbol sizes at point locations to identify areas that have been repeatedly burgled. Care needs to be taken when using this approach as large symbols can obscure smaller symbols or lead to inaccurate interpretation of the location of the symbol. Reproduced by permission of Ordnance Survey on behalf of HMSO. © Crown copyright 2005. All rights reserved. Ordnance Survey Licence number 100044021

6.4 Geographic boundary thematic mapping

Crime events, mapped as points, can be aggregated in a GIS to geographical administrative or statistical boundaries such as census output areas, block groups, police beats or election polling districts. These counts of crimes by geographic area can be used to create thematic maps (also termed choropleth maps) that display the distribution of crime across the study area (Eck *et al.*, 2005). Using the robbery and residential burglary point data displayed in Figure 6.1, Figure 6.3 displays the pattern of the count of these crimes, aggregated to a defined set of geographic areas. Based on these aggregations of the crime data, many people would identify areas A, B and C as the high crime hotspot areas for these two different crime types. However, others may have reasons to identify other quite distinct areas or disagree over the size and the shape of the areas that have been chosen. Again, it can be difficult to determine who is correct.

Thematic mapping is a very popular technique, yet there are a number of geographical tyrannies associated with this technique that

(a)   (b)

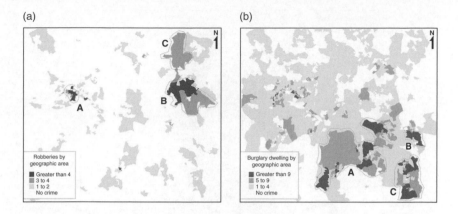

Figure 6.3 (a) A thematic map of the number of robberies per geographic area and (b) a thematic map of the number of residential burglaries per geographic area, labelled with their three possible main hotspot areas

can lead to misinterpreting the geographical distribution of the underlying crime data.

Geographic areas of various sizes and shapes, when aggregated and thematically shaded can be misleading. For example, natural tendency is for the map reader's attention to be drawn to the large areas that are boldly shaded (MacEachren, 1995; Monmonier, 1996). Thematic mapping is a process that shades the whole of a region and can often be too coarse to represent the detailed spatial patterns of actual crime events. Indeed, from a reading of Chapter 4, it is highly unlikely that crime events are uniformly spread throughout a region. This becomes clear when examining Figure 6.4, which shows the hotspot area that was marked as area C in Figure 6.3b. At this more detailed level of inspection it is possible to see that the distribution of the crime data across these geographical areas is not evenly spread, and that there are large areas where there are no crimes. The thematic shading gives the impression that crime is spread over the whole area. Although commonly done by crime mappers, most cartographers would consider it unacceptable to map raw counts of crime to areal units. The appropriate approach, when dealing with polygons of different sizes, is to divide the number of crimes by some appropriate denominator, such as the number of houses (for burglary) or the number of residents in an area (for robbery). This is discussed in more detail in a later section in this chapter, and in section 12.5.

A second problem that affects all types of thematic mapping is the MAUP (Openshaw, 1984). This is where the results of any geographic aggregation process, such as the count of crimes within a set of geographic boundaries, may be as much a function of the size, shape and orientation

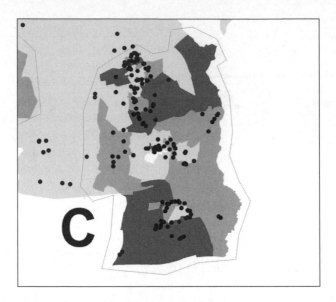

Figure 6.4 Thematic maps may mislead and not truly represent in detail the locations of where crimes cluster. This map shows that there are large areas where there are no crimes, yet the thematic shading gives the impression that crime is spread over the whole area

of the geographic areas as it is of the spatial distribution of the crime data. In essence, when thematically mapped, different boundaries may generate different visual representations of where the hotspots may exist.

A third point to consider is the class boundaries that can be set for the thematic map. In general, classes should be organised so as to be as easy to understand as possible. In many cases it is useful to apply a custom range, where each range break is logical. The map should be the central message, and if the audience question the legend that has been used, then this can distract attention away from the message that is being presented with the map. A logical range setting (e.g. 0, 1–5, 6–10, 11–15, Greater than 15) can be easier to understand than the majority of other settings (see Harries, 1999 and section 12.5.1 for more information on class boundaries).

6.4.1 The ecological fallacy

A final consideration with thematic maps occurs at the interpretation stage. Many unwary crime mappers make statements about the residents of areas, statements that fall foul of the ecological fallacy. The ecological fallacy can occur when an inference is made about an individual based on aggregate data for a group. For example, an analyst may examine the aggregate income for a neighbourhood of a city and discover that the average household income for the residents of that area is $30 000. To state that the average income for residents of that area is $30 000 is true and accurate. The ecological fallacy can occur when the analyst then states, based on this data, that people living in the area earn about $30 000. This may not be true at all, and may be an ecological fallacy. Close examination of the neighbourhood might show that the neighbourhood is actually composed of two housing estates – one of a lower socio-economic group of residents and another of a higher socio-economic group. The residents in the poorer part of town earn about $10 000 while the more affluent citizens earn about $50 000. When the analyst states that individuals who live in the area earn $30 000 (the mean rate), this did not account for the fact that the average in this example is constructed of two disparate groups, and it is likely that not one person earns $30 000.

This does not mean that identifying associations between aggregate figures is necessarily defective, and it does not necessarily mean that any inferences drawn about associations between the characteristics of an aggregate population and the characteristics of sub-units within the population are absolutely wrong either. What it does say is that the process of aggregating or disaggregating data may conceal the variations that are not visible at the larger aggregate level and that crime mappers should be careful in how they interpret aggregate values. This care is also extended to how the aggregation of data at different geographic levels produces different relationships to that which exists at the individual level – a geographical tyranny referred to as *ecological correlation* which we describe in more detail in Chapter 7.

6.5 Grid thematic mapping

A technique that can help overcome the problems of varying sizes and shapes of geographic areas is to use a uniform grid, where each cell (a quadrat) is of the same size and shape. Each grid cell can have a crime count associated to it, which can then be thematically mapped. The mapped value could also be a density value calculated from the count and cell area (Eck *et al.*, 2005).

Choosing an initial grid cell size is difficult. An initial guide can be to use one that is approximately the distance in the longest extent of the map, divided by 50. For example, if a study area's longest extent is 10 km in distance, an initial trial grid cell size to use would be 200 m.

same data, diff technique =)
diff - HS

Most GIS software include tools for the creation of these types of grids. Once the grid is created, a point-in-polygon operation can calculate the number of crime points in each grid cell. The grid can then be thematically shaded in relation to the count of crime points within it. Figure 6.5 shows the thematic grid maps of the robbery and residential burglary sample data, and areas A, B and C have been marked as the likely crime hotspots. When compared to the three respective hotspot areas identified in Figure 6.3, differences between the areas that were selected are revealed. Both maps use the same data, but the techniques have identified certain different distinctions.

Grid thematic mapping does tend to better represent the spatial pattern of crime when an appropriate cell resolution has been set, in terms of determining the location, size and relative scale of hotspots (Eck *et al.*, 2005). However, grid thematic mapping does suffer from certain similar problems to all thematic mapping in that it can still be affected by the MAUP. A coarse series of grid cells may hide some of the spatial patterning detail within the cell and that inappropriate class boundaries for the thematic map can produce unhelpful or misleading results. Many crime mappers also often comment that they do not like the appearance of grid maps, in that the 'blockiness' can be distracting. Improving the granularity by reducing the cell size does help to identify some of the high crime areas in more detail, but can often only have the effect of losing much of the visual patterning of crime that a slightly coarser grid offers (Figure 6.6). It also comes at a cost of larger file sizes and processing time.

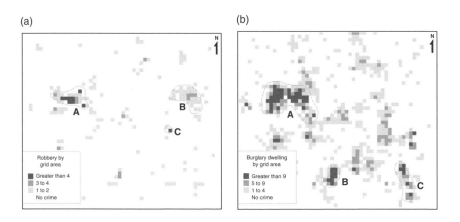

Figure 6.5 (a) A thematic map of the number of robberies per grid cell and (b) a thematic map of the number of residential burglaries per grid cell, labelled with their three possible main hotspot areas

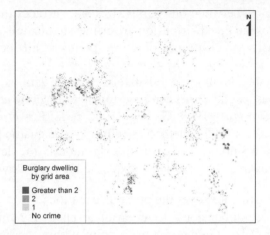

Figure 6.6 By improving the resolution of the grid cells, it may be possible to identify in more detail some of the high crime areas, but this can often have the affect of losing much of the general visual interpretation of crime distributions. This figure shows the distribution of burglary dwellings, counted per grid cell, using a grid cell that is a fifth of that used in Figure 6.5. The general pattern of the distribution of burglary dwelling is similar, and it does help to more finely identify some of the high crime areas, but loses some of the visual impact to assist in identifying hotspots

6.6 Continuous surface smoothing methods

An increasingly popular method for visualising the distribution of crime and identifying hotspots is one that creates a smooth continuous surface to represent the density or volume of crimes distributed across the study area (Chainey *et al.*, 2002; Eck *et al.*, 2005). These types of methods are commonly referred to as interpolation techniques and include inverse distance weighting and kriging. These types of techniques use an intensity or population value taken from sample locations to estimate values for all locations between sample sites. An example of this is a continuous surface map that is used to represent the distribution of rainfall, where the values between rain measurement points (the sample sites) are estimated from a function that considers the rainfall readings at these points and the distribution of these points (Eck *et al.*, 2005).

With crime data we do not necessarily have sample sites where there is an intensity value, and neither are we trying to estimate the number of crimes that may have occurred between these crime point locations. We should therefore avoid methods that aim to create estimated intensity values in the gaps between our points. Instead, surfaces that we wish to create to represent the distribution of crime should tell us something about the density or clustering of crime points at all locations in our study area.

A suitable method for visualising crime data as a continuous density surface is quartic kernel density estimation (Ratcliffe and McCullagh, 1999; Williamson *et al.*, 1999; Chainey *et al.*, 2002; Eck *et al.*, 2005). Eck and colleagues describe the method as follows: 'The quartic kernel density method creates a smooth surface of the variation in the density of point events across an area. The method is explained in the following steps:

- a fine grid is generated over the point distribution;
- a moving three-dimensional function of a specified radius visits each cell and calculates weights for each point within the kernel's radius. Points closer to the centre will receive a higher weight, and therefore contribute more to the cell's total density value (Figure 6.7); and
- final grid cell values are calculated by summing the values of all kernel estimates for each location' (Eck *et al.*, 2005).

The cell values that are generated typically refer to the number of crimes within the area's unit of measurement (e.g. crimes per square kilometre).

The quartic kernel density estimation method is available in most GIS software as extensions (e.g. HotSpot Detective or Vertical Mapper for MapInfo and Spatial Analyst for ArcGIS) and is also available in CrimeStat. It typically requires two parameters to be entered before it can be applied against crime data. These are the grid cell size and bandwidth (also known as the search radius or 'interval'). Of these, bandwidth is the parameter that will lead to most differences in output when it is varied.

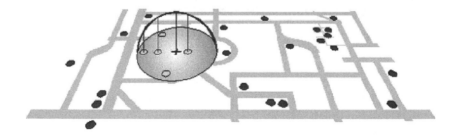

Figure 6.7 A search radius (or bandwidth) is selected, within which intensity values for each point are calculated. Points are weighted, where incidents closer to the centre contribute a higher value to the cell's intensity value of the cell. Source: Ratcliffe (1999a)

6.6.1 Bandwidth selection

Choice of suitable bandwidth has seen much discussion, both from the statistics field and also from spatial epidemiology, where kernel density estimation has been widely applied (Stone, 1974; Hogg, 1979; Bowman, 1984; Silverman, 1986; Cliff and Haggett, 1988; Sheather and Jones, 1991; Bowman and Azzalini, 1997). For example, one method makes use of a Moran's I correlogram to examine how spatial autocorrelation changes with distance. The correlogram shows the shape of the distance decay as well as the distance upon which it approaches the global Moran's I. Using this method, the shape of the distance decay can indicate the type of kernel to select, while the 'sill' (the point at which the correlogram approaches the global Moran's I) can indicate an appropriate bandwidth (Cliff and Haggett, 1988; Levine, 2004). However, this method cannot be applied on point data because the calculation of Moran's I is dependent on data being aggregated to geographic units. For point data, Brimicombe (2004) suggests values for the bandwidth to be 6, 9 or 12 times the median nearest neighbour distance, while Bailey and Gatrell (1995, pp. 86–87) explain that 'the value of kernel density estimation is that one can experiment with different values [of the bandwidth], exploring the surface...using different degrees of smoothing in order to look at variation in [the surface] at different scales'.

One method for choosing the bandwidth value for kernel density estimation on crime data, suggested by Williamson *et al.* (1999), was to vary the bandwidth relative to different order values of the mean nearest neighbour distance. These different orders refer to the mean nearest neighbour distance between each point and either its closest nearest neighbour, its second closest nearest neighbour or its nth closest nearest neighbour (often referred to as different orders of K). These nearest neighbour distances for different orders of K can be calculated using software such as CrimeStat (Box 6.1). Using a bandwidth value that relates to the spatial distribution of the crime points has the advantage of retaining a spatial component in the calculation of a density estimation surface, rather than selecting an

---

Box 6.1 Using CrimeStat to produce kernel density estimation surfaces

CrimeStat offers functionality to create kernel density estimation surfaces and also allows the user to generate mean nearest neighbour distance values for different orders of K.

Continued on page 158

Continued from page 157

Mean nearest neighbour distances

- Open the crime data in CrimeStat and set the necessary parameters, values and units in the 'Data setup' tab. The study area's coverage area should also be entered in the 'Measurement Parameters' tabbed area.
- Choose the 'Spatial description' tabbed area and 'Distance Analysis'.
- Click the 'Nearest neighbour analysis (nna)' check box and set the 'Number of nearest neighbours to be computed' to 20. This will generate mean nearest neighbour distance values for 1 to 20 K orders. Note these K-order distances.

Kernel density estimation

To create a kernel density estimation surface in CrimeStat requires the user to initially specify the extent and cell resolution of a grid that covers the entire point data coverage.

- Select the 'Data setup' tabbed area and 'Reference file'. Ensure that 'Create Grid' button is selected and enter the maximum and minimum x and y extents of the coverage of the sample area.
- Choose a 'Cell specification' and next to the 'By cell spacing' prompt enter a cell size value that is suitable for your data. As a guide, you could follow the guidelines on cell size that are referred to in section 6.6.2.
- Choose the 'Spatial modelling' tabbed area and select 'Interpolation'. Choose the kernel density estimation 'Single' option. Choose the following parameters: 'File to be interpolated' – 'Primary'; 'Method of interpolation' – 'Quartic'; 'Choice of bandwidth' – 'Fixed Interval'; 'Interval' – this is the bandwidth, therefore experiment by entering a value of a K-order mean nearest neighbour distance for this crime dataset; 'Output units' – select a suitable unit; 'Calculate' – 'Absolute Densities'; and 'Output' – save the result to a specified GIS format (e.g. ArcView 'SHP').
- Click 'Compute'.

CrimeStat will then generate a file that can be opened in a GIS and thematically shaded to produce a kernel density estimation surface similar to those shown in this section.

arbitrary number as the parameter value. However, using a K-order mean nearest neighbour distance approach still requires a decision as to which K order to apply, meaning that its selection could still be quite arbitrary.

Following on from Bailey and Gatrell's suggestion (1995), a useful rule to follow is to consider the geographic scale at which crime hotspots need to be identified and choose a K-order mean nearest neighbour distance that will best reflect this scale. Low K orders are best applied when the crime mapper wishes to visualise crime patterns in fine detail, for example to identify specific localities (e.g. street corners) that have high crime levels. *depends* Larger K orders will more heavily smooth point data to provide a more *as* general view of the distribution of crime, and could be used for more *purpose* strategic purposes such as identifying neighbourhoods that require strategic crime reduction investments. It may be clear from this that more research is required from the crime mapping field, research that identifies the affect the bandwidth has on the accuracy of hotspot mapping output and the interpretation of the images, rather than just determining bandwidth based on what looks the best. In this section we will demonstrate the impact that bandwidth size has on the visual output generated from kernel density estimation.

Using the kernel density estimation method also requires specifying the cell size of the fine grid that is initially generated across the crime series distribution. Large cell sizes will result in more coarse-looking maps but are suitable for large scale output, while smaller cell sizes will offer a finer level of granularity more akin to a continuous surface, but may generate large file sizes. Where the user is unsure over the cell size to use, we suggest following the methodology of Ratcliffe (1999b) where cell size resolution is the result of dividing the shorter side of the minimum bounding rectangle (i.e. the shortest of the two extents between the maximum x and minimum x, and maximum y and minimum y) by 150. Although an arbitrary value, it works as a reasonable starting point for most crime mapping applications.

Figure 6.8 shows six different kernel density smoothing surfaces for the robbery and burglary dwelling crime data. The cell size that was applied was calculated by following the Ratcliffe (1999b) methodology, as explained above. The figures demonstrate the differences in map output in relation to different bandwidths and the smoothing impact that large bandwidths have on the results. Each bandwidth was determined by calculating the mean nearest neighbour distances for different orders of K. K orders of 1, 5 and 10 were applied to the kernel density estimation routine used to create these surface maps for robbery and burglary dwelling data. Note that because the mean nearest neighbour distances are different for each individual crime dataset, there are differences in the level of smoothing that are applied between the two sample datasets at the same orders of K. In other words, the K-10 distance used as the cell bandwidth

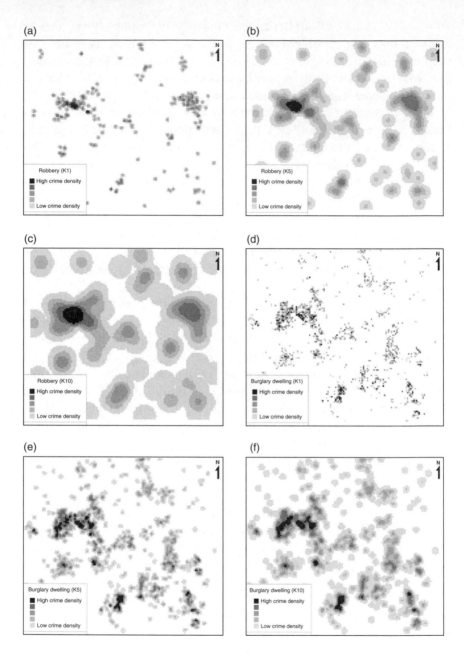

Figure 6.8 Kernel density smoothing maps of robbery (a, b and c) and residential burglary (d, e and f) generated using bandwidths of K order 1 mean nearest neighbour distance (MNND) (a and d), K order 5 MNND (b and e) and K order 10 MNND (c and f). The maps illustrate that different levels of smoothing result from the use of different sizes of bandwidth

for the robbery data is different to the K-10 distance used in the burglary set. This difference reflects the different spatial patterns of the crime distributions. The mean nearest neighbour distances for the residential burglary data were approximately half that of the robbery data.

The rise in popularity of kernel density estimation as a way to visualise spatial patterns in crime data has been largely due to its availability in software packages, and the visual appeal in the outputs that can be generated to represent crime hotspots. A kernel density estimation output allows for an easier interpretation of where crimes cluster in comparison to point, geographic area thematic, and grid thematic maps. The maps area thematic, and also reflect more accurately the location, relative scale and spatial distribution of crime hotspots in comparison to these other methods. The kernel density estimation method also considers concentrations of crime at all event levels, rather than grouping some crime events into clusters and discounting others – an issue that was identified in some of the early crime hotspot mapping techniques such as STAC (Chainey *et al.*, 2002; Eck *et al.*, 2005). The kernel density estimation method also creates grid cell value outputs that can be compared from map to map (assuming the same parameters are employed). Finally, the method has the advantage of deriving crime density estimates based on calculations performed at all locations (Levine, 2004) and retains some practical flexibility in map design (Eck *et al.*, 2005).

6.6.2 Issues in the use of kernel density surface estimation

Certain problems do exist with the kernel density estimation method. First, kernel density estimation is a smoothing technique where the level of smoothing is determined by the bandwidth. This can result in surfaces that smooth over and into areas where no crime has happened and where no crime point data exists (such as just outside a study area), and hence exaggerate the distribution of the crime problem. The issue over which class boundaries to choose to represent the different thematic thresholds also still presents itself as a problem. Many of those who use the technique often fail to question the validity or statistical robustness of the map that is produced, being caught instead in the visual lure of the image (Eck *et al.*, 2005). As a result, little regard is given to the legend thresholds that are set. For example, a map showing the distribution of crime as a kernel density estimation surface can have a variable number of hotspots depending on the ranges selected by the map designer to show spatial concentrations of these point events (Eck *et al.*, 2005). This is demonstrated in

(a)   (b)

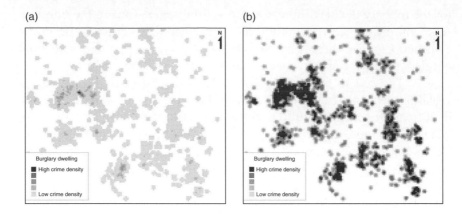

Figure 6.9 Different threshold settings produce different results. Figure (a) appears to show that the crime problem is not too bad because very few areas appear in the high crime density category (darkest shading). Figure (b) appears to show a widespread crime problem with many areas in the high crime density category. Each map was produced using the same crime data and differs only by the threshold values that were chosen to represent the different thematic shades

Figure 6.9. The source data of residential burglary remain the same in each map, yet Figure 6.9b appears as though there is a worse crime problem. In essence, by setting a lower threshold for the highest colour value on the map, the crime analyst can give the impression that there is a higher crime problem in an area than there may be in reality.

      Chainey *et al.* (2002) suggested the application of incremental multiples of the grid cells' mean to help standardise thematic thresholds. Whilst this approach can help determine useful thematic thresholds, the approach does require some careful calculations for it to be applied. What is still missing is a method that can statistically define those areas that are hotspots. The kernel density estimation method can produce some excellent results in identifying the location, size, shape, relative scale and orientation of hotspots, but care must be taken in the selection of thematic range settings. After the following case study, the next sections explore statistical methods of hotspot detection.

Case study: Mapping hotspots of thefts of vehicles in Camden, London

Camden is a borough in inner London. As part of an analysis of vehicle crime in the area, hotspots of thefts of vehicle crime were mapped using kernel density estimation. The kernel density estimation parameters (cell size

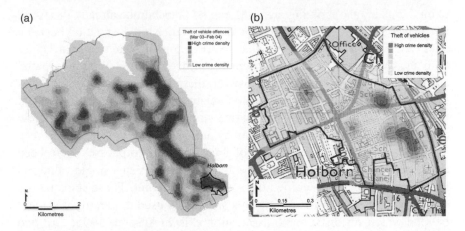

Figure 6.10 Using kernel density estimation to produce crime hotspots. (a) Theft of vehicle crime hotspots in the London Borough of Camden. This map provides a general view of hotspots across the whole area, using a cell size resolution of 40 m and a bandwidth of 310 m. Figure (b) shows the hotspots within the hotspot of Holborn. This figure used data only for this area of Camden, a cell size resolution of 5 m and a bandwidth of 25 m. This figure reveals three main areas in the Holborn hotspot where crime is most concentrated. Reproduced by permission of Ordnance Survey on behalf of HMSO. © Crown copyright 2005. All rights reserved. Ordnance Survey Licence number 100044021

and bandwidth) were selected based on the application of the mapping output. Initially a cell size of 40 m and a bandwidth of 310 m were chosen to view hotspots across the whole study area. This revealed several hotspots of vehicle thefts across Camden (Figure 6.10a). The parameters were then modified after crime data points were selected just for those areas within and in the vicinity of the main hotspots so that more detailed spatial patterns in these areas could be explored. For example, Figure 6.10b shows the detailed hotspots of thefts of vehicle crime in the Holborn area of Camden, created using a cell size of 5 m and bandwidth of 25 m (the Holborn area can be seen at the bottom right of Figure 6.10a). This application of kernel density estimation revealed three main crime problem areas in Holborn which allowed for the commencement of further analysis that helped explore and better understand why these areas were persistent problem crime areas.

6.7 Local Indicators of Spatial Association (LISA) statistics

From a crime mapping perspective, spatial association statistics test whether the number of point events in an area is similar to the count of point events in neighbouring areas. Moran's *I* (a technique already discussed in Chapter 5) is one such test. It explores the spatial autocorrelation between data variables (i.e. where events that are close together have similar values than those that are further apart) and can determine if positive spatial autocorrelation is said to exist. However, these global statistical measures may offer little insight into the location, relative scale, size, shape and extent of hotspots, and often only summarise an enormous number of possible disparate spatial relationships in crime data.

In recent years a number of local statistical processes have been developed, processes that identify the association between a single value and its neighbours up to a specified distance from the point. These statistics are suited to the identification of hotspots and can be used to identify distances beyond which no discernable association exists (Anselin, 1995; Ord and Getis, 1995; Getis and Ord, 1996; Ratcliffe and McCullagh, 1999). In addition, a problem with the mapping techniques that have been discussed so far in this chapter is that none of them can define with any statistical significance those areas that are suspected of being hotspots. Local statistics can offer additional statistical robustness to support the suspicion that certain areas can be defined as hotspots. These LISA statistics include Local Moran's *I*, Local Geary's *C* and the Getis and Ord Gi and Gi* statistics (Ord and Getis, 1995; Getis and Ord, 1996).

All these LISA statistics assess the local association in data that is mapped, but do so in different ways. The Local Moran's *I* is based on covariance, the Local Geary's *C* is based on measuring differences between points and the Gi and Gi* statistics compare local averages to global averages. It is the Gi and Gi* statistics that have received most attention in recent years by crime mappers (for example, Ratcliffe and McCullagh, 1999; Chainey *et al.*, 2002; Eck *et al.*, 2005) because as a method they fit neatly with the definition of a hotspot – they identify those areas where the local averages (e.g. concentrations of crime) are significantly different to the global averages (i.e. in comparison to what is generally observed across the whole study area).

6.7.1 Gi and Gi* statistics

The use of the Gi and Gi* statistics is best demonstrated with an example.[1] We will use an area subdivided into regions, shown as a 16×16

---

[1] Here we use a similar example methodology as Getis, A. and Ord, J. (1996). Local spatial statistics: an overview. In *Spatial Analysis: Modelling in a GIS environment*, ed. by P. Longley and M. Batty. London: GeoInformation International.

1	1	1	5	0	0	0	1	0	0	0	0	0	0	3	2
0	3	0	0	6	1	0	1	1	0	0	0	0	0	1	3
5	0	0	0	0	1	9	5	0	0	3	0	0	1	0	1
1	4	0	2	0	5	0	0	0	1	1	0	0	0	0	2
1	0	2	3	0	3	6	0	1	2	0	0	0	1	5	0
3	5	0	4	0	0	0	2	1	2	1	1	0	0	1	0
0	0	1	1	8	1	6	6	2	2	0	1	0	1	2	0
0	2	2	2	4	6	12	9	2	2	3	6	2	0	0	2
0	0	3	8	5	1	2	1	1	1	5	0	0	0	2	2
1	2	4	2	1	0	1	0	1	3	0	0	2	3	0	2
4	4	1	0	0	1	1	1	0	2	1	4	2	1	6	4
1	1	0	0	0	0	0	0	1	4	5	2	2	6	1	0
0	0	0	2	0	0	1	0	2	6	1	3	0	4	0	0
1	1	0	0	0	0	0	0	0	2	0	0	13	0	0	0
0	0	0	1	1	0	0	0	1	4	6	0	2	0	0	0
0	8	2	6	0	0	0	4	3	1	4	7	0	0	0	0

Figure 6.11 A 16×16 matrix of cells, each holding a value representing the number of crimes within their respective areas. In this example the distance between each grid centroid is 125 m

cell grid shown in Figure 6.11. Each cell can be identified by its centroid point (positioned in the centre of each cell) and each cell has a count of the number of crimes in that area (Figure 6.11). In our example, the centroid distance from each cell to another is an arbitrarily chosen value of 125 m.

Consider the point positioned in the eighth row of the eighth column in Figure 6.11. This point (which we shall call i) has the value 9. In the first instance we specify a null hypothesis that site i is not the centre of a group of unusually high values centred on i and its surrounding cells (neighbours). We can use standard terminology and call each neighbour j, such that our null hypothesis states that the total crime count in i and a group of j cells up to a distance d from i is not significantly higher than anywhere else on the grid.

This null hypothesis states that the sum of values at all the j sites within a radius d of i is not more (or less) than one would have expected by chance given all the values in the entire study area (both within and beyond the distance d). If local spatial autocorrelation exists, it will be exhibited by a spatial clustering of high or low values. This can be determined by Gi or

Gi* statistics. When there is a clustering of high values, the Gi and Gi* values will be positive. Low values will yield a negative Gi or Gi* value.

Anselin (1995), Ord and Getis (1995), Getis and Ord (1996), and Ratcliffe and McCullagh (1999) provide details of the equations for Gi and Gi*. We will not worry about looking at them in this chapter, but instead will focus on how we can apply these statistics to crime data. The difference between the two statistics is that the Gi* statistic also includes the value of the point i in its calculation. Gi excludes this value and only considers the value of its nearest neighbours (all of the j cells) against the global average. Gi* is the more popular of the two statistics because it considers all values within d.

Geographical Information Systems software such as ArcGIS and MapInfo do not routinely offer these types of spatial statistical analysis functionality. A useful tool for calculating Gi and Gi* statistics is a freeware Excel-based tool called Rook's Case, developed by Michael Sawada from the University of Ottawa (Sawada, 1999). The reference also lists the URL for sourcing Rook's Case. Using Rook's Case requires the crime point data to initially be aggregated to a fine grid in a GIS. These data can then be imported into Excel in the format required for Rook's Case. The output generated from Rook's Case can then be exported back into the GIS by linking to the polygon-based grid that was initially used to generate aggregate grid cell values. Box 6.2 helps to further explain this process.

Box 6.2 Using Rook's Case to generate Gi* statistics

Download Rook's Case from: http://www.uottawa.ca/academic/arts/geographie/lpcweb/newlook/data_and_downloads/download/sawsoft/rooks.htm

Follow the online information from this website for installing Rook's Case into Microsoft Excel. Help files supplied with Rook's Case also add to the steps provided in this box.

- In your GIS generate a grid, similar to that used in section 6.5 of this chapter, and populate this grid with the count of the crimes per grid cell. We recommend that you start with a small grid (maximum 50×50 cells) to initially test the speed and functionality of Rook's Case.
- Generate a table from your GIS that lists each grid cell's centroid x and y coordinate pair and the respective count of crimes within

Continued on page 167

Continued from page 166

each cell. It is also useful to allocate an identity number to each cell to ease a later process of linking the Gi* results to this grid (e.g. label each cell with a unique number 1, 2, 3...n). Export this table in a format suitable for opening in Excel. We will call this table 'Crime grid data'.

- Open Excel. If you have installed Rook's Case by following the installation directions you will now see a toolbar appear that is called ROOKCASE.
- Open the file 'Crime grid data'. For ROOKCASE to work the data need to be arranged in the format as a list of x, then y, then the crime count for each grid cell. Remove any other information such as the identity number and column headers. These can be re-added later.
- Click on the ROOKCASE button. You will see the ROOKCASE dialog box appear.
- Click on the 'Local Spatial Auto' button. Select 'Getis-Ord Gi and Gi*'.
- Enter appropriate 'Lag distance' and 'Lag values' (see section 7.1 for guidance on these parameters).
- Click the compute button.
- Your 'Crime grid data' spreadsheet will be populated with the Gi and Gi* statistics for each point, and for each lag. The Gi* statistic is listed under the 'z-Gi*(d)' and the Gi as 'z-Gi(d)'. If you are only interested in the Gi* results for your specified lags it may be worthwhile removing all other new columns. It may also be worthwhile replacing the identity number for each cell in the first column in Excel.
- Save this file and import into your GIS. The Gi* results can be linked back to the original grid by performing a join between the identity numbers or by using the x and y coordinates.
- Once the grid has been populated with the Gi* results, these results can be thematically mapped, or mapped to reveal only those cells that have statistically significant Gi* values (see Figure 6.15).

The Gi* statistic requires two parameters: the lag distance and the number of lags to apply. A lag distance is the radius of a moving circle that visits each grid cell. In many ways it is similar to the bandwidth measure that is used for kernel density estimation, except that a suitable value is easier to determine. The lag distance should be at least the distance between each grid cell, and preferably at least the distance of the radius from each cell's centroid that has a coverage that will consider all of its immediate neighbours. The distance between the grid cells in the example 16×16 cell grid in Figure 6.11 is 125 m. A suitable lag distance to apply could be 178 m as the furthest distance to one of the eight immediate neighbours is 178 m (i.e. $\sqrt{(125^2 + 125^2)}$). As we will test for local clustering we should determine an appropriate distance. To ensure that all eight of the immediate neighbours were included within the first lag, the lag distance was set to 180 m. Increasing the number of lags allows a crime mapper to explore how far spatial association exists from each point. A lag of 5 for the 16×16 matrix will calculate Gi and Gi* values within a distance (d) of 180 m, 360 m, 540 m, 720 m, 900 m.

Table 6.1 shows the Gi* statistics for the cell positioned in the eighth row of the eighth column with the crime count value of 9. Higher positive values of Gi* indicate greater clustering of high values. At a lag of 1 the Gi* statistic is positive. This high and positive Gi* statistic indicates that there is positive local spatial association between this cell and its neighbouring cells. That is, this particular cell and the eight cells immediately surrounding it and forming a 3×3 matrix have a high total count and that the high total in these cells is greater than the global average. The Gi* statistic at a lag of 2 is also positive but not as high as the Gi* statistic at a lag of 1. The Gi* value remains similar up to a lag of 4 (i.e. up to a distance of 720 m from the cell in question), but then reduces considerably to 0.3 at a lag of 5. For those cells that have low values (i.e. low crime counts) and which are also surrounded by cells of low values, the Gi* statistic would be negative.

To legitimise the Gi* results it is possible to determine a threshold measure that defines whether these values are statistically significant. Ord and Getis (1995) suggest a Bonferonni-type test to generate measures of significance. (Bonferonni is a statistical procedure that performs multiple tests to determine levels of significance in a data sample.) Ord and Getis

Table 6.1 Gi* statistics for the cell positioned in the eighth row of the eighth column of the 16×16 grid. At a lag of 1, within a distance of 180 m from this cell, there is a high positive level of local spatial association. This reduces at a lag of 2, remains stable to a lag of 4, but then reduces close to zero at a lag of 5

Gi* at 0 to ≤ 180 m (Lag 1)	Gi* at 0 to ≤ 360 m (Lag 2)	Gi* at 0 to ≤ 540 m (Lag 3)	Gi* at 0 to ≤ 720 m (Lag 4)	Gi* at 0 to ≤ 900 m (Lag 5)
4.2	2.4	2.3	2.6	0.3

(1995) publish a comprehensive table for Gi* and Gi at the 90, 95, 99 and 99.9% significance levels. A common significance level to use is 95%. Figure 6.12, derived from Ord and Getis (1995), summarises the Bonferonni-generated 95% significance levels for a range of sample sizes.

The 95% significance level for the sample of 256 records is approximately 3.55. This means that there is significant statistical evidence of crimes clustering (i.e. there is a hotspot) at the cell positioned in the eighth row of the eighth column of the 16×16 grid, up to a lag 1 distance. This means that the cell is at the centre of a cluster of high cells, a cluster that would only occur by chance if the values in the 16×16 matrix were randomly scattered around the matrix less than 5% of the time. This type of result provides a distinct advantage over the previous techniques that have been explored in this chapter for identifying and mapping hotspots: it identifies those areas that can be statistically defined as 'hot'. Figure 6.13 shows those cells that display statistical evidence of local spatial association above the 95% significance level at lag 1. In essence, these are the statistically significant hotspots in this grid of cell-based aggregated crime data.

The Gi and Gi* statistics can be applied against simple crime counts or where the grid's cells contain other values that relate to the underlying crime data, such as kernel density estimation values for each cell (Ratcliffe and McCullagh, 1999). Figure 6.14 shows the robbery and residential burglary crime hotspots, used as examples earlier in this chapter, both as the original hotspot surfaces and as defined by the Gi* statistic at a 95% significance level. The Gi* results are compared to the kernel density estimation surfaces produced using a K order 5 bandwidth. In both Gi* crime maps the hotspot areas correspond to the high crime density areas, but add robustness to the kernel density estimation maps by identifying those areas where there is a statistically significant local spatial association of high crime areas. That is, these are the areas that can be statistically defined as being hot!

A problem sometimes associated with the kernel density method is that in some locations it may smooth across or into area where there is no recorded crime, because it is not constrained to the high detail of the underlying geography of the crime point distribution. Crime mappers sometimes have to explain to police officers why there is an apparent crime level (i.e. shading) in an area where no crime is possible. An advantage of the Gi* technique is that as it compares local averages against global averages. Therefore areas where it is impossible for certain crimes to happen (e.g. for residential burglary this would include reservoirs, rivers or any areas where there was no residential housing) can be identified in a GIS

Number of grid cells in datasets (n)	5	10	50	100	500	1000
Gi* 95% significance level values	2.3189	2.5683	3.0833	3.2889	3.7134	3.8855

Source: Ord and Getis (1995)

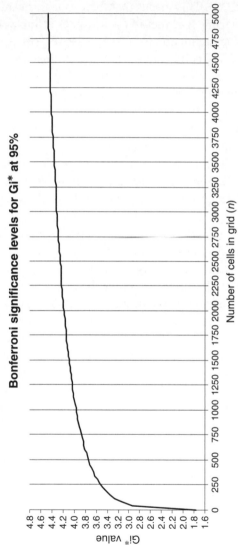

Bonferroni significance levels for Gi* at 95%

Figure 6.12 Bonferroni-generated significance levels for Gi* at 95%. For example, for a grid that contains 500 cells, for the Gi* value to be significant to the 95% level requires it to have a minimum value of 3.7134

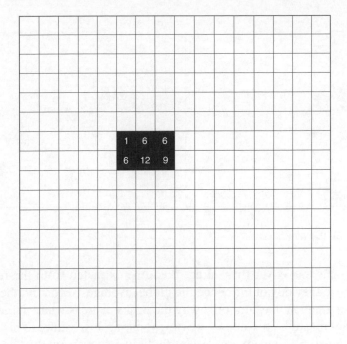

Figure 6.13 Cells in the 16×16 matrix grid that show statistical evidence of local spatial association above the 95% significance level at a lag distance of 180 m. The distance across each cell is 125 m

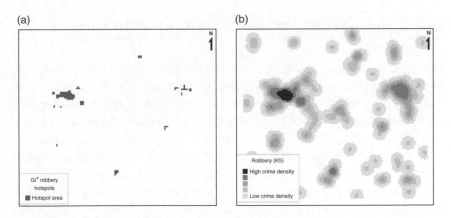

Figure 6.14 Gi* hotspot areas compared to kernel density estimation results. (a) Gi* hotspots at the 95% significance level for robbery, (b) a kernel density estimation surface for robbery, (c – on page 172) Gi* hotspots at the 95% significance level for residential burglary, and (d) a kernel density estimation surface for residential burglary

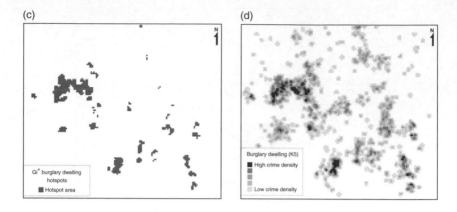

Figure 6.14 (Continued)

and the cells that cover these areas could be extracted from the full grid coverage so they do not influence the global average. Secondly, the kernel density estimation may smooth away the peaks in areas where large bandwidths aggregate high values with neighbouring low values. This can be seen in Figure 6.14 where areas identified as hotspots from the Gi* results are not grouped in the 'high crime density' threshold on the kernel density estimation maps. This, however, may be a small price to pay for the increased readability of a kernel density estimation map, or can be corrected for with a Gi* derived output.

6.8 Considering the underlying population

So far in this chapter we have reviewed the limitations of point mapping and suggested kernel density estimation as a replacement. We have then exposed the limitations in kernel density estimation surface and proposed the Gi* statistic as a way to circumvent the lack of statistical rigour in kernel density estimation surfaces. It can be argued that there is one remaining flaw that exists in what appear to be good techniques such as kernel density estimation and the Gi* statistic for identifying crime hotspots. The distribution of these hotspots could be influenced by an underlying population. For example, residential burglary hotspots could be influenced by the distribution of the density of housing. Hotspot analysis typically focuses on identifying high crime areas: those areas that experience a high volume of crime and where there is a high level of crime concentration, relative to the distribution of crime across the whole region of interest. This is certainly practically useful as it helps to direct attention to where crime happens the most. However, in some crime reduction scenarios (such as programme evaluation) the population at risk must be included in any calculation, and hotspot methods should equally be able to identify hotspot areas that experience high rates of crime.

Residential burglary rate maps are a common product of the crime mapper, where data on the number of residential properties are typically sourced from the census, linked to geographic areas and calculated against residential burglary counts within each of these geographic areas. Figure 6.15 shows a residential burglary rate map generated for the sample area used earlier in the chapter. These types of maps are however still subject to the problems associated with the MAUP. Until recent years it has been difficult to improve upon these problems because the underlying population data have only been available at aggregated geographic unit level such as a census output area or census tract. Property gazetteers that have been created in a number of countries (as referred to in Chapter 3) provide a data source of all residential dwellings, listed with high resolution geographic coordinates for the area the gazetteer covers. This type of data

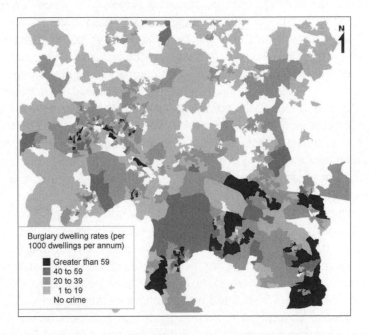

Figure 6.15 Residential burglary rate map thematically shaded by geographic areas. Maps of this type are helpful in understanding problems of risks of being a victim of a crime, but are still subject to the problems associated with the MAUP

resource significantly improves the quality of the denominator used to identify areas of high burglary rates (Chainey, 2001).

Case study: Identifying street crime risk hotspots in the West End of London using pedestrian counts

Robbery and theft from the person (commonly referred to as *street crime*) are problems in the United Kingdom's town centres. Identifying where these crime problems exist is typically performed by generating kernel density estimation hotspot maps that identify the concentrations of these crimes. This type of hotspot analysis has been useful in focusing resources to specific areas to help tackle some of the issues of street crime, but these 'volume' maps hide the relative levels of *risk* that people experience on the street.

Most attempts at calculating and mapping the rates of street crimes use census population data as the denominator. These census data are available at a geographic aggregate unit level (e.g. output area, tract or block). These rate calculations for robbery and theft of the person typically only exaggerate the crime levels in town centres where these types of crimes tend to happen, but where the resident population is relatively low.

An innovative study in the London Borough of Westminster has explored patterns in the risk of these types of crime by using new techniques that model pedestrian traffic so that accurate counts of pedestrians at the street level can be calculated. Westminster covers the popular West End district of London, including famous streets and localities such as Oxford Street, Leicester Square and Piccadilly (Figure 6.16). The detailed pedestrian counts for this area are calculated against similarly precise street crime data from London's Metropolitan Police to produce informative measures that identify those areas where the likely risk of being a victim of robbery or theft of the person is high.

The analysis of these street crime risk hotspot maps is helping to better inform the design of appropriate responses to tackle these types of crime. In particular, the analysis has helped to understand that the street crime responses to an area of high crime volumes and high risk require different attention and design to those responses required in areas of high crime volume and low risk, and low crime volume but high risk.

Several techniques that consider the *at risk* population that are more advanced than simple rate calculations are also now available to crime analysts. These include the risk-adjusted nearest neighbour hierarchical clustering routine and dual kernel density estimation that are available in

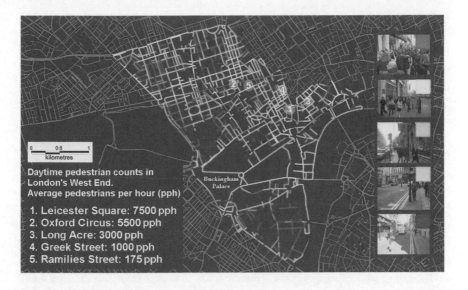

Figure 6.16 Modelled daytime pedestrian counts per street segment in the West End of London, shown thematically with a photographic representation of the on-street population. Source: Chainey and Desyllas (2004). Reproduced by permission of Ordnance Survey on behalf of HMSO. © Crown copyright 2005. All rights reserved. Ordnance Survey Licence number 100044021 (see Plate 2)

CrimeStat (Levine, 2004), the spatial scan statistic (Kulldorff, 1997) and the Geographical Analysis Machine (GAM) (Openshaw *et al.*, 1987). The spatial scan statistic is a likelihood-based method that compares points within a fixed circular or cylindrical search window that visits each point to those points that lay outside each search window. Where the likelihood function is greatest is where the cluster is most likely to exist. This can also be applied with a baseline variable so that the likelihood is relative to the underlying population. The risk-adjusted nearest neighbour hierarchical clustering routine applies a search window over a data distribution that changes in size in relation to the baseline variable (the underlying population) – for areas where there is a large population the search radius will be small, whereas the opposite will be true where there is a low population. The method seeks to identify clusters of crime points that are more concentrated than would be expected on the basis of the underlying population. In other words, it identifies clusters with high risk. These two techniques offer new opportunities for crime analysts to explore how the underlying population influences the distribution of clusters. We next describe in more detail the two other methods that also offer these spatial analysis possibilities; dual kernel density estimation and GAM.

more concentrated relative to underlying pop.

175

6.8.1 Dual kernel density smoothing

Dual kernel density is a technique that can compare crime data against an underlying point-based population dataset to produce a surface map that is normalised without being constrained by thematically shading the geographic boundaries (Oberwittler and Wiesenhütter, 2004). For example, with the use of a property gazetteer it is possible to map residential properties at a very precise level and use these data as the denominators against burglary dwelling data to produce a continuous surface map that identifies those areas with high crime rates. Dual kernel density is not typically available in most GIS software, but is available in CrimeStat. Figure 6.17 shows the dual kernel density map produced using a property gazetteer of dwellings for the study area, calculated against the residential burglary data for this same area. In comparison to the kernel density estimation map of the same data (Figure 6.8), the residential burglary risk surface map identifies several different areas. The same problems that have been described in regard to kernel density estimation still apply to this type of map, particularly in terms of the need for care in deciding upon the thematic class boundaries that distinguish the areas that are 'hot'. However,

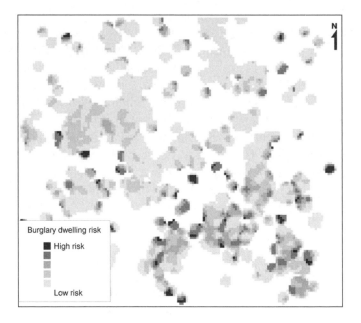

Figure 6.17 A dual kernel density burglary dwelling risk map generated using a property gazetteer of residential dwellings as the secondary variable

the Gi* statistic can be applied against these outputs to determine those areas that can be statistically defined as 'hot', similar to how the Gi* statistic can be applied to kernel density estimation cells (Ratcliffe and McCullagh, 1999).

6.8.2 The Geographical Analysis Machine

A second technique that makes use of an underlying population variable with event point data to identify hotspots is the GAM (Openshaw *et al.*, 1987). GAM was initially developed to identify clusters of cancer (Openshaw *et al.*, 1988), but is equally applicable to crime data (Jefferis, 1999). Its practical strength over dual kernel density estimation is that it tests for, and identifies, clusters of crime that are statistically significant.

The GAM works by generating a circle that visits every location over the area of interest, retrieving information about the point data and the population variable that is mapped within each circle, and applying a statistical assessment over whether or not the incident rate in each circle is unusually high (or low). This process is then repeated a large number of times using circles of different sizes. In each case the circles overlap to allow for edge effects and to provide a degree of sensitivity in the analysis. GAM's use with crime data has often been restricted due to the limited availability of the software and the lack of availability of denominator data that match the quality of the geocoded crime data. With the increasing availability of property gazetteers and other richer forms of geographic information (e.g. pedestrian count data) this means that suitable high precision denominator data are now becoming increasingly available for applying GAM to crime. GAM is also now available to use as a Java application (http://www.ccg.leeds.ac.uk/smart/gam/gam.html). However, few examples of its use with crime data have been published.

6.9 Predictive crime mapping

The use of hotspot mapping as a simple form of forecasting future trends in crime and predicting where crime will happen next has been common in the crime mapping field. The technique has the advantage of being variable, so 'where next' could mean in the next 24 hours for the application of targeting operational police patrols, or within the next 12 months for strategic crime prevention planning purposes. Now, however, this use of retrospective crime data to determine where to target policing and crime reduction resources is being questioned. Research into crime prediction techniques suggests that forecast methods that use retrospective data, particularly those that attempt to use data for the same month from the previous year, are not particularly accurate (Gorr and Olligschlaeger, 2002; Groff and LaVigne, 2002). In addition, when large volumes of retrospective data are used to create hotspot maps, very little consideration is given over whether the data volume is suitable for determining what will actually happen 'next'. For example, if 12 months of crime data are used to produce a hotspot map, then all the methods discussed in this chapter apply an equal weight to a crime that happened yesterday and a crime that happened 365 days ago. Bowers *et al.* (2004) suggest a prospective hotspot mapping method, drawing from techniques that have been previously used in epidemiology. This method is based on research that demonstrates that the risk of burglary is communicable, with properties within 400 m of a burgled dwelling being at significantly greater risk of victimisation for up to two months after the initial event. Crimes that have occurred in the past are therefore weighted in relation to their communicable risk, applying greater weight to those crimes that have happened more recently and which happened close to other similar events. They suggest that their methodology is 30% more accurate than existing hotspot mapping techniques such as kernel density estimation and thematic mapping of boundary areas.

Crime prediction and forecasting is a field that has received increasing attention in recent years, although the field is still very much in its infancy. Several techniques have emerged for predicting crime patterns. These include univariate techniques, methods that use multivariate leading indicators, point process modelling and artificial intelligence methods.

Univariate methods use a previous value of one variable to predict what will happen in the future, and whilst they can be straightforward to apply, they can also be the least accurate if only simple methods are applied (Gorr and Olligschlaeger, 2002; Gorr *et al.*, 2002). Leading indicators methods use multiple variables that tend to vary with crime, rather than just crime itself, to predict future crime patterns. This technique is showing great promise for large areal units (Gorr and Olligschlaeger, 2002), but they do require a significant level of expertise on the part of the user, both in terms of understanding environmental criminology theory and multivariate spatial modelling (Groff and LaVigne, 2002). A third technique is referred to as point process modelling (Groff and LaVigne, 2002). This technique is based upon the preference shown by offenders in how they offend (i.e. their rational choices – see Chapter 4), and combines multivariate models that attempt to explain behaviour with components from kernel density estimation and a geostatistical interpolation technique called kriging.

Another group of techniques that have recently emerged for use for predicting crime are artificial neural networks (Olligschlaeger, 1997; Corcoran and Ware, 2003). These techniques are fed with past crime events which have variables that help explain where and when crime occurs (i.e. the areas and times that are vulnerable to crime). The neural network is trained based on this information, and iteratively learns about the ambient conditions that influence the space and time patterns of crime, to predict where and when future crimes will occur.

Groff and LaVigne's (2002) review of crime mapping prediction techniques identified that this is a field that is still very much in development, meaning that many of the techniques still only exist in prototype versions and continue to be researched and refined. Groff and LaVigne also remarked that any new methods that are developed should be founded on strong theoretical grounds, commenting that several of the emerging techniques were lacking in their guiding theory. For example, they say that, 'the accuracy of the multivariate methods depends upon the appropriateness of the variables included in the model. Since the identification of appropriate variables is grounded in theory, one without the other will only have limited utility' (Groff and LaVigne, 2002). Clearly, crime mappers of the future will not only have to be technically competent, but will also still require a background in environmental criminology theory.

6.10 Summary

Hotspot mapping, like all forms of mapping, is a naturally iterative and experimental process (i.e. the first map that is generated may not be the one that is finally used). Care needs to be taken in the selection of parameters to ensure that what is being identified as a hotspot is accurate. Hotspot mapping also requires good cartographic skills, and although it is not a contest to find whose map looks the prettiest, sound cartographic principles are important if the message that the map conveys is easy to interpret by its reader.

This chapter has explored the application of a number of techniques for identifying hotspots. The discussion was not to necessarily identify the optimum technique but to explain and understand the strengths and weaknesses in those techniques that are in common use by crime mappers. Several more advanced techniques have also been presented, and whilst there are others beyond the realm of this book, our focus has been to explore methods which are currently available for use in policing, crime reduction and crime prevention.

Further reading

Eck, J., Chainey, S.P., Cameron, J., Leitner, M. and Wilson, R. (2005). *Mapping Crime: Understanding Hotspots*. Washington, DC: National Institute of Justice.

This booklet complements this chapter. It adds to the discussion on the need for strong theoretical principles in identifying crime hotspots, guides the readers through methods for mapping crime and understanding hotspots and demonstrates the application of these methods in a range of GIS and other software tools.

Monmonier, M. (1996). *How to Lie with Maps*. Chicago: University of Chicago Press.
MacEachren, A.M. (1995). *How Maps Work: Representation, Visualization and Design*. New York: Guilford Press.

The two books are an excellent reference for the crime mapper that wants to find out more about the power of maps and how to represent spatial features on a map.

References

Anselin, L. (1995). Local indicators of spatial association – LISA. *Geographical Analysis*, 27(2), 93–115.
Bailey, T.C. and Gatrell, A.C. (1995). *Interactive Spatial Data Analysis*. Harlow: Longman.
Bowers, K.J., Johnson, S.D. and Pease, K. (2004). Prospective hot-spotting. *The British Journal of Criminology*, September 2004, 44, 641–658.
Bowman, A. (1984). An alternative method of cross-validation for the smoothing of density estimates. *Biometrika*, 71, 353–360.
Bowman, A. and Azzalini, A. (1997). *Applied Smoothing Techniques for Data Analysis: The Kernel Approach with S-Plus Illustrations*. Oxford: Oxford University Press.
Brimicombe, A.J. (2004). On being more robust about hotspots. Paper presented at the 7th Annual International Crime Mapping Research Conference, April 2004. http://www.ojp.usdoj.gov/nij/maps/boston2004/papers.htm.
Chainey, S.P. (2001). Combating crime through partnership: Examples of crime and disorder mapping solutions in London, UK. In A. Hirschfield and K. Bowers (eds) *Mapping and Analysing Crime Data*. London: Taylor & Francis.
Chainey, S.P. and Desyllas, J. (2004). Measuring, identifying and analysing street crime risk. Presentation at the 2004 UK National Crime Mapping Conference, London: University of London. http://www.jdi.ucl.ac.uk/news_events/conferences/index.php.
Chainey, S.P., Reid, S. and Stuart, N. (2002). When is a hotspot a hotspot? A procedure for creating statistically robust hotspot maps of crime. In *Innovations in GIS 9*. London: Taylor & Francis.

Cliff, A.D. and Haggett, P. (1988). *Atlas of Disease Distributions.* Oxford: Blackwell Reference.

Corcoran, J. and Ware, A. (2003). Crime hot spot prediction: A framework for progress. In D. Kidner, G. Higgs and S. White (eds) *Innovations in GIS 9: Socio-economic Applications of Geographic Information Science.* London: Taylor & Francis.

Eck, J., Chainey, S.P., Cameron, J. and Wilson, R. (2005). *Mapping Crime: Understanding Hotspots.* Washington, DC: National Institute of Justice.

Getis, A. and Ord, J.K. (1996). Local spatial statistics: An overview. In P. Longley and M. Batty (eds) *Spatial Analysis: Modelling in a GIS Environment* (pp. 261–277). Cambridge, England: GeoInformation International.

Gorr, W. and Olligschlaeger, A. (2002). Crime hotspot forecasting: Modelling and comparative evaluation. Final report to the National Criminal Justice Reference Service (NCJRS).

Gorr, W., Olligschlaeger, A. and Thompson, Y. (2002). Short-term forecasting of crime. *International Journal of Forecasting*, 19(4), 579–594.

Groff, E.R. and LaVigne, N.G. (2002). Forecasting the future of predictive crime mapping. In N. Tilley (ed.) *Analysis for Crime Prevention* (Crime Prevention Studies Volume 13). Monsey NY: Criminal Justice Press.

Harries, K. (1999). *Mapping Crime: Principle and Practice.* United States National Institute of Justice. http://www.ojp.usdoj.gov/nij/maps/pubs.html.

Hogg, R.V. (1979). Statistical robustness: One view of its use in applications today. *American Statistician*, 33, 108–116.

Jefferis, E. (1999). A multi-method exploration of crime hot-spots: A summary of findings. Crime Mapping Research Centre intramural project. Washington, DC: National Institute of Justice.

Kulldorff, M. (1997). A spatial scan statistic. *Communications in Statistics: Theory and Methods*, 26, 1481–1496.

Levine, N. (2004). *CrimeStat Version 3 Users Guide.* Washington, DC: National Institute of Justice. http://www.icpsr.umich.edu/NACJD/crimestat.html.

MacEachren, A.M. (1995). *How Maps Work: Representation, Visualization and Design.* New York: Guilford Press.

Monmonier, M. (1996). *How to Lie with Maps.* Chicago: University of Chicago Press.

Oberwittler, D. and Wiesenhütter, M. (2004). The risk of violent incidents relative to population density in Cologne using the dual kernel density routine. In N. Levine (ed.) *CrimeStat Version 3 Users Guide.* Washington, DC: National Institute of Justice.

Openshaw, S. (1984) The modifiable areal unit problem. *Concepts and Techniques in Modern Geography*, 38, 41.

Openshaw, S., Charlton, M., Craft, A.W. and Birtch, J.M. (1988). Investigation of leukaemia clusters by the use of a geographical analysis machine. *Lancet*, I, 272–273.

Openshaw, S., Charlton, M., Wymer, C. and Craft, A.W. (1987). A mark I geographical analysis machine for the automated analysis of point data sets. *International Journal of Geographical Information Systems*, 1, 335–358.

Olligschlaeger, A. (1997). Artificial neural networks and crime mapping. In D. Weisburd and T. McEwen (eds) *Crime Mapping and Crime Prevention*. Monsey NY: Criminal Justice Press.

Ord, J.K. and Getis, A. (1995). Local Spatial Autocorrelation Statistics: Distributional issues and an application. *Geographical Analysis*, 27, 286–306.

Ratcliffe, J.H. (1999a). *Spatial Pattern Analysis Machine Version 1.2 Users Guide*.

Ratcliffe, J.H. (1999b). *Hotspot Detective for MapInfo Helpfile version 1.0*.

Ratcliffe, J.H. and McCullagh, M.J. (1999). Hotbeds of crime and the search for spatial accuracy. *Geographical Systems*, 1(4), 385–398.

Sawada, M. (1999). ROOKCASE: An Excel 97/2000 Visual Basic (VB) Add-in for Exploring Global and Local Spatial Autocorrelation. Bulletin of the Ecological Society of America, 80(4), 231–234. http://www.uottawa.ca/academic/arts/geographie/lpcweb/newlook/data_and_downloads/download/sawsoft/rooks.htm.

Sheather, S.J. and Jones, M.C. (1991). A reliable data-based bandwidth selection method for kernel density estimation. *Journal of the Royal Statistical Society* (Series B), 53, 683–690.

Silverman, B.W. (1986). *Density Estimation for Statistics and Data Analysis*. London: Chapman and Hall.

Stone, M.A. (1974). Cross-validatory choice and assessment of statistical predictions. *Journal of the Royal Statistical Society* (Series B), 36, 111–147.

Williamson, D., McLafferty, S., Goldsmith, V., Mallenkopf, J. and McGuire, P. (1999). A better method to smooth crime incident data. *ESRI ArcUser Magazine*, January–March 1999, 1–5. http://www.esri.com/news/arcuser/0199/crimedata.html.

7
Mapping Crime with Local Community Data

Learning Objectives

This chapter explores how crime data can be complemented with non-police data to explore the picture of criminal behaviour and how criminality can be tackled. In this chapter you will learn that reducing crime is not just something that the police should do in isolation, but in collaboration with partners active in a number of other areas. This partnership approach also extends to the local community and how they can participate in supporting policing and crime reduction efforts.

The chapter reviews the sources, content and uses of partnership data. This sharing of data between the police and local partners also comes with its challenges. In this chapter you will learn about these challenges, including those that relate to data exchange, legislation, data management, how they can be overcome and the typical resources required for effective crime mapping in a partnership framework.

The chapter also presents examples of how geographic data can be combined, whether it is for dissemination purposes or for analysis of relationships, and case studies that show partnerships at work.

GIS and Crime Mapping   Spencer Chainey and Jerry Ratcliffe
© 2005 John Wiley & Sons, Ltd

7.1 Introduction

Problems of crime are not just something for the police to resolve. To deal with the issues of crime, contributions need to be made from other parts of government and the community. The police view of crime can often only be a limited one. The police can only react to the information and intelligence they are given, and within the limits of their responsibilities. If a local crime problem originates from the conditions on a local housing estate, the ineffectiveness of the correction systems, poor inclusion levels at local schools and landscape design flaws that only help to encourage crime rather than curb it, then there can be frustration on the part of police working alone. Given the problems detailed, what can the police do on their own if their actions are to bring demonstrable and sustainable reductions in crime? This chapter will show that working with local agencies in partnership can enhance the value of police action.

The growth of partnership approaches to crime reduction originated from recognition that the problems of crime require a multi-agency response. Partnerships between police forces and their local government organisations, education authorities, justice system, health, fire and ambulance services allow a more informed view of criminal behaviour to be developed. Data such as census statistics, deprivation indices, land use profiles, housing tenure, noise nuisance, graffiti, vandalism, truancy, offender statements, parolee information, drug and alcohol abuse and arson incidents offer a way to explore the possible causes and links to crime and how an effective response can deliver real and sustainable reductions in crime. In this chapter we do not cover community policing, the definition and approach of which we leave to explain in Chapter 11. However, as all data and intelligence are derived from people and their activities, local services and the fabric of an area, the community acts as the essential source to report, comment, capture and gather information that helps decide how crime problems can be tackled. How this type of data can be shared and used is the challenge we aim to answer in this chapter.

7.2 What are crime reduction partnerships?

In many parts of the world, the responsibilities for reducing crime are being organised through local partnerships. For example, the local city government may be the body responsible for providing housing services, and in so doing, capture information on the occupancy and condition of these houses. Using such data, the police and city council can explore if a crime problem is particular to housing provided and managed from the state, and if so, with the local city council, target a response to these houses to help reduce the associated crime problem.

The activity of crime reduction partnerships has been formalised in England and Wales since the 1998 Crime and Disorder Act. The Act established a statutory duty for the police to work with their respective local government bodies to reduce crime. In other parts of the world, these partnerships are not as formal, but the spirit of working together to solve crime problems exists in many countries. The 2003 Brazilian Government's Public Security Plan (Brazil Ministry of Justice, 2003) calls for the contribution of local municipal bodies to work with the country's tiers of police (municipal, military and federal) to combat the nation's crime problems. In the United States the spirit and activity of partnership working continues to develop, although it is also in the US where in many places the crime reduction role is still seen to be the exclusive domain of policing. In these different parts of the world, much of the growing need for collaboration stems from recognising that the police picture of crime, and ability to respond to crime issues, can only ever be a limited one and that it requires contributions from many other local partners to fully recognise the levels of crime in an area. This may also require a responsibility to consult widely with the local public to make sure that the partnership's perception matches those of the local communities.

In England and Wales, 375 crime reduction partnerships exist, covering the map of local authority services in these two countries. Every three years each partnership is required to produce an audit that measures and analyses their local crime and disorder problems. From the audit, a strategy is devised that describes and prioritises how the local crime problems will be tackled, setting targets against which partnership performance will be measured. The audit, and its subsequent plans for crime reduction, is made public and consultation with the public ensures that the measures and priorities are matched with the public experience and their expectations. The strategy is then monitored and reviewed over its three-year cycle to ensure it is flexible to new priorities that may emerge and is on target for delivery.

Audits and strategies for crime reduction may not be produced in other parts of the world as formally as they are in England and Wales, but within these crime reduction partnerships there is a common need to share information to accurately picture, explore and understand the local crime issues, and design and devise the multi-agency response to addressing crime problems. This requires an appreciation of the data that are available from these partners, their uses and how they can be sourced.

7.3 Mapping and the benefits of partnership working

One of the main challenges in bringing together and joining up information from a range of partners is to work out how these data can be integrated. Geography and mapping often act as the common denominators that run through these disparate datasets – each dataset usually containing some form of locational reference. This may be an address, a location or a defined geographic boundary area (e.g. a census geographical unit). For example, a crime can be referenced to a property or a location; an offender that is processed through the justice system can be linked to their place of residence; a city council which grants the licenses for alcohol and entertainment venues can map these bars and nightclubs by their address; the levels of exclusion or truancy of children from school can be linked to the school and also a child's home address; and the demographic profile of an area can be described within its census boundary areas. Using mapping as the basis to integrate these data provides a number of useful opportunities. These include the following:

- As a powerful information media that communicates large data volumes.
- Presenting information on a map helps to provide clarity in areas that need attention, areas that might otherwise be hidden in other information formats such as text, graphs or tables. Maps help to invoke discussion by identifying more easily the areas that need some form of crime reduction targeting.
- Presenting the crime picture in a map that shows the problems that may exist in certain areas makes it easier to identify other partnership activities that operate in these high crime areas. For example, an area may be identified as a burglary hotspot, and mapping and disseminating this information for circulation within the local partnership may identify that this hotspot coincides with an area that is soon to receive significant funds to support regeneration. These regeneration funds could be leveraged to better support the crime reduction response.
- Crime hotspot analysis is often questioned by those working close to crime reduction as being a process that reveals what they may already know. Research by Ratcliffe and McCullagh (2001) found that uniform police officers were not familiar with the location of crime hotspots for a number of different types of crime. In most situations, particularly across a partnership made up from a number of different agencies, the hotspots of crime may be the knowledge of only a few, and not general knowledge to a partnership of many. This could also be the case for hotspots of offender activity, deprivation, high truancy and other socio-economic problems. Knowledge

of where hotspots are thought to exist may also be anecdotal, based on historical information, insensitive to changing and newly emerging patterns, and unreliable. This may result in targeting responses to the wrong places. Mapping this data helps to provide clarity to all.

- Mapping crime and other partnership data provides statistical evidence for validating why resources are targeted to particular areas.
- Maps present a picture that prompts discussion, better enabling all partners to contribute, rather than following the view of one single partner. This includes questioning the robustness of the information that is presented and exploring ways in which complementary data from the partnership can be added to help better inform local decision-making processes.
- Integrating the data allows for a better informed and holistic view of the issue and how it can be responded to in a multi-agency manner.
- Mapping provides a basis to help monitor and measure partnership-targeted actions.
- A map can help to raise the profile of data that are missing or incomplete and raise caution to any errors. It is important that those who use maps are given some indication of their reliability. This level of inquiry can promote a culture of data improvement that can raise the quality of information that can be used to help identify and diagnose the problem.

Case study: Comparing the perception of where crime happens with where crime actually happens

A number of staff at a London-based crime reduction partnership were sceptical over the role that mapping could bring to support its efforts. They viewed it as an unnecessary investment which would only reveal what they already knew. They claimed to know where the robbery hotspots were, where burglary was most concentrated and where the worst crime-ridden places were to park cars. To test these perceptions a survey was conducted of 21 people active in making contributions to the partnership. These included the senior police chief responsible for liaisons between the police and its partners, the police crime analyst, the manager of the local Crime and Disorder Reduction Partnership team, three community safety officers based in this team and partnership representatives from education, housing, justice, health, youth offending and action against drugs. Using the methodology of Ratcliffe and McCullagh (2001), each was asked to precisely mark the places on three different maps of their area where they felt the respective robbery, burglary dwelling and vehicle crime hotspots were located. The results were then aggregated on to a map and compared to the spatial distribution of the previous six months of actual recorded robbery, burglary dwelling and vehicle crimes.

Stark differences were seen between the three sets of maps. For robbery, which was meant to be a top priority for the partnership, the location of the main hotspot was missed by the majority of all partners. Statistically, the two hotspot maps were only 7% similar (i.e. only 7% of the actual areas that were hotspots were identified by the partnership). For burglary of dwellings, none of the actual hotspot areas were identified (a statistical match of 0%) and for vehicle crime the match was 6% (Figure 7.1). Further questioning of the partnership about their perceived understanding of where vehicle hotspots existed revealed that many felt the car park adjacent to the building in which they held their meetings was a crime hotspot. Actual crime data revealed this was not the case, and that their perceptions were largely influenced by the fact that the partnership representative from the local justice team once had his car broken into during a partnership meeting.

This study was useful in demonstrating the potential application of mapping for providing timely and factual information on crime to support the partnership in making informed and accurate decisions on where to target crime reduction resources.

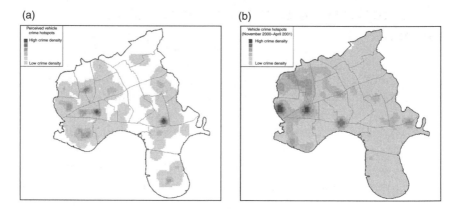

Figure 7.1 Perception versus actuality. Figure (a) shows the perceived location of vehicle crime hotspots as identified by the area's local crime reduction partnership. Figure (b) is the actual location of vehicle crime hotspots. Statistically, only 6% of the hotspot areas in (b) were identified by the partnership

7.4 Partnership data

There are many types and sources of data that can be used by partner-ships to help build a more comprehensive picture of crime beyond that which is captured by the police. These data include non-crime data that can be used to help understand the generators behind what causes crime to happen, and include physical, social and economic measures of the local landscape. This section does not attempt to review all the possible data that could be used, but instead points to the core datasets typically used by partnerships – what they tend to consist of, the spatial scale at which they are usually available, where they could be sourced and how they could be used. These core datasets are categorised into three groups relating to their typical partnership requirement level, importance and application in supporting crime reduction efforts. We term these three categories:

1. Primary
2. Secondary
3. Tertiary.

In many ways these groups act as the information-sharing priorities for partnerships. However, we do refer the reader back to Chapter 3 when we discussed the importance of asking certain spatial questions about the application of data to avoid unnecessary 'shopping trips', trips that may result in collecting data that are unnecessary and not fit for purpose. Information sharing helps to provide the factual detail that partnerships require in order for them to make informed and robust decisions about where to target their resources. The crime mapper should be aware that partnerships can quickly suffer from information overload and wasted effort if the purpose for which they need certain data is not accurately defined and its purpose not clearly identified.

A separate consideration to these partner-derived data is the primary map-base source required to access cartographic and boundary maps in a GIS format. In some countries these types of maps are available to all in government (central, regional and local) via the country's national mapping agency. However, in those countries where these agencies do not exist, the production of detailed cartographic products may be the responsibility of the local municipal government or public utility companies. The partner-ship may provide a new means of access to this mapping data for partners who had previously hit this barrier. This can be a resource for crime mappers who encounter problems affecting the development of their digital mapping applications or who had been unable to afford mapping products that are created by commercial companies.

7.4.1 Primary datasets

There are typically two core primary datasets that are used by partnerships. These are:

1. Police recorded crime
2. Census of population.

In the first instance it is vital to know where crime is happening and certain characteristics of these crimes, hence police recorded crime data are required. Census data that describe the population and can include data on the physical and socio-economic conditions in areas help to put these crime data into context. Census data can also be used as the denominator for calculating crime rates (e.g. using population data to calculate the number of crimes per 1000 population in an area). Table 7.1 lists descriptions, typical content and uses of these data.

7.4.2 Secondary datasets

Secondary datasets offer added richness to that which has already been sourced. These data include:

- Disorder and anti-social behaviour information relating to:
  - Noise nuisance
  - Abandoned vehicles
  - Graffiti and vandalism
  - Police calls for service
- Offender records
- Victim records
- Neighbourhood statistics
- Transport Police crime data.

Table 7.2 lists descriptions, typical content and uses of these data. Police calls for service (command and control data) can identify where incidents happen and reveal levels of disorder that are not captured in crime recording systems. Other agencies may also collect anti-social behaviour data on noise nuisance, abandoned vehicles, graffiti and vandalism that add to the picture of delinquent behaviour and criminal activity. Data from the

Table 7.1 Primary partnership datasets

Data source	Description	Typical content	Partnership uses	Maximum spatial precision available
Police	Crime recording information system	• Offence details including crime type classification, date and time committed, and the location of the offence • Offender information including age, gender, ethnicity and home address details • Victim information including age, gender, ethnicity and home address details or company details When a crime is committed to property, this recording system also records the details of the property affected (e.g. vehicle make and model) and the items stolen or damaged	Mapping hotspots of crime, creating profiles of offenders and victims, and analysing if patterns or trends exist in crime activity	Exact address or location
National Statistics Agency or local government	Census of Population	• Population statistics (e.g. age, gender, ethnicity) • Housing statistics (e.g. dwellings, tenure, type of housing) • Economic statistics (e.g. level of employment, occupation, hours of work) • Social statistics (e.g. health, education)	To provide detail on the population and physical and socio-economic characteristics of an area	Census output geography (e.g. in the United Kingdom the smallest census boundary area covers approximately 115 households)

Table 7.2 Secondary partnership datasets

Data source	Description	Typical content	Partnership uses	Maximum spatial precision available
Local government body and/or Police	• Noise nuisance complaints • Abandoned vehicles • Graffiti and vandalism • Police calls for service	• Incident type • Location of incident • Date and time of incident • Result of incident (e.g. if a warning notice was served on the tenant for the noise dispute, or if a disorder incident resulted in a criminal event) Other, specific to particular dataset • Response time (noise complaint, police response) • Make, model and age of cars • Tagging description of graffiti • Value of damage	Mapping of incident hotspots, providing a picture of disorder and anti-social behaviour	Exact address or location
Corrections/ Probation and Police	Offenders	• Personal details about the offender (e.g. age, gender, home residence) • Offence history (offence types and persistency) • Problems related to offending (unemployment, drugs, alcohol, mental health) • Motivations for offending • Emotional well-being, attitudes and behaviour • Sentencing decisions • Intervention programmes • Re-offending risk assessment (e.g. used to help identify certain risks that the offender may pose to themselves or others and for helping to decide on actions for person management of the offender)	Creating profiles of offenders and understanding what makes them offend	Exact address

Source	Dataset	Contents	Purpose	Geography
National Statistics Agency or local government	Neighbourhood Statistics	• Deprivation and affluence statistics (poverty, income, access to services, housing conditions) • Income and other welfare benefits • Educational attainment • Training provision • Hospital admissions • Commercial environment (e.g. commercial and retail areas)	Providing detail on the population and physical and socio-economic characteristics of an area, in complement to the census	Census output geography
Transport Police	Crimes on public transport and at stations	• Offence details including crime type classification, date and time committed and the location of the offence • Offender information including age, gender, ethnicity and home address details • Victim information including age, gender, ethnicity and home address details or company details	Mapping hotspots of crime at stations and on transport routes, creating profiles of offenders and victims, and analysing if patterns or trends exist in crime activity, complementing the policing and disorder picture	Exact address or location

Police on offenders can be richly complemented by data from the corrections or probation service with details they may hold about an offender's history (e.g. their drug habits).

In some countries a separate agency police the transport network. Data on crimes committed on the network or at stations can add to the picture of criminal activity that happens at all other areas. These types of data provide an example of other law enforcement agencies that may exist locally (e.g. University campus police) and from whom other incident data could be sourced.

Neighbourhood statistics are a growing form of information developed and distributed by government statistics services. In addition to census data, these types of statistics turn administrative data collected by government bodies into useful statistics that describe the characteristics and conditions of neighbourhoods. These include aggregate data (e.g. by census areas) on benefits, education, skills, training, health and housing, and can often include crime data for neighbouring police areas. In the United Kingdom and in the United States, neighbourhood statistics are being supported with coordination at a national level. For example, the US National Neighborhood Indicators Partnership (NNIP) collaborates efforts between the Urban Institute and local government partners to develop and use neighborhood-level information systems to support local policymaking and community building (see http://www.urban.org/nnip/). In the UK, neighbourhood statistics are the responsibility of the Office for National Statistics (www.neighbourhood.gov.uk). In certain other places in the world, neighbourhood statistics are captured and coordinated at the regional or local level.

7.4.3 Tertiary datasets

Classification as a tertiary dataset does not mean that these last sets are unimportant. These data provide information that adds value and extends the comprehension of local crime problems. These data include:

- Housing
  - Housing stock owned by the government
  - Register of Social Landlords (e.g. Housing Associations)
- Education
  - School pupil educational attainment
  - School exclusions
- Fire service incidents

- Health
  - Hospital admissions from non-accidental injuries
  - Drug and alcohol treatment register
- Land use
- Transport network and termini.

Tertiary datasets can also include local data that are similar to those available from the national census or neighbourhood statistics sources, but which are of higher resolution than that offered though national means. For example, rather than using aggregate statistics on pupil educational attainment that could be sourced from a national body, these data may be available locally in a format that is precise to individual pupils. Table 7.3 lists summary descriptions, typical content and uses of these data.

These lists of primary, secondary and tertiary datasets do not aim to be exhaustive of all data that could be available from a partnership, but do offer a flavour of what can be sourced, and how it can be used to help those embarking on sharing information between a local partnership to prioritise their data requirements.

Case study: Crime And Disorder Information Exchange (CADDIE), Sussex, England

Material supplied by Tim Hemsley, CADDIE Project Manager, West Sussex Police

In the county of Sussex in England a number of local agencies have partnered together as stakeholders to reduce the area's crime and disorder problems. These include the police, the local government authorities and the other emergency services – fire and ambulance. The key requirement of the partnership is to ensure that all the agencies involved have relevant, accurate and timely information to help them tackle the fear of crime, reduce crime and disorder and improve the quality of life in the community.

In response to this a web-based information solution, named CADDIE, coordinated by a team of information officers and analysts from the partnership and funded by the partnership, has been developed. CADDIE is a mechanism for sharing crime and disorder data between the local authorised partners. Initially piloted in the Crawley and Horsham areas of West Sussex, CADDIE provides a solution to many of the problems associated in the United Kingdom with sharing data and information across

Table 7.3 Tertiary partnership datasets

Data source	Description	Typical content	Partnership use	Maximum spatial precision available
Local government (Housing services)	• Housing stock owned by the government • Register of other social landlords (e.g. Housing Association properties)	• Address of property • Rooms and amenities in property • Void or occupant property • Condition of property • Value of property • Benefit provision to occupants of property	Comprehensive detail on social housing provision	Exact address
Local government (Education services)	• School pupil educational attainment • School exclusions	• Pupil performance and ability at key stages of education • Truancy record • Exclusion record • Linked to school or pupil residence	Detail on educational attainment and exclusion history of schools and their pupils. Used to identify schools that are underperforming and areas where attainment and exclusions could be linked to young offending	Exact address
Fire service	Fire incidents	• Incident type (e.g. arson to property) • Location of incident • Date and time of incident • Response time to incident	Provides picture of fire incidents that are linked to crimes (e.g. criminal damage, arson) and if similar areas draw the same demand on responses required by other emergency services	Exact address or location

Health service	• Hospital admissions from non-accidental injuries • Drug and alcohol treatment register	• Record and details of casualties that sustained injury from a criminal assault (e.g. glass related injuries from a pub fight) • Record and details of impact of drug or alcohol abuse on offenders and/or victims	Provides a picture of the associated health risks and health costs to crime incidents	Exact address or location
National or local government, cadastre agency or commercial organisation	Land use	• Location and descriptions of land use in an area (retail, entertainment, types of industry, housing, open space)	Providing detail on the physical, social and industrial characteristics of an area, in complement to the census	Exact address or location
National or local government (Transport department)	Transport network and termini	• Location, coverage and frequency of the transport network (e.g. trains, metros, buses) • Location and characteristics of transport hubs, stations and bus stops • Passenger volumes	Providing detail on people's movements around space and the physical characteristics of stations and other transport exit and entry points	Exact address, location or route

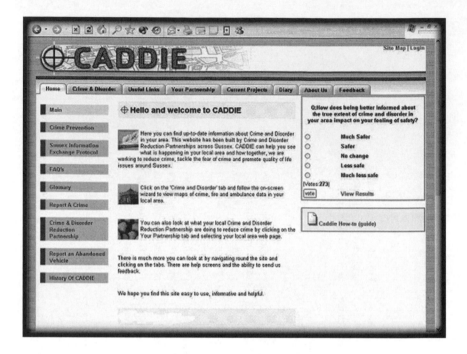

Figure 7.2 The public-access home page to CADDIE www.caddie.gov.uk

Crime and Disorder Reduction Partnerships (CDRPs). Since its conception in 2001, it has developed into a holistic approach for reducing crime, improving performance and delivering best value through partnership working. The system is now available to partners across the whole of Sussex and now also extends to Kent, the county that neighbours Sussex to the east.

The first step in bringing data into CADDIE is for the data to be cleaned, de-personalised and geocoded to ensure they meet data privacy rules and are quality assured. Data currently included in CADDIE are police crime records, census data, fire incidents, and ambulance calls for service. This dataset is up to date and is presented in a simple, user-friendly format allowing partners quick access to the information they require to monitor and react to crime and disorder problems in the county. CADDIE also provides public access to certain levels of information (Figure 7.2). These include:

- thematic maps describing the prevalence of crime and disorder by census ward;
- feedback on the crime reduction initiatives that are in operation in local areas;
- advice and information on crime and disorder;

- diary pages listing forthcoming public events;
- community safety issues in the area;
- polls and questionnaires relating to crime and disorder issues; and
- links to other useful web sites.

Partners that are stakeholders in the county's effort for reducing crime and disorder also have access to additional information by logging on to a secure site from the public front end. The information available includes:

- *Maps* – providing a facility to 'self build' hotspot maps of crime and disorder incidents at the touch of a button.
- *Management reports* – providing a facility to create reports of crime and disorder hotspots, which include geographical, social and demographic data from the 2001 census.
- *Reporting of incidents* – providing a facility to input previously unreported incidents of anti-social behaviour and disorder.
- *A library* – including the Audit of crime and disorder – the local strategy document that sets out the partnership's plans and targets for reducing crime and disorder, action plans, terms of reference, diary of events, contacts list and minutes of meetings.
- *Analytical support* – access to analytical reports and results to studies carried out by the analysts team that supports CADDIE.

CADDIE is based on the recognition that no single agency can tackle the problems of crime and disorder in the community. Combining the information, knowledge and expertise from a partnership is leading to the better recognition of how local partners can work together to reduce crime, planning the effective allocation of resources to where they are needed, and the ability to regularly audit, monitor changes and evaluate the progress against their targets for reducing crime.

Further information on CADDIE is available at www.caddie.gov.uk.

7.5 Information sharing

Information sharing can be a greater challenge than people initially perceive it to be. Evidence from those involved in information sharing suggests that barriers are often put in the way due to requirements to adhere to data privacy rules, silo attitudes that are against information sharing, issues with the handling of sensitive information, poor handling and management of data, questions (or embarrassment) over data quality and a lack of investment in resources to support information sharing (Chainey, 2004). Experiences from working with partnerships suggests that if information sharing is to be effective the following four factors need to be in place:

1. Data are easy to access from their source so that information can be quickly and effectively used.
2. Data that can be accessed are up to date to ensure that decision-making is based on the most timely picture of crime and other activities.
3. Data are comprehensive, in that they provide the detail required to inform those that make decisions from the data.
4. Data are fit for purpose. In other words a dataset is of the quality required to ensure that the picture it presents is a reliable one.

These principle factors also adhere to the general processes required for managing and organising information required and generated in crime mapping for policing and crime reduction applications. These processes and structures are discussed in more detail in Chapter 13.

7.5.1 Problems experienced with data sharing

Different partnerships have different requirements, but for partnerships to be productive and well-informed there is a need to generate a routine to how information is shared and used. Many partnerships often suffer by only operating on an ad hoc data exchange basis that results in information being too little and too late for it to have a real effect. Effective partnership information sharing requires a regular, if not continuous, process of information supply in order for partners to be routinely informed with timely information about crime and disorder patterns.

The lack of any routine and lack of defined processes for sharing information between partners can create a number of difficulties:

- There is a lack of consistency in what types of information are requested, meaning that data suppliers perform new processes each time a request is made. This has an obvious time demand that affects how quickly a request can be turned around.
- The lack of structure in reporting processes means that information is very rarely served in the timely format it is required. This leads to a reactive rather than proactive approach to tackling local crime and disorder problems.
- The over-reliance on one partner providing information can create stresses and tensions in an information-sharing relationship, particularly

when the data supplier receives little in return. This over-reliance can also mean that only a single view of crime is being seen by the partners, and not the multi-agency view that is needed to complete the picture and help identify possible opportunities for tackling causation.

- The only information that is exchanged is aggregated data or reports containing statistics, and not raw data. Whilst individual partners may perform analysis on their own data, the lack of sharing raw data can prevent the types of multi-agency data analysis that the partnership, in part, was designed to do. Without a broader view of the crime problem, partnership analysts are unable to provide a full service, such as analyses that could assess crime reduction initiatives the partners have designed and targeted, correlation analysis that explores the relationship between data from the same area, or analysis that helps to diagnose why areas experience particular problems and how these can be overcome.

In addition for the need to install some routine, there are also several other barriers to information sharing that are useful to consider and need to be appreciated.

Resourcing and funding

- Partnership working is hard work and requires time and effort to ensure that trust and respect between partners is built, and whatever effort each put in, they will receive worthwhile benefits in return.
- There is often a lack of resources (especially people) to bring all the partnership information together and analyse it. Adequate resources need to be put in place.
- Skills in information sharing and geographical analysis are rare commodities. Finding people with both will help ensure the success of any information partnership.
- Issues of staff management: Who pays for any centralised staff that manage and analyse data? Are all partners expected to contribute equally, and if not how does one divide the costs? These issues need to be resolved fairly and be proportional to partner requirements.

Adhering to data (and spatial) privacy rules

- The processing requirement to depersonalise data can be resource intensive. Because a time resource is required to depersonalise data, the excuse often given is that people do not have the time to carry out this duty – therefore access to data is denied.

- Understanding problems at the local level may require personalised data to be used, but data privacy rules, or confusion over these rules, may prevent these data being shared.
- If legislation is in place that permits the sharing of personal data, these laws are often not well understood by the practitioners, and to avoid any risk of breaking the law, legislation provides a barrier behind which partners hide, refusing access to data that could support crime reduction, rather than adopting a 'can do' attitude and working through the confusion.

Data quality

- Poor data quality will lead to poor or badly informed decision-making.
- Lack of quality data for small areas will fail to identify the problems at the community and neighbourhood level.
- Lack of geocoded data will mean that the value of GIS will not be fully realised.

Paper records

- For some records and data of interest, this data may not exist in electronic format and will need to be digitised. The time requirement for this should not be underestimated.

Information management, data access and central coordination

- The lack of a coordinated strategy and a coordinated resource for information sharing and analysis will result in poor information management, poorly structured partnership analysis and little opportunity for strategic information improvements to data quality.
- The low priority often given to data management will mean that any problems associated with data will continue to remain.
- Poor communication between the right people in agencies will often fail to breakdown the communication barriers in a partnership, frustrate access to information and discussions over its content and use.
- Access to too much data in an uncoordinated, poorly managed way could result in information overload.
- If the partner agency is not joined up internally within a partnership, it will most likely be ineffective in working with others outside the partnership.
- Inconsistency in exchanging data makes it more difficult to handle and manage data.
- The lack of application of common data standards means that data can be inconsistent and difficult to cross-reference.

Culture against information sharing

- *Information is power* – sharing this information releases the power individuals, departments and agencies have over the data. It should be recognised that some may be unwilling to surrender control of data.
- *The silo mentality* – 'it is our data and you are not using it' – continues to act as one of the main barriers to information sharing.

Coordinated analysis and information interpretation

- The ability to misinterpret patterns and give false correlations could damage partnership relations and make any targeted actions irrelevant.
- Exchanging data makes data custodians nervous over how they will be used and what results will come from analyses, especially if data are used in a manner for which they were not designed.

Technology issues

- High costs prevent development, upgrades and system maintenance.
- Incompatible systems make it difficult for data to be joined together.
- The lack of quality recording practices, robustness and reliability fails to instil confidence in data held on information systems.
- The need for a system that is sustainable and has a long life span for it to be compatible and proportional to its resourcing inputs.

Any information-sharing partnership requires careful nurturing. All of the challenges will not be solved in one go. Many of the challenges result from cultural barriers that often require examples to demonstrate how information-sharing can work to the level of precision required yet remain compliant with any legislation. Examples can also help to show that partnership working can be practical and cost effective, improve decision-making, improve the reporting process and show a return on investment in terms of a reduction in crime.

7.5.2 Guidelines for information sharing

In a post-mortem of a failed GIS (Openshaw *et al.*, 1990), the GIS that was the subject for discussion was one of the earliest crime mapping and information-sharing partnerships ever established between a police agency and several local partners. In their review, Openshaw and his colleagues listed a number of failings and mistakes they encountered when setting up a police GIS, and they also listed a number of helpful suggestions. Several years on, the suggestion list still looks very familiar to those struggling with implementing and using crime mapping in partnerships. With some adaptations from what has been experienced since, these points help guide those that are aspiring to succeed in sharing information in a GIS and spatial database framework. Many of these points relate to ensuring good information is available to support good analysis. The list below is an adaptation from Openshaw *et al.* (1990) and Chainey (2001):

- Clearly identify data sources, data collection methods, data manipulation procedures, types of analysis to be performed and system/analysis outputs.
- Develop a data archive that has the flexibility to be added to with continual updates of complete and consistent data sets.
- Develop a metadata facility describing information that is made available for data exchange.
- Introduce a virtually automated and accurate process for geocoding partner information, where the extraction, cleaning and any necessary depersonalisation of information is carried out as part of this software process.
- Introduce a simple method and well-structured procedure for handling and delivering information between partners.
- Establish a high level of communication between data partners and agency contacts. In particular, identifying a one-point contact (designated officer) in each partner agency for data access and data queries.
- Introduce compatible processes for data transfer between data providers and their partners.
- Introduce analytical techniques that can be consistently used by all partnership members so that conflicting results do not arise.
- Document information-sharing procedures to ensure consistent processes are being applied for data exchange and ensure these are readily available to partnership members and new staff that replace those that leave.
- Review the appropriate choice of technology, analysis tools and dissemination software that are appropriate for the partnership and make changes where required.
- Review the necessary communication channels required within each partner agency and across the strategic level of the partnership to ensure that information and analysis outputs are used.
- Produce reporting templates for management and operational staff, and if the resources permit, automate these processes so that the analysts can concentrate on analysing data, and are not just used for producing administrative statistics and reports for management.

- Review the funding and other resourcing required to enable changes and updates to be made to the system or training of user(s).
- Ensure that where information or data is used within the work of the partnership in, for example, reports, charts and maps, the data suppliers are sent copies of the final piece of work. Where data are to be made publicly available, the data custodian must give formal approval.

To assist the management of this information, experience from partnerships engaged in sharing information has identified that a centralised information hub brings a number of practical, operational and strategic benefits. These include:

- consistency in how data are sourced, handled, processed (including cleaning, geocoding and depersonalising), managed and how dissemination is coordinated;
- the ability to perform routine reporting and analysis on behalf of the partnerships;
- the ability to apply independent quality assessments on data that are contributed and explore ways to improve data that are of strategic importance to the partnership;
- to be the source of information management and analysis expertise;
- providing support and guidance in understanding the format, content and limitations of data use;
- to meet specific requests from the partnership for analysis products; and
- to coordinate and produce consistency in the required management reports.

One such hub is the East Valley COMPASS Initiative, which brings together data from a wide variety of organisations in both the City of Redlands and the county of San Bernardino, California. COMPASS stands for Community Mapping, Planning and Analysis for Safety Strategies. Data are shared between such traditionally disparate organisations as the state corrections department, the local public works department and academic organisations. For more information, see http://www.eastvalleycompass. org/hs/compassinfo.htm.

7.5.3 Handling the requirements of data privacy

One of the biggest issues with sharing data is how to handle data privacy (Box 7.1). Personal and private details of a person can relate to not only the person's name and details about them, but also the address at which

Box 7.1 The principles of data privacy

To help understand some of the needs for data privacy and how to achieve compliance through information sharing, it is useful to consider these seven data protection principles. These principles are based on the UK's data privacy laws, but are similar to data privacy principles established in many other countries (e.g. the UK's principles draw heavily from the European Union's Directive on Data Protection, therefore are also similar to principles in many other EU member countries. Other countries, such as Australia, also describe similar principles):

1. There must be a legitimate basis for the processing of personal data and information processing must be fair and lawful.
2. Disclosure of personal data must be compatible with the purpose(s) for which the personal data was obtained.
3. Personal data must be adequate, relevant and not excessive to the purpose(s) for which it is processed.
4. Personal data must be accurate and where necessary kept up to date.
5. Personal data processed for any purpose(s) must not be kept for longer than is necessary for that purpose(s).
6. Personal data must be processed in accordance with the legislative rights of data subjects (e.g. human rights).
7. Appropriate technical and organisational measures must be taken against unauthorised or unlawful processing of personal data and against accidental loss or destruction of, or damage to, personal data.

they live. The requirement to depersonalise data on an individual requires the removal of details that allow a person to be uniquely identified. In the context of crime mapping, handling this spatial data privacy is not simply a question of making the point object that represents a person's home address a larger dot! Many crime mappers have made the mistake in the past that in an attempt to depersonalise the address information of an offender on a map that is published to the public (e.g. a point object representing the address of a male sex offender), the point object is simply made a bigger object. The object may remove the ability to singularly identify the offender, but in so doing also identifies any neighbours under the point object as potentially being the sexual offender. This results in pointing the accusatory finger to all males that live under the dot on the map!

A common view also exists that data privacy, and any laws that a national government may impose on it, is a hindrance to information sharing and to the sharing of intelligence. There will always be tension between the need for information and the need to respect personal rights. Consistent implementation of data privacy principles in an agreed information-sharing protocol framework (e.g. a document that describes what can be shared to who it can be shared, and how the publishing of this data is handled) and the use of 'privacy enhancing technology' represents an effective way of balancing these needs.

In 1999, the requirements for sharing crime maps and spatial data were the topic of a US Crime Mapping Research Center (CMRC) roundtable. The result of this roundtable was the creation of guidance on confidentiality, data sharing and related security issues pertaining to crime mapping (Wartell and McEwen, 2001). Similarly, the British Home Office released 'Data exchange and crime mapping' (Radburn, 2000), as a practical guide to help those in the UK implement the challenges posed by spatial data privacy. Many of the points raised in these guidance documents, although designed primarily for their respective audiences, are pertinent to the international audience and complement many of the points raised in this chapter. Further details of this guidance are provided in the 'Further reading' section of this chapter.

7.5.3.1 Handling spatial privacy through degrading spatial precision

Issues with spatial data privacy can be avoided when data are spatially degraded to a more coarse level of precision. For example, rather than mapping to a property, a suitable sanitised resolution to position the event would be to a suitable geographic reference on the street. Data could also be aggregated to a more coarse geography (e.g. a count of crimes for each police beat). For analysis purposes, the spatial precision still needs to be fine enough to ensure that any subsequent spatial analysis is not so coarse that it actually hinders the identification of crime problems at the community or neighbourhood level. For many partnership applications, the first stage of the analysis of a problem does not require personal data (e.g. the exact address of where a crime happened). In the United Kingdom, spatial data privacy is often handled by depersonalising address details to the fine granular level of the postcode. In the US, this can be achieved by mapping data to the block or block group rather than the actual address. A UK postcode, on average, contains 14 addresses. Depersonalising the address detail in a crime requires matching the address to its postcode, using the postcode centroid as the spatial coordinates that represent the crime and removing the fine address content from the crime record. For example, the address fields in a crime record at 125 Mandrake Road, London, SW17 7PX become Mandrake Road, London, SW17 7PX after a depersonalising process has been performed. For postcodes that contain fewer than four addresses, the geographic coordinates assigned to the crime are those of the next nearest neighbouring postcode that has greater than three addresses within it. This method of depersonalising the spatial content of crime data retains the detail in the record that is not subject to being depersonalised (e.g. the crime type, time and date of the crime), rather than losing it if it was aggregated to a count of crimes in an area.

Figure 7.3 shows the difference in these two levels of spatial data representation. Figure 7.3a is a hotspot map created using personal address data and Figure 7.3b is a hotspot map created using data that has been depersonalised, compliant to strict UK data protection laws, to the postcode level. Certain subtle visual differences do exist, but broadly the spatial patterns have the same effect for focusing and initiating the direction of future stages of analysis.

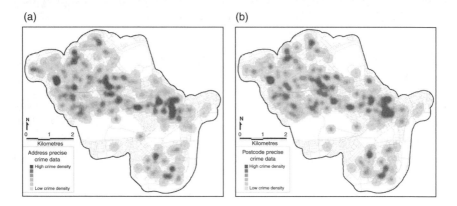

Figure 7.3 A spatial comparison between crime hotspots produced from (a) personal data geocoded to the exact address and (b) depersonalised data geocoded to postcode level. The crime data are taken from an area in central London to demonstrate that depersonalising data in this manner retains the granularity of spatial precision required for partnership-type geographical analysis. Reproduced by permission of Ordnance Survey on behalf of HMSO. © Crown copyright 2005. All rights reserved. Ordnance Survey Licence number 100044021

In countries that do not have postcodes or similar address aggregation and coding systems such as the nine figure zip code, depersonalising the address content of a crime record to the street segment could offer a level of granularity that retains spatial privacy. There is, though, a need for research to be conducted in this field of spatial data privacy and spatial depersonalisation to describe the impact that this aggregation has on subsequent analyses.

7.5.3.2 Handling the exchange of personal data

To help explain what can and what cannot be shared it is useful to consider the following three points. These points are based on UK practice but are useful reference for other countries.

1. *Is this exchange of data justified by the public interest?* When people abuse the privilege of living in a free society by committing crime, the community in effect passes legislation empowering the partners within a crime and disorder reduction partnership to stop such actions in the public interest.

2. *Is this data exchange lawful under data protection?* Data protection legislation may often not allow the sharing of personal data without the permission of the individual involved, even if they are a serious offender. The need to fight crime and disorder can result in a reinterpretation of the legislation, or may require a legislative modification as in the case of the UK's Data Protection Act, to enable the sharing of personal data if action is required to prevent a crime or if it is agreed the action is in the public interest. Data protection legislation and the exchange of information tread a fine line between protecting the individual and protecting society. The key behind this fine line is whether by sharing information about people, the rights of those individuals are still being respected, and that any action that requires the use of personalised data is in the public interest.

3. *Is this data exchange lawful under human rights legislation?* Human rights legislation usually gives everyone the right to respect for their private and family life, their home and their correspondence. However, this right can only usually be exercised in accordance with the law and in the interests of a democratic society, for example, for national security, public safety, crime and disorder, and the protection of health. Sharing of personal data is allowed if the public interest demands it.

In other words, so long as information shared under this power is directly linked to the purpose or objectives of the legal requirements of agencies to provide services to reduce crime and while complying with any legally enforceable provisions of data protection legislation, the sharing of personal data can be legal.

In practice, the justification for sharing personal data must be evidence-based – this is essential to maintain public confidence that partnerships do not adopt a 'big brother' coverage and share data as a matter of course. For example, a crime hotspot map produced from depersonalised crime data (e.g. postcode or census output area) might indicate the need to examine in more detail the causes and the people involved. Based on the previous evidence of a hotspot, personalised information for this area can now be requested and used by the partnership. However, it also follows that data used to identify the hotspot must be accurate to ensure the privacy rights of individuals and to prevent the risk of mistaken identity.

7.5.3.3 A practical example

Consider the following scenario: A spatial analysis of residential burglary using depersonalised (postcode level) crime data has revealed a number of burglary hotspots of properties. Research indicates that burglary hotspots spatially coincide with areas where properties are repeatedly burgled (Johnson *et al.*, 1997; Bowers and Johnson, 2005). The local crime partnership wish to act on this burglary problem by identifying those properties that are suffering from being repeatedly burgled by making them safer and reducing the personal impact this crime has on the people that are being repeatedly victimised.

Burglary repeat analysis requires precise property addresses. Depersonalised information is not precise enough to reveal the properties that are being repeatedly burgled. The partnership can seek permission to access the personal crime details (i.e. the personal address details) for this, and only this, analysis, based on the evidence that hotspots of burglary exist and the statutory legal duties of the partnership. If properties that are repeatedly burgled are identified, the partnership can also request other personalised data from partners (e.g. housing data from the municipal government) to help identify if patterns exist in the types of properties that are repeatedly burgled. This exercise is not a data 'fishing trip' because clear evidence exists about the crime and its location, and it is in the public interest to tackle it. The personal data that are used must though not be kept for longer than is necessary for this analysis purpose. This means that once the analysis has been completed, the personal data must be destroyed and not used for any other purpose.

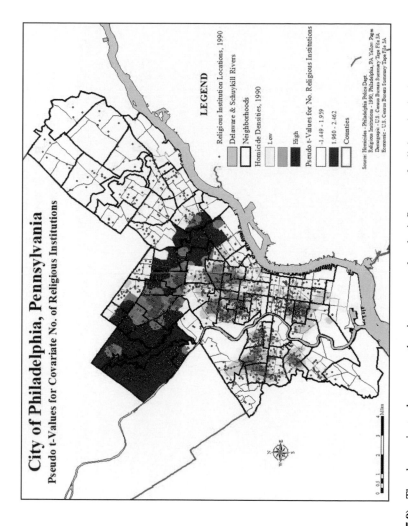

City of Philadelphia, Pennsylvania
Pseudo t-Values for Covariate No. of Religious Institutions

LEGEND

- Religious Institution Locations, 1990

Delaware & Schuykill Rivers

Neighborhoods

Homicide Densities, 1990

Low

High

Pseudo t-Values for No. Religious Institutions

-1.449 - 1.959

1.960 - 2.462

Counties

Source: Homicides - Philadelphia Police Dept.
Religious Institutions - 1990, Philadelphia, PA Yellow Pages
Demographic - U.S. Census Bureau Summary Tape File 3A
Economic - U.S. Census Bureau Summary Tape File 3A

Plate 1 (Figure 5.9)   The change in *t*-values across the city suggests that the influence of religious institutions on the incidence of homicide is stronger in some parts (the darker shaded areas) than others

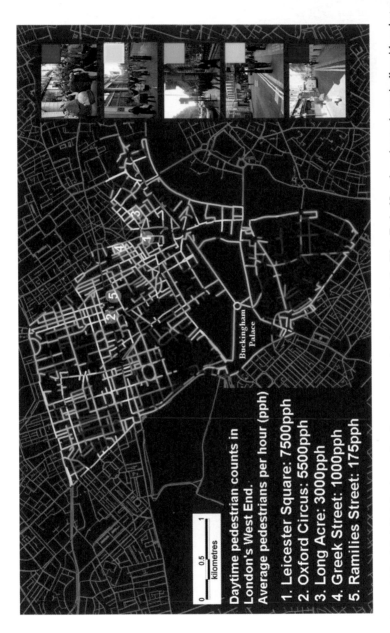

Plate 2 (Figure 6.16) Modelled daytime pedestrian counts per street segment in the West End of London, shown thematically with a photographic representation of the on-street population. From Chainey and Desyllas (2004). Reproduced by permission of Ordnance Survey on behalf of HMSO. © Crown copyright 2005. All rights reserved. Ordnance Survey Licence number 100044021

Daytime pedestrian counts in London's West End.
Average pedestrians per hour (pph)

1. Leicester Square: 7500pph
2. Oxford Circus: 5500pph
3. Long Acre: 3000pph
4. Greek Street: 1000pph
5. Ramilies Street: 175pph

Buckingham Palace

kilometres
0   0.5   1

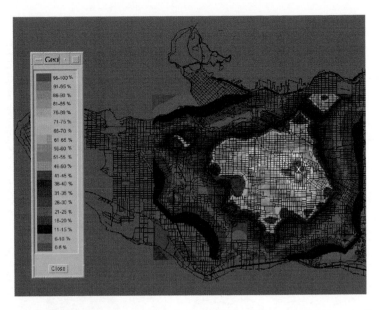

Plate 3 (Figure 10.5) An example of a jeopardy surface produced as a result of a geographic profile. The area of highest probability of where an offender lives is shown by the dark red and orange areas The white cross on a blue circular background positioned at this location is where the offender actually lived. Source: Rossmo (2000)

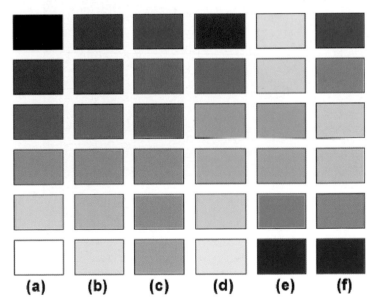

Plate 4 (Figure 12.7) Six columns demonstrating different colour plans for thematic mapping

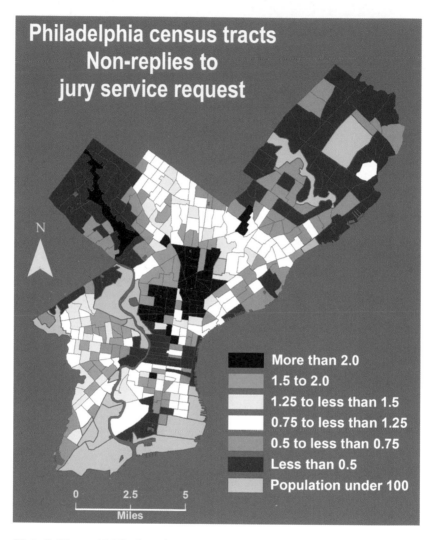

Plate 5 (Figure 12.12) Location quotients of non-replies to jury service requests, Philadelphia, 2002

Case study: The Amethyst Crime and Disorder Information Hub, Cornwall, England

Material supplied by Phil Davies, Information Manager, Project Amethyst

Amethyst is a web-based solution for crime mapping and data exchange. There are two areas to this website:

1. A publicly accessible area that provides district crime statistics and maps. It displays crime information in an easily understood format, and aims to provide a clearer picture of district crime and as a result reduce the perceived fear of crime (Figure 7.4).
2. A secure area for partners – Key partners supply data to Amethyst, which is then cleaned and mapped in a GIS. The partners include:

   - Devon and Cornwall Police Constabulary
   - Cornwall County Council
   - Local district councils
   - Cornwall Fire Brigade
   - Devon and Cornwall Probation Service
   - Cornwall Drug and Alcohol Action Team
   - Cornwall Education Department
   - Cornwall Health Service
   - Cornwall Youth Offending Team
   - Cornwall Social Services.

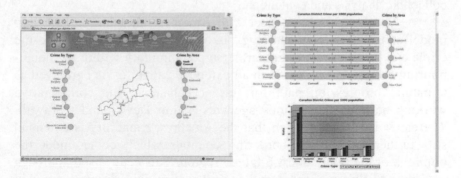

Figure 7.4 Crime and disorder data disseminated to the public from the Project Amethyst website. www.amethyst.gov.uk

Amethyst, and its associated information hub, has been providing valuable online information and analysis services to the crime and disorder reduction partnership community across Cornwall since July 2002. This centralised facility has proven to be of immense value for visualising and identifying patterns of crime and disorder within Cornwall; in turn it has enabled Amethyst partner organisations to target priority areas much more cost-effectively, and allocate people and technical resources that support the partnerships efforts for crime reduction.

Amethyst is more than just an IT solution – it is a holistic approach to crime and disorder data sharing. The public-facing website focuses on reducing the fear of crime by presenting a clear picture of what crime and disorder is happening in local areas. The site does not only display crime information, but also details initiatives and interventions put in place by the crime and disorder reduction partnership to tackle local crime and disorder issues, helps raise awareness on crime prevention tips, encourages a safer community, plus offers useful links to other sites.

Authorised partners have access, through a secure login, to a dynamic geographical mapping tool that provides detailed cross-partnership data, which are relevant, accurate and timely and all exchanged under a strict data sharing protocol. This central point for crime and disorder information vastly reduces duplication and provides a quick online one-stop-shop to data that once took many days to obtain.

Amethyst also provides analytical support to the crime and disorder reduction partnership. Through this collaborative working, regular reports are provided to the partners, forming the basis of a rolling audit. Crime audits in the past have been too focused on police data alone. The new audits now draw on quality-assured police and secondary data collection, analysis and evaluation.

In recognising the success of Amethyst, the same partnership model and underlying technology is being extended to embrace the crime and disorder reduction partnership community within the neighbouring county of Devon. Such an extended capability is perceived as a natural progression path for further enhancing the collaborative working between the various agencies within Devon and Cornwall. Thereafter it is the intention that the Amethyst capability is extended still further to embrace bordering counties which come under the umbrella of the South West England region. This centralised approach is demonstrating the value and importance of joined-up working in identifying problems of crime and disorder at both a local level and cross-border levels.

7.6 Combining data from different geographic units

In Chapter 5 we discussed a number of spatial statistical routines that allow the crime mapper to explore statistical relationships between different datasets. In this chapter we have presented a range of data that are maintained from police and non-police sources which could be shared to help better understand local crime problems. The use of geography as a powerful common denominator to integrate data has also been presented, but a problem which may exist for these spatial statistics to be applied is that much of these data may be held in different spatial objects. For example, crime data may exist as point objects and census data exist in polygon object form, providing an aggregate measure of an attribute within defined census boundaries. For a relationship to be explored between the crime data and a census variable requires these data to be represented in the same geographic object form, such as aggregating the crime data as a count of the number of crimes or as a rate to the census geographic boundary unit. Figure 7.5 demonstrates this concept, aggregating crime records from point data to counts of crime for each of the census geographic units.

In certain situations the geographic variables may both exist as points which do not spatially coincide. For these data to be compared, they again need to be represented in a geographic object form that spatially coincides with the variables. For point data this would require both data to

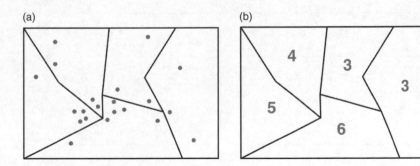

Figure 7.5 For a relationship to be explored between data that are represented in different geographic object form requires these data to be combined to a common object geography. (a) shows crime data represented as points within census geographic boundary areas that hold a variable for housing tenure. For the relationships of crime and housing tenure to be explored, these crime data are aggregated as a count within each of their respective census geographic polygons (b)

be aggregated to a common geographic polygon object that could be either an administrative boundary such as census areas or to an arbitrary geometric grid. Figure 7.6 shows two sets of points that do not spatially coincide. An arbitrary spatial grid can be overlayed across these points and used as the geometric grid to which these point variables are aggregated (as either a count or a rate).

A third scenario is where one spatial variable may be represented as a point and the second is represented as a line segment. For example, the points could be street robberies and the line segment variable is the count of pedestrians along a street segment. The same rules described above need to be applied where these data need to be represented in a common geographic object form that spatially coincides for any relationship between the two variables to be explained. This will require the points to be either aggregated to the line segments (Figure 7.7) or both objects aggregated to a polygon unit.

LeBeau (2000) used this method effectively when aggregating drug arrest locations (points) to the line segments of the streets where they took place in Charlotte-Mecklenburg, North Carolina (USA). LeBeau mapped the data by varying the thickness of the line relative to the number of arrests on each street. The display (Figure 7.8) shows the highest drug arrest street segments and blocks.

The final scenario is one in which two variables in polygon object format are to be combined or compared, but the polygons that represent the variables do not exactly spatially coincide. This requires the data to be

(a)

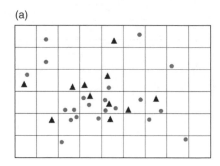

(b)

0,0	1,0	0,0	0,0	0,1	1,0	0,0	0,0	0,0
0,0	1,0	0,0	0,0	0,0	0,0	0,0	1,0	0,0
1,1	0,0	0,1	0,1	1,1	0,0	0,0	0,0	0,0
0,0	0,0	2,0	2,1	3,1	1,1	1,1	0,0	0,0
0,0	0,1	2,0	1,0	0,1	0,0	1,0	0,0	0,0
0,0	0,0	1,0	0,0	0,0	0,0	0,0	1,0	0,0

Figure 7.6 (a) Two sets of point data do not coincide (one set is displayed as circular dots, the other as triangles). For these variables to be compared they can be aggregated to an arbitrary geometric grid that overlays the point data's coverage; (b) these point data can then be represented as counts for each of these geographic polygons. In this example the first digit represents the count of circular symbol points in each grid polygon cell and the second digit represents the count of the variable that is displayed as triangle point symbols in the grid cell polygon

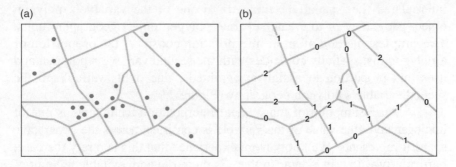

Figure 7.7 (a) To explore the spatial statistical relationship between two variables that are represented as points and street segments would require these data to be represented in a common geographic object form that spatially coincides. (b) This could involve aggregating the point variables as counts to the line segments (i.e. the count of crimes on the section of street on which they occurred)

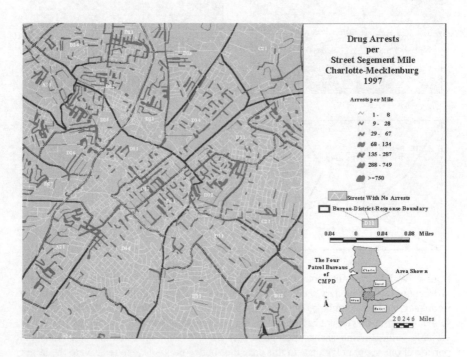

Figure 7.8 Drug arrests represented thematically along street segments in Charlotte-Mecklenburg, North Carolina. Source: LeBeau (2000)

215

interpolated (i.e. spatially estimated) to one of the variable's polygon object geography or to create separate polygon object geographic units. The simplest method of areal interpolation considers the proportion of a polygon that spatially coincides with the second variable's polygon and uses this proportion to assign a weighted value of its variable to the second variable's polygon geography (Figure 7.9).

An assumption of many areal interpolation techniques is one of homogeneity – the value of the variable is consistent across the geographic space it represents. We know from experience that this is rarely the case, particularly with crime data, so there is therefore an inevitable amount of error in the values that result from an areal interpolation. A number of more sophisticated areal interpolation techniques have been developed that attempt to more accurately interpolate the underlying spatial values.

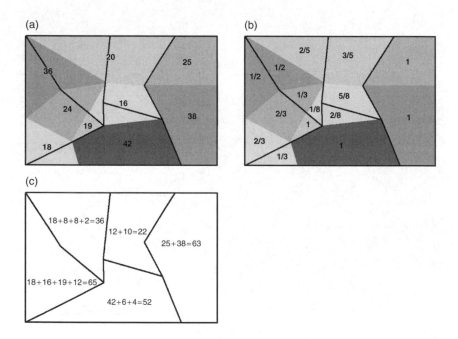

Figure 7.9 The areal interpolation of spatial polygon objects. In this figure there are two different polygon boundary sets: one is shown as the shaded regions and contains a crime frequency, and the boundaries of the target polygon set are shown as black lines. (a) Two sets of polygons that do not have exact spatially coinciding boundaries are shown with the variable value for the shaded polygons. (b) Simple areal interpolation calculates the portion of one set of polygons – in this case the shaded polygons – in relation to the larger polygons. (c) The portion of each shaded polygon area is used to estimate the value of the variable from each of these polygons that should be assigned to the other areal units

These require more advanced spatial statistical routines that are beyond the scope of this book, but can be reviewed from a large source of other literature (Goodchild and Lam, 1980; Green, 1989; Flowerdew *et al.*, 1991; Langford *et al.*, 1991; Flowerdew and Green, 1994; Bailey and Gatrell, 1995; Coombes, 1995; Fisher and Langford, 1995; Openshaw and Rao, 1995; Gregory, 2000; Sadahiro, 2000).

The main points for the crime mapper to consider when aggregating or performing areal interpolation methods are as follows:

- Any aggregation will mean a loss of spatial detail from the original data and could impair any analysis of spatial patterns.
- Areal interpolation assumes that spatial variables are homogenous. This is usually quite unrealistic. However, spatial statisticians have over the years developed new methods that attempt to limit the effects of this assumption.
- Areal interpolation methods of polygon units ultimately estimate the resulting variable values, therefore introducing an inevitable amount of error into these final statistics. This error and statistical uncertainty does need to be considered as it is propagated when these areal units are further combined or used in statistical calculations (Burrough and McDonnell, 1998; Mowrer and Congalton, 2000; Hunsaker *et al.*, 2001; Longley *et al.*, 2001).

With all these conditions and problems to worry about, the crime mapper may feel that areal interpolation is a technique to avoid. On occasions it is inevitable that a crime mapper will have to perform areal interpolation, but considering the above and taking care of how data are interpolated can help to avoid the main pitfalls.

Geographic administrative areas are becoming more harmonised (e.g. in the United Kingdom, efforts have been made over the last five years to harmonise police beat areas with census geographic areas and areas of local government administration). If the crime mapper ever has to create or recommend new geographic areas (e.g. to represent the area covered by a proposed crime reduction initiative), it is useful to consider the geographic boundary units that already exist and use these as the building blocks for this new geography (e.g. using the smallest census geographic units as the building blocks for the area of the crime reduction initiative). If the spatial building blocks are of fine resolution, this can helpfully avoid the new areas becoming awkwardly constrained to impractical arcas, and also means that future areal interpolation can be avoided so that the aggregation of variables for subsequent analysis is more straightforward and reliable.

Ecological correlation

In chapter 6 we described the concept of the ecological fallacy – when an inference is made about an individual based on aggregate data for a group. In terms of data that are aggregated it is useful to make a distinction between the ecological fallacy and another geographical tyranny referred to as ecological correlation. Ecological correlation applies to two issues – the false interpretation of a correlation between variables; and a fallacious inference.

A false interpretation of the relationship between two variables may occur when the proxies of variables are correlated rather than the variables themselves. For example, many analysts often find a negative correlation between income levels and crime. That is, areas where income is low have the highest levels of crime. However, the relationship is not direct – low income people do not necessarily commit crime because they have a low income. The correlation exists because low income neighbourhoods are associated with other social conditions that, in turn, are predictors of crime. Care, therefore, needs to be taken in the interpretation of the relationship between geographic variables.

A correlation between two variables is possible at any geographical level – individual, Census block, output area, ward, block group, city, district – yet the relationship displayed between one level of geography may be different to the relationship displayed between the same variables at a different geographic level. The fallacy in a correlation arises by trying to make an inference about the correlation that is displayed at one level of geography and directly applying its interpretation to another level of geography. For example, Levine (personal communication) found that at an individual level there was a negative correlation between income and the likelihood of divorce (i.e. low income persons are much more likely to become divorced). Applying this correlation to a higher level of geography would have been incorrect because at the larger aggregation level of the State the opposite relationship was true – there was a positive correlation (i.e. areas that on average have persons on a high relative household income are where divorce rates are higher). Both were valid and strong correlations, but because they were different an ecological correlation exists.

Similar to the problem of the ecological fallacy, this does not mean that identifying relationships between aggregate figures will necessarily produce inaccurate results, and it does not necessarily mean that any inferences drawn about relationships between the characteristics of an aggregate population and the characteristics of individuals are absolutely wrong either. Studying correlations between areal units can help to reveal trends between data (for example, the correlation between high levels of repeat burglary victimisation and deprivation, Johnson *et al.*, 1997). What it does say is that different relationships may exist at different levels of geographical aggregation, that crime mappers should be careful in how they interpret the relationships drawn from aggregate correlations, and that analysis should be performed consistently at a geographical level to avoid a fallacy.

7.7 Summary

Agencies other than the police are playing an increasing role in supporting the efforts to reduce crime. Geography can play a key role in enabling the joining up of data that are available across an information-sharing partnership, and provide the platform for helping to ensure a consistent, comprehensive, legislative-compliant and accurate management resource for tackling local crime problems and disseminating information. These important roles include:

- Data integration – to integrate disparate datasets across partnerships.
- Data sanitising – the different levels (or precision) of geographic referencing provide a useful means for depersonalising records but without compromising certain types of analysis.
- Data analysis and exploring of relationships – geographical enquiry could be used to reveal the conditions or characteristics within a particular area (e.g. crime hotspots, location of licensed premises, offender patterns or demographic profiles) and how they can be tackled.
- Management reporting – timely area profile reports can be generated to support a strategic view of local issues, and when presented in map form can help partners to understand patterns. These reports can also help to monitor the partnership's area-based initiatives and help to prompt discussion on further strategy implementation.
- Consulting across departments, with other organisations and with the public – map output can help to inform, better enable contributions to consultation, make sense of the decision-making process that has led to the choice of a local initiative, and also be used to present the options in the design of the crime reduction process.
- Partnership and public access to crime and community safety data, service information and local initiatives – enquiry using a geographic reference (e.g. a postcode) can provide the gateway for accessing information about local conditions and the programmes being targeted to combat local issues.

Partnership working and how geographic information is used by partnerships has its challenges, but by following the experiences and lessons learnt from others and considering certain key principles, the activity of crime mapping in partnerships can flourish and provide an essential base on which informed decision-making for crime reduction can take place.

Further reading

LaVigne, N. and Wartell, J. (2001). *Mapping Across Boundaries: Regional Crime Analysis*. Washington, DC: US National Institute of Justice. http://www.ojp.usdoj.gov/nij/maps/pubs.html.

'Mapping Across Boundaries' addresses the obstacles and develops answers in developing regional crime mapping. The report is a primer for police agency personnel and students of mapping who want to enhance crime control and prevention efforts, and discusses how cross-boundary mapping can better reveal hotspots of crime that occur along jurisdictional boundaries or identify serial crimes by offenders operating in neighbouring jurisdictions. The report provides guidance through case studies on a range of regional mapping models – from central archiving systems to ambitious multi-agency consortia with common database structures and GIS platforms.

Wartell, J. and McEwen, J.T. (2001). *Privacy in the Information Age: A Guide for Sharing Crime Maps and Spatial Data*. Washington, DC: US National Institute of Justice. http://www.ojp.usdoj.gov/nij/maps/pubs.html.

In July 1999, the US CMRC hosted a two-day Crime Mapping and Data Confidentiality Roundtable. The purpose of the roundtable was to generate discussion and initial guidance on issues of confidentiality, data sharing and related security issues pertaining to crime mapping. Participants of the roundtable included representatives from law enforcement, the research community, the legal profession, the GIS field, victim advocacy and the media. The report 'Privacy in the Information Age: Guidelines for Sharing Crime Maps and Spatial Data' documents the outcomes from the roundtable.

Radburn, S. (2000). *Data Exchange and Crime Mapping: A Guide for Crime and Disorder Partnerships*. London: Home Office. http://www.crimereduction.co.uk/technology01.pdf.

This guide discusses many of the challenges and offers practical tips on data exchange and crime mapping. Its topics include how data can be used to assist crime and disorder reduction, advice on how to comply with data protection laws, certain basics of crime mapping and how mapping and information-sharing systems can be set up within a partnership.

Office of the Deputy Prime Minister–Social Exclusion Unit (2000). Report of Policy Action Team 18: Better Information. London: The Stationery Office. http://www.socialexclusion.gov.uk/publications.asp.

The problems of information sharing and lack of good data to help identify problems, diagnose them and be in a position to use data to monitor and assess if responses are effective were neatly accounted in the review the British Government completed in 2000 on 'better information'. In total there were 20 Policy Action Teams that mainly focused on specific experiences of social exclusion (e.g. problem housing). Policy Action Team 18 focused purely on information.

References

Bailey, T.C. and Gatrell, A.C. (1995). *Interactive Spatial Data Analysis*. Harlow: Longman.

Bowers, K. and Johnson, S.D. (2005). Domestic burglary repeats and space-time clusters: The dimensions of risk. *The European Journal of Criminology*, January 2005, 2, 67–92.

Brazil Ministry of Justice (2003). Public Security Plan. Brasilia: National Secretariat of Public Safety. www.mj.gov.br/senasp.

Burrough, P.A. and McDonnell (1998). *Principles of Geographical Information Systems*. Oxford: Oxford University Press.

Chainey, S.P. (2001). Combating crime through partnership: Examples of crime and disorder mapping solutions in London, UK. In A. Hirschfield and K. Bowers (eds) *Mapping and Analysing Crime Data*. London: Taylor & Francis.

Chainey, S.P. (2004). Crime mapping use in Crime and Disorder Reduction Partnerships in the UK: Results from an interactive survey conducted at the National Community Safety Network Conference, Cardiff, Wales. www.jdi.ucl.ac.uk.

Coombes, M. (1995). Dealing with census geography: Principles, practices and possibilities. In S. Openshaw (ed.) *Census Users' Handbook* (pp. 111–132). Cambridge: GeoInformation International.

Fisher, P.F. and Langford, M. (1995). *Modeling the Errors in Areal Interpolation between Zonal Systems by Monte Carlo Simulation Environment and Planning A*, 27, 211–224.

Flowerdew, R. and Green, M. (1994). Areal interpolation and types of data. In A.S. Fotheringham and P.A. Rogerson (eds) *Spatial Analysis and GIS* (pp. 121–145). London: Taylor & Francis.

Flowerdew, R., Green, M. and Kehris, E. (1991). Using areal interpolation methods. In *Geographic Information Systems*. Papers in Regional Science, 70, 303–315.

Goodchild, M.F. and Lam, N.S.-N. (1980). Areal interpolation: A variant of the traditional spatial problem. *Geo-Processing*, 1, 297–312.

Green, M. (1989). Statistical Methods for Areal Interpolation: The EM Algorithm for Count Data. North West Regional Research Laboratory, Research Report 3: Lancaster.

Gregory, I.N. (2000). An evaluation of the accuracy of the areal interpolation of data for the analysis of long-term change in England and Wales. Proceedings of the 5th International Conference on GeoComputation University of Greenwich, United Kingdom 23–25 August 2000. http://www.geocomputation.org/2000/GC045/Gc045.htm.

Hunsaker, C., Goodchild, M.F., Friedl, M. and Case, T. (2001). *Spatial Uncertainty in Ecology: Implications for Remote Sensing and GIS Applications*. New York: Spinger-Verlag.

Johnson, S.D., Bowers, K. and Hirschfield, A. (1997). New insights into the spatial and temporal distribution of repeat victimisation. *British Journal of Criminology*, 37(2), 224–241.

Langford, M., Maguire, D. and Unwin, D.J. (1991). The areal interpolation problem: Estimating population using remote sensing in a GIS framework. In I. Masser and M. Blakemore (eds) *Handling Geographical Information: Methodology and Potential Applications* (pp. 55–77). New York: Longman.

LeBeau, J. (2000). *Demonstrating the Analytical Utility of GIS for Police Operations: A Final Report*. Washington, DC: NIJ (Crime Mapping Research Center).

Longley, P., Goodchild, M., Maguire, D. and Rhind, D. (2001). *Geographic Information Systems and Science*. Chichester: John Wiley & Sons.

Mowrer, T. and Congalton, R. (2000). Quantifying Spatial Uncertainty in Natural Resources: Theory and Applications for GIS and Remote Sensing.

Openshaw, S. and Rao, L. (1995). Algorithms for Re-aggregating 1991 Census Geography. *Environment and Planning A*, 27, 425–446.

Openshaw, S., Cross, A., Charlton, M. and Brunsdon, C. (1990). Lessons learnt from a post mortem of a failed GIS. 2nd AGI Conference, October 1990, Brighton.

Radburn, S. (2000). *Data Exchange and Crime Mapping: A Guide for Crime and Disorder Partnerships*. London: Home Office. http://www.crimereduction.co.uk/technology01.pdf.

Ratcliffe, J.H. and McCullagh, M.J. (2001). Chasing ghosts? Police perception of high crime areas. *British Journal of Criminology*, 41(2), 330–341.

Sadahiro, Y. (2000). Accuracy of count data transferred through the areal weighting interpolation method. *International Journal of Geographical Information Science*, 14, 25–50.

Wartell, J. and McEwen, J.T. (2001). *Privacy in the Information Age: A Guide for Sharing Crime Maps and Spatial Data*. Washington, DC: US National Institute of Justice. http://www.ojp.usdoj.gov/nij/maps/pubs.html.

8
Mapping and Analysing Change Over Time

Learning Objectives

Crime is a dynamic event. The incidence of crime is neither stationary over space nor stationary over time. While GIS provide practitioners and researchers with a remarkable tool for mapping crime over space, the development of temporal analysis tools has lagged behind the development of spatial methods. This chapter explores ways to examine temporal and spatio-temporal patterns of crime across different temporal scales and aims to equip the reader with not only a range of tools but also an understanding of the value that a temporal understanding can contribute to the crime reduction effort. An appreciation of the temporal characteristics of a crime pattern can go a long way to preventing and reducing crime.

8.1 Introduction

Much of this book has until now concerned itself with the production and application of single crime maps: fundamental tools in the description of criminal behaviour and crime patterns. When determining a crime reduction strategy, understanding the changing mosaic of criminal behaviour over time becomes equally as important. First, it is useful to know if an observed crime pattern is steady in space over time or moves from place to place. Offenders occasionally curtail their activities in one area and find new

GIS and Crime Mapping   Spencer Chainey and Jerry Ratcliffe
© 2005 John Wiley & Sons, Ltd

targets in response to police activity, a process referred to as 'displacement' (Barr and Pease, 1990; Hesseling, 1994). Secondly, once a crime reduction strategy has been implemented, it can be useful to evaluate the strategy in order to determine if there has been a reduction in crime in the target area as well as to test for the existence of any displacement. This type of work requires the crime mapper to have at their disposal a range of analysis tools that are applicable to exploring patterns not just spatially, but also temporally. The development of temporal crime mapping tools has tended to lag behind the expansion of spatial techniques. This chapter will introduce a number of techniques that can map crime patterns in the temporal dimension.

While most people readily grasp the three dimensions of space, time is an altogether more difficult aspect to work with. We cannot touch it, nor can we see it, yet we know just how concrete and real it is when assignments are due, deadlines loom or we are late for a train. Crime information is usually recorded with day and time attributes. The degree of precision of the temporal information determines the temporal resolution available for analysis. Temporal resolution can be best described as the minimum difference between two independently measured temporal values that can be distinguished by the analytical method applied (Langran, 1992). More complex analysis is possible with call for service data, often recorded to the second or minute, than with the movement of glaciers which have a temporal resolution measured in years. The temporal resolution determines the minimum level of temporal analysis that can be conducted, but does not inhibit longer time analysis.

Data recorded to the minute could be analysed on a minute-by-minute temporal basis, though it may not reveal much useful intelligence. However, the same data could easily be analysed on an hourly or daily basis, allowing variation during the course of a day to be seen. Crime data that only record the day of the offence limit the minimum temporal resolution to daily frequency counts at best.

Time can be displayed using a simple notation – the letter t with a subscript number that is used to indicate a time component. For example, t_0 is used to signify either the first in a series of temporal values or the value at the present time. Moving forward into the future increases the number (t_1, t_2, $t_3 \ldots t_n$), while past values are shown with negative values (t_{-1}, t_{-2}, $t_{-3} \ldots t_{-n}$). Vasiliev (1996) identified five categories of temporal information (moments, duration, structured time, time as distance, space as clock) of which the first four are most relevant to crime analysis. To these relevant four, we add another that is particular to crime mapping – the time span. In the following section, these categories will be shown against a timeline, a method for visualising time.

8.2 The timeline

The timeline is a useful tool with which to visualise the different temporal categories. We tend to think of time as having distinct periods that end at the commencement of the next period (days, hours). Periods of time, such as weeks or seconds, are convenient tools to help people communicate information. Real time is not constructed of distinct blocks of time that run together, but is a continuum, and we can visualise this continuum as a timeline that stretches into the past and into the future.

8.2.1 Temporal categories

Figure 8.1 (first row) shows a timeline for five hypothetical burglaries carried out by one offender. If offences are recorded temporally as moments, then they would appear on the timeline (second row) as individual events with no duration, happening at particular points along the timeline. The actual duration (third row) of each offence differs depending on how long the offender spends in each property. Both of these can be measured relative to familiar structured time units, such as hours (fourth row). If we could

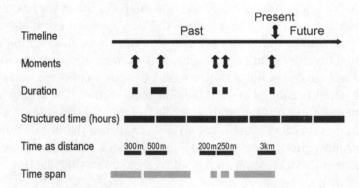

Figure 8.1 Timeline for five hypothetical burglaries. The five events are shown as moments along the timeline, though the duration shows that some burglaries took a longer time to commit than others. The structured time shows the fixed interval durations of hours during the day. Time as distance shows how far the offender travelled from home address to burglary site and how long it took to get there and back. The last offence was committed with the aid of a car, so the time to cover a much larger distance is relatively shorter. The final row indicates the time span of the offence start and end times as reported to police, shown in grey to indicate the range of possible times of the offence (in order to distinguish the actual offence duration, shown in black). The duration is the best indication of actual offence time, but is rarely available to crime researchers

225

interview the burglar, we may discover the patterns of behaviour (fifth row). These would show that the burglar travelled on foot to most offence locations, returning home after each one. Most offences were within a few hundred metres of the burglar's home, but travelling on foot has a significant time penalty which influences the range of possible targets as well as the number of burglaries that could be committed. The last offence was committed with the aid of an accomplice's car that enabled the burglar to travel to a location much further than before. The final row shows the time span as reported to police. In the first two examples and in the fifth offence, the homeowners were away from the home for at least an hour. They were also away during the offences with short time spans, but their property was unguarded for a shorter time. We describe these temporal categories in more detail in the following section.

8.2.1.1 Moments

A moment is usually expressed as a simple time expression, where a crime event is known to have happened at a particular time or is attributed with an estimated time. An event usually denotes a change in state of an object (Peuquet and Niu, 1995). In the case of a residential burglary, the object (the residential address) can be thought of as changing from a state of unburgled to burgled. When mapped as a moment, the event has just one temporal attribute – the time it happened. These types of event are easy to map. For example, a map of all calls for service to a police station in one day can show the location of each incident with the time of the call mapped next to the point. At some scales the point is not necessary, and the time stamp is enough to show the time and location. The ability to map the temporal pattern is determined by the temporal resolution available. If the crime recording system is only able to record the hour of the call, trying to query this system to map all incidents in a 10-minute period will be unsuccessful. Moments are usually depicted along a timeline (Figure 8.1) as points with no length.

8.2.1.2 Duration

Duration is an indication of the length of time that it takes for an event to occur. For example, an intelligence officer might use surveillance data in court to show that a burglar took six minutes to steal from a property. Duration can also become a search parameter indicating the temporal extent of a query. With a map of all calls for service in one day, the day in question becomes the duration for the query that generates the map data.

Events with a measurable duration are usually depicted along a timeline (Figure 8.1) as lines with a length determined by the duration.

8.2.1.3 Structured time

Structured time is a basic mechanism of human activity that enables people to effectively communicate temporal change. Although time is a continuum, it is helpful for us to be able to describe when and how long events take place in terms of seconds, minutes, hours, days, weeks, months and years. There are of course smaller periods of structured time than seconds, and longer periods than years; however, the range from minutes to years is the general area of concern for crime mappers. It is worth noting that conversations involving structured time often implicitly involve some rounding of time. For example, if an intelligence analyst says that a known burglar has been active for a couple of months, then it is often the case that the offender has actually been active for a slightly longer or shorter period of time. Structured time is depicted on timelines as lines with equal length and spacing (Figure 8.1).

8.2.1.4 Time as distance

Time can be used as a measure of distance. A light year – the distance light travels in one year – is not of much use to a crime analyst, but from a crime theory perspective it can be useful to consider time as a spatial factor in the movement of offenders and the resulting patterns of crime. If, for example, an offender is required to report to a police station in New York every evening, the offender is unlikely to be responsible for a pattern of rapes in San Francisco. Space is something that offenders often have to cross to reach their target, and it takes time to cross that space. The concept is also useful for some operational policing functions. The placement of police stations and vehicles can be optimised for emergency response by mapping the location of hotspots of calls for service and factoring in the time it would take to travel to the location.

8.2.1.5 Time span

Time span is a temporal category that is fairly unique to crime mapping. It has evolved from the realisation that the actual moment and duration of many crime events are not known. These temporal categories only come to light, if ever, when offenders are caught and interviewed. The reality of modern law enforcement is that many offenders go undiscovered and the only available information at the time of the crime report is when the victim

227

last saw their property in a good condition, and when they returned to find their property missing or damaged. Many police databases record these variables as the *from* date and time and the *to* date and time. They can also be called the *start* and *end* date and time. The period of time between the *from* and *to* times is called the time span, and it represents the duration within which the crime event happened. Within the time span, the event occurred as either a moment or a duration. Offences which have a recorded time span but do not have a distinct offence time often create analytical difficulties. The distinguishing characteristic of the time span compared to other temporal categories is the ability to reflect the degree of uncertainty of an event. Even though this reflects a lack of knowledge about a criminal incident, it does not mean that the crime analyst does not now have enough information to understand patterns of criminal activity. The section of this chapter on aoristic analysis has a method for analysing such offences.

8.3 Temporal resolution and querying a temporal database

When mapping crime incidents, the scale of temporal resolution becomes relevant. The more precise the data, the more precise the mapping potential. A crime database that only records the date of offences can be queried in terms of calendar time, but not in terms of clock time, when the time of the offence on the day in question is recorded. Recording the time of offences or calls for service dramatically increases the range of analysis options.

For example, when mapping all calls for service to one police station in one day, the query (or search) duration of the map is the 24-hour period. On such a map, each incident could be plotted as a point with a fairly precise temporal representation. Display options for this type of map include colour coding each point to represent each hour of the day, or replacing the point symbol with text to show the time of the call (for example, 1245 or 12:45 pm). Other possibilities include classifying the point symbol by police shift. The incidents become moments with a time span of a single minute, and the day becomes the search duration. However, if the temporal resolution of the map changes, so do the subsequent map characteristics. If we change the search parameter to a year, a map of all calls for service in one year that shows each point as a time of occurrence will more than likely swamp the display with numbers or symbols. With such a map, the temporal resolution of the moments becomes untenable, and a more

Figure 8.2 Chronological temporal classes and thematic temporal classes shown along the timeline

appropriate cartographic solution for displaying the calls for service must be found. More often than not, point mapping becomes inappropriate at this resolution and kernel density estimation surface maps (see Chapter 6) or other mapping techniques become necessary.

Before mapping any change in crime patterns over time, it is necessary to interrogate a crime database in order to be able to select the correct subset of the data. There is little point in mapping 10 years of crime when the task only requires a map of new offences since last week. A given search duration of crime events will result in the return of a number of incidents from a crime database. This is the most basic form of a temporal query and is one with which most crime researchers are familiar: find all burglaries that happened in 2003, or find all robberies that happened from midnight to 6 am. This raises two possible cartographic options for simple crime mapping. First, it is possible to map offences in a chronological fashion so that offences that happen close to the same time as other crimes will be in the same class. A second option is to map offences that happen around the same shift or time each day. This mapping approach is more thematic than chronological. These two options are shown as a timeline example in Figure 8.2.

When performing a temporal query on a crime database it is necessary to anticipate the type of question to be asked. Simple questions regarding 'what happened' and 'when' are easy to determine (though 'when' can be difficult for crime that is recorded as a time span and requires special treatment – we address this later in this chapter). More advanced spatio-temporal questions relate to questions of change. Peuquet (1994) identified three classes of spatio-temporal query:

1. One class of query addresses change in an object or feature. For instance, has an offender moved house in the last year? Where was the suspect living six months ago?

2. A second class of question addresses the nature of a spatial distribution over time. For example, where is the highest intensity hotspot for burglary this year compared to last year? Which police district has the highest number of drug arrests since March?

3. A third class examines temporal variation in multiple phenomena. Questions of this nature could include asking if nearby major sporting events are associated with an increase in local car crime, or are assaults associated spatially and temporally with bar-closing times.

To address these more complex queries it is necessary to consider the range of possible temporal relationships between crime events. This becomes especially important when evaluation of crime reduction strategies and police operations takes place because it is often necessary to demonstrate the spatio-temporal location of crimes before, during and after an intervention in order to determine the success (or not) of the strategy. It is also a factor with crime events recorded with *start* and *end* (or *from* and *to*) times, as will be explained in a later section.

Peuquet (1994) recognised seven temporal relationships, as shown in Figure 8.3. Some of these can be converted into crime-specific illustration queries, as the following examples show:

- Select all crimes that occurred prior to the police operation (X before Y).
- Select all crimes that happened on Tuesday and select all crimes on Wednesday (X meets Y).
- Select all robberies that happened as the convenience store closed (X ends Y).
- Select all burglaries with a start time before 6 pm and an end time after 6 pm (X overlaps Y).
- Select all murders that happened during 2003 (X during Y).

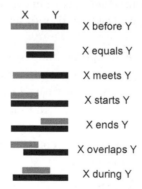

Figure 8.3 Temporal relationships. Source: Adapted from Peuquet (1994, p. 455). Reproduced by permission of Blackwell Publishing Ltd

Designing the appropriate temporal query is a prerequisite to spatio-temporal mapping. Once completed, there are different techniques that can be used to address different mapping problems.

8.3.1 Temporal questions

There are generally three important temporal questions that policy-makers, decision-makers and crime researchers ask. First, they are concerned with the degree of change of a temporal pattern: How has the distribution of offending changed since last month? Where are new crime patterns appearing? These questions address issues of pattern fluctuation and are answered by comparison of one distribution with another.

The second significant question is the problem of temporal magnitude: Which areas show an increase in crime? Where is the largest increase in crime? These questions can be a little trickier because they are not simply a matter of mapping the change, but also a matter of the magnitude of the crime in an area. For example, a police district which has one robbery a month will show a 100% increase if robberies go up to two a month. This 100% increase appears to be substantial; however, it pales next to a district that only has a 5% increase in robbery, but which moved from 100 robberies a month to 105.

The final type of question is becoming more common with the current move towards intelligence-led policing and problem-oriented policing: Has a crime reduction strategy made a difference? Again, this requires the crime mapper to compare past and present crime distributions, but also to seek a determination that any measured change is significant. The remainder of this chapter explores different spatio-temporal mapping methods that help to address all of these questions.

8.4 Comparing two distributions

Univariate maps chart the spatial extent of one variable, captured over a single period in time. Many of the maps that are discussed in this book are univariate maps – in other words they deal with one variable: crime – and show patterns of crime over a single period (e.g. between January and December in any one year). They can also be useful in addressing some of the questions raised in the previous section by simply showing and comparing two maps side by side.

Figure 8.4 shows two maps of Philadelphia police districts. The first indicates the number of shootings in each district from July to December

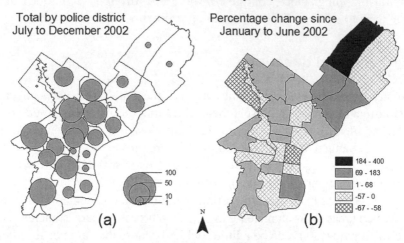

Figure 8.4 A proportional symbol map is used to indicate the (unstandardised) raw count of shootings in each police district for the last six months of 2002. The choropleth map enhances this display by showing the percentage change in the figures from the six months at the start of 2002. Notice that although the map on the right shows a police district with a huge increase in shootings, the proportional symbol map indicates that this area still has a low number of shootings. Both maps are necessary to fully understanding the data

2002. The second shows the change in the number of shootings in each district compared to the preceding six months (January to June 2003). If the map on the right had been viewed on its own, there would be some concern regarding the most northerly police district where the highest percentage increase in shootings was registered. However, this is an increase that is relative to the number of shootings in the earlier part of the year. Both maps together show that although this police district did register a substantial percentage increase, the volume of actual shootings in this area is relatively low. These two maps are able to indicate both where the current 'hot' districts are for shootings (Figure 8.4a) and the degree of changes within the districts from the previous temporal period (Figure 8.4b).

The maps in Figure 8.4 are aggregated to police districts. These would suit the district captains and the senior executives but, for a senior detective responsible for a number of districts, this may not convey sufficient spatial detail. A detective may be more interested in the spatial patterns in certain streets and neighbourhoods. The actual point distributions of incidents could be portrayed in separate maps for the two six-month periods, with the maps shown side by side (Figure 8.5). However, most users

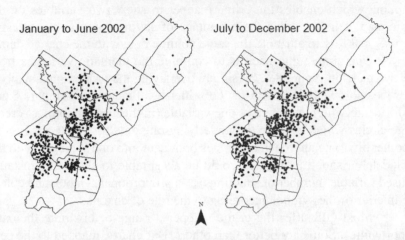

Shootings in Philadelphia, 2002

Figure 8.5 Two point–pattern maps of Philadelphia shootings, January to June 2002 and July to December 2002. Although the dots are an increase in spatial accuracy over Figure 8.4, the point clouds are arguably more difficult to interpret

would find the two comparative dot clouds difficult to interpret and compare, and may revert to gleaning information from Figure 8.4, even though the spatial detail of the dot maps are more precise.

A further option is to use a method such as kernel density estimation (as discussed in Chapter 6) to show continuous surface distributions of shootings for both time periods side by side. The use of this method would help to visually identify any changing spatial distributions. However, three particular problems do emerge: first, the temporal magnitude component could be lost in small areas on the edge of hotspots; secondly, it may be difficult to visually signal and attribute a rate of increase or decrease between some parts of the map; and thirdly, the parameters and thematic classes between the two maps would need to be controlled to the same values in order for relative comparisons to be made. These problems aside, kernel density estimation comparison maps would most likely be an improvement over Figure 8.5.

8.4.1 Bivariate choropleth maps

Side-by-side maps are easy to create and can be powerful tools. The use of two univariate maps to show volume and change, or past and present, increases cartographic clarity by simply and clearly portraying one component of the temporal question. An alternative approach is bivariate mapping which enables the crime mapper to show two variables on the same map. This can have the advantage of saving space and increasing the user's ability to attribute the same volume data with the change data.

One simple approach is to combine the thematic variables from the two maps in Figure 8.4, by simply overlaying the proportional symbols over the choropleth surface. Most GIS offer this functionality (Figure 8.6a). Unfortunately high numbers of one variable (in this case shootings) create large circles which obscure the underlying rate data for small areal units. Another bivariate approach combines both data into one symbol. With the Philadelphia shooting data, it would be acceptable to show the unstandardised variable (number of shootings) as a proportional symbol, and colour the interior of the symbol according to the rate of change.

Most GIS offer the crime mapper a range of bivariate thematic opportunities. Some are better than others. Bar charts, mapped to the centroid of a polygon, can be used to show past and present crime levels in an area. They can even be employed to show the change over a number of time periods (as long as the periods have the same temporal resolution). However, these types of maps are often not easy to read and, as a result, lack impact. Although functional, they often convey too much information and the

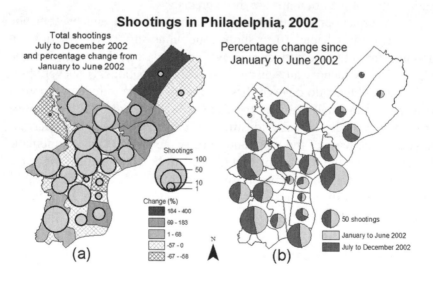

Figure 8.6 Volume and change in number of shootings shown on single maps. Map (a) shows proportional symbols over a choropleth surface, while map (b) shows proportionally sized pie charts

overall significance of the map is lost in the detail. Pie charts can show the magnitude of total offending over two (or more) time periods. They work best by aiming to show the total offending in an area over a longer period, such as a year, with smaller periods of structured time indicating the volume of offending in pie pieces (Figure 8.6b).

There are other visual variables that can be manipulated for bivariate maps. Pilots and meteorologists are familiar with point symbols that use different orientation and symbology to depict wind speed and direction on surface analysis charts. These can be adapted for crime mapping purposes, though caution should be employed. Although these types of maps are quite clever, they are difficult to interpret for a law enforcement and crime reduction audience not used to such symbology. An example of such a map would be to depict the standardised variable (shooting count) of an area as a proportional symbol at the centroid of the area, and adjust the orientation according to the increase or decrease over the previous year. Although this cartographic functionality is not available in most GIS software, it can be achieved with a little thought.

8.5 Mapping temporal change with graphs

Offending is inextricably linked to time in a number of ways which can be analysed. Analysts can explore the time of day of offending, the day of the week, patterns in the season or school holidays. All of these are different structured time periods along the timeline. One of the simplest charting tools is a temporal histogram. A histogram is a chart that shows the frequency of a variable. It is a tool that is available in statistical software such as SPSS, as well as Microsoft Excel[1] (it requires the installation of the free Analysis ToolPak, an add-in that is activated from the 'Tools' menu).

Histograms can be applied to a number of non-temporal variables. For instance, an analyst can use the histogram function to count all break-and-enter offences in a crime database, broken down into modus operandi (assuming this data is recorded with a numerical code), or by the value of the goods stolen. Histograms can be applied to any numerical or date format data in Excel.

Figure 8.7 shows a screen capture from an Excel spreadsheet. In this example, the analyst has searched a crime database and found 40 shooting

---

[1] This discussion and the following example employ MicroSoft Excel 2002. The menu structure and availability of analysis tools may vary in previous and subsequent versions.

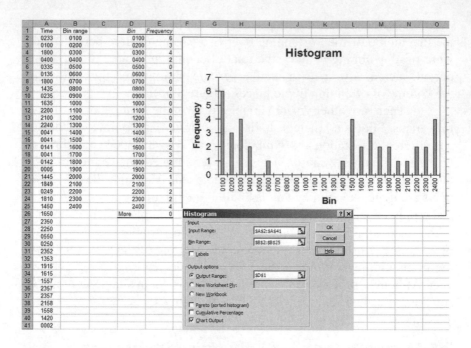

Figure 8.7 Screen capture from an Excel spreadsheet, showing the hour of 40 shooting incidents in the range A2 to A41. These are captured from the crime database. The bin range (cells B2 to B25) is entered by the analyst. Once the histogram function of Excel is activated, cells A2:A41 become the input range and B2:B25 become the bin range. In this example, an output range of D1 is chosen over the default option of a new spreadsheet, and the chart output has also been selected, as shown by the dialog box at the bottom of the image. The results are shown in the chart and in the range of cells D1 to E26

incidents that occurred in one police district. These have been entered into the cell range A2 to A41. They are in numerical format (so the analyst can use the histogram tool) but have had leading zeros added so that the 24-hour format of the time is obvious. Histograms require a range of values into which the software will sort the crime data. In this case, the analyst wants to sort the shooting times into the hour in which they occurred. The bin range has been entered into cells B2 to B25. The histogram function is activated by selecting 'histogram', once having clicked on 'Tools', and then 'Data analysis...'. Once the histogram function of Excel is activated, cells A2:A41 become the input range and B2:B25 become the bin range. In this example, an output range of D1 is chosen over the default option that would show the result in a new spreadsheet (though this is also a perfectly acceptable option). The chart output has also been selected. All of these choices are shown in the histogram dialog box at the bottom of Figure 8.7.

The results of the histogram function (selected with the options shown in the dialog box) are shown in the chart and in the range of cells D1 to E26. The zero value next to a label of 'More' confirms that there were no records that had a value under 0000 or over 2400. Temporally, this would be impossible, but this does act as a handy safeguard against data entry error. Histograms work with interval data by rounding values upwards. In other words, six records have a value of up to 0100 (in effect, from 0000 to 0100), three records have a value from 0101 to 0200, four records have a value from 0201 to 0300 and so on. From this we can quickly see that there were no shootings recorded from about dawn through to midday, with a peak in the hours around midnight.

The example in Figure 8.7 uses hour of the day as the variable. It is equally possible to map offences onto a histogram of weeks or months. One point to watch for when creating histograms of monthly crime data is the impact that the uneven calendar can have on results. Months, while familiar to all audiences, are not strictly structured time, as some months are longer than others. This can hinder accurate interpretation of monthly histograms. When mapping offences to months, it is better to standardise the values by charting a rate of offence per day.

Histograms of longer time periods allow an analyst to examine the change in crime volume over long time periods. It is easily possible to chart the distribution of offences over years, by changing the bin range in Excel or SPSS. Figure 8.8 shows the weekly burglary rate for the Australian

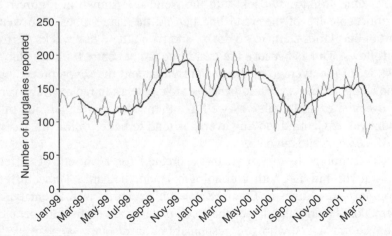

Figure 8.8 Temporal histogram of weekly reported burglaries in the Australian Capital Territory, from January 1999 to mid-March 2001. The bold black line represents an eight-week moving average trend. Source: Ratcliffe (2001, p. 5)

Capital Territory from January 1999 to mid-March 2001. As the data are shown for weekly periods, a structured time of exactly seven days, no correction for varying time periods is needed, as would be necessary if monthly data were shown. The values are therefore directly comparable.

A problem that can occur with a small temporal analysis period (short structured time) or when the number of records in the data is small is the spikiness of the histogram. This occurs because there are insufficient values in each class to properly overcome the normal fluctuation in the crime rate or because a short temporal search duration is employed. Although the general trend in the burglary rate can be seen from the grey line in Figure 8.8, there is still some variability evident from week to week. One solution is to replace the raw figures with a trend line of some description, as is possible with most charting computer packages.

The broad black line in Figure 8.8 shows an eight-week moving average. With this type of trend line, each value represents the average of the eight weeks of burglary counts up to that point. This is why the trend line does not start until the eighth week – there were insufficient values to start the line before that week. The trend line has the advantage of smoothing out some of the minor fluctuations that are caused by the natural variation in the crime rate. Trend lines permit the analyst to see the overall trend in the behaviour of the crime distribution over time, and are often a better choice when disseminating charts to other users. When charting in Excel, right mouse click over a line of values, and choose the 'Add Trendline...' menu option.

One slight drawback with the trend line shown in Figure 8.8 is the noticeable lag of the trend line behind the data values. The weekly count of burglaries starts to go down, and then after a few weeks the trend line follows. This is because the trend line is not centred. Each value is a construct of the average of the present value and the seven preceding it. Some computer packages allow the user to create a centred moving average that uses the existing value as well as other values on either side of the present value. Centred moving averages tend to better follow the rise and fall of the key trends in data.

Temporal histograms can be created for a variety of different temporal resolutions, both chronological and thematic. Three different histograms are shown in Figure 8.9, each depicting a different type of information. Hourly and day-of-the-week patterns are useful for operational police commanders who are responsible for resource allocation. The chronological range of events is useful for a senior commander who might want to know the overall trend over time and if a significant problem has been developing or dissipating.

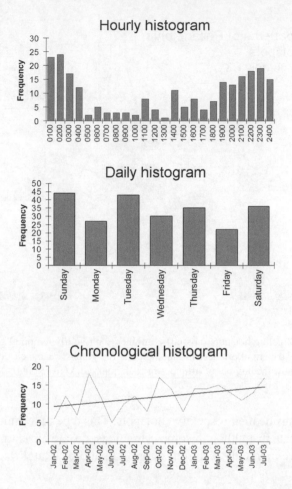

Figure 8.9 Three histograms that show a wealth of information regarding shootings in one police district. The hourly pattern shows that shootings build throughout the evening coming to a peak in the hours around midnight. The daily pattern indicates that Sunday and Tuesday are the peak days for shootings, and the bold (linear) trend line in the chronological histogram shows that the frequency of shootings appears to be gradually increasing over time

Temporal histograms lack spatial information, but can be applied to subsets of spatial data to create spatio-temporal comparisons. One way is to aggregate offences by police district and then to generate a temporal histogram for each police district. These can be charted together to show the changing pattern of crime across a city. Another way is to combine hotspot areas or polygon areal units with temporal information. An example of this is Figure 8.10 where the temporal patterns of shootings have been

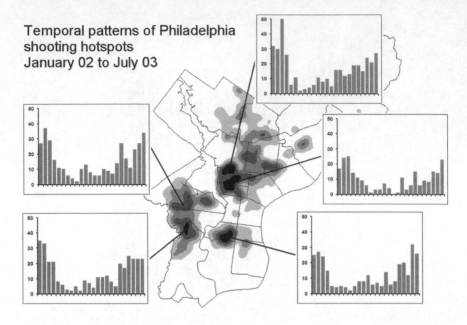

Temporal patterns of Philadelphia shooting hotspots January 02 to July 03

Figure 8.10 Shooting hotspots (spatial) combined with hourly temporal graphs to show spatio-temporal pattern of offences in Philadelphia. Bar graphs are associated with different hotspots and show the number of offences in each hour block, from 0000–0059 through to the period 2300–2359

mapped alongside their respective hotspots. This type of technique has been used to indicate that although some crime hotspots are close to one another, they can have markedly different temporal patterns (Ratcliffe, 2002).

8.6 Using animation

So far we have only considered static images; however, modern computing speeds allow more intricate mapping techniques for the visualisation and description of temporal crime patterns. Two such possibilities are temporal GIS and animation. Temporal GIS are computer programs that are capable of performing temporal queries and mapping the results 'on-the-fly'. At present, these tend to be fairly specialised and available in only custom-built programs to which few crime analysts have access. Over time, this situation is bound to improve and programs will come down in price. Rudimentary temporal functions are, though, beginning to appear in more mainstream GIS, and any GIS that is able to export maps into an image file can be used to create animations that are both impressive to view and informative to audiences. Animation provides for an opportunity to examine spatio-temporal patterns in new and innovative ways (Dorling and Openshaw, 1992) and are particularly good as descriptive mechanisms and in presentations as briefing tools.

Frame-based animation involves the creation of a series of single 'snapshot' (Langran, 1989) images, each taken for a particular period of time, which are then displayed in rapid succession (Peterson, 1995). For example, Figure 8.11 shows shootings in Philadelphia displayed as 12 snapshot image frames, each depicting a different month. The spatio-temporal pattern of shooting hotspots can be seen to change over time (issues with the different number of days in a month aside). It is, though, important to recognise that a snapshot does not provide a detailed picture of the changing behaviour of crime patterns between snapshots; the images tell an analyst what has changed from snapshot to snapshot but does not convey how or why the patterns changed. For example, an animation made up of frames of crime each one year apart will show the general change in patterns from year to year, but not the way that change varied over the year. However, this type of snapshot framing is akin to small multiples (Tufte, 1990), repeating the same basic design structure from image to

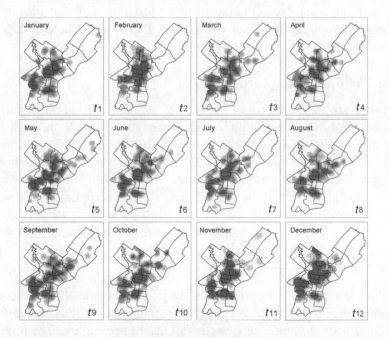

Figure 8.11 Monthly snapshot images of Philadelphia shooting hotspots for 2002

image. This approach has the interpretative advantage that once a user has decoded the initial image, they can easily apply the same understanding to the subsequent images, increasing readability.

To create frames for an animation and to knit them in sequence requires a number of steps and certain map design principles to be considered.

8.6.1 Saving a map image

The first stage when constructing an animation is to collect the map images. This process can be automated (by those with some programing experience) or conducted manually. Each image is displayed on screen and then saved as an image file (e.g. JPG or TIF format are the most acceptable formats). TIF images have little compression which means that they are high quality images but at the expense of file size, while JPG pictures have much smaller file sizes but sacrifice a degree of image quality. Animations can be quite computer processor intensive; So unless higher resolution pictures are required, JPG format will usually be fine.

8.6.2 Making the animation

The next stage involves creating a file or program that will display the images at the correct frame rate or duration. There are a couple of different ways to do this.

8.6.2.1 Windows Explorer

Recent versions of Windows Explorer have the capacity to display image files as a slideshow. Open Windows Explorer and navigate to the folder that holds the image files. On the 'View' menu it is possible to select 'Filmstrip' when image files are present. If the left-hand task pane is visible, there is one option called 'View as a slide show'. This function is free and easy to run, as long as the image files are named in a correct alphanumeric order. Unfortunately, there is little control over the slideshow function, and this approach is therefore not recommended for more than basic exploratory work.

8.6.2.2 Microsoft PowerPoint

Although not specifically designed for the purpose of animations, Microsoft PowerPoint can be used to generate rudimentary animations, as long as a limited number of frames are viewed. This can be accomplished by inserting map image files in sequential order into separate PowerPoint slides.

Once the maps are in the correct order (24 slides for an hourly image), the next stage is to adjust the slide transition. Select all of the slides and then click 'Slide Show' and then 'Slide Transition...'. Select a subtle transition method such as 'Appear' or 'Dissolve' and then click the 'Automatically after' box, choosing a time interval of one or two seconds. This will ensure that it is not necessary to continually click the mouse to advance the frames when viewing the presentation.

The trick with images in PowerPoint is to ensure that each image is sited at the same place on each slide. This can be achieved by selecting the picture and then clicking 'Format' and then 'Picture...' on the menu bar. The Format Picture dialog has tab options for size and position, and if these settings are the same for all images (assuming that all of the pictures are the same original size) then the animation should flow without jumping around the screen.

8.6.2.3 Animation software

The Internet is awash with freeware and shareware programs that will compile a collection of images into an animation file. These are best explored on a trial-and-error basis to find the program that works best for the image size and number of frames. It is certainly possible to generate passable animations at little or no cost. If you have image processing software, it is worth exploring the options that are often bundled with the packages. For example, Paint Shop Pro now comes with a program called Animation Shop which can create respectable animations in a reasonable time.

If money is no object, then specialised software is available. These programs would be considered fast and good, but not cheap. Adobe Premiere is able to smoothly link animations to live action video and produce the result in a host of different audio-visual formats. It can also offer the option of selecting from a range of different transitions between frames. Animations are time-consuming and it is therefore a decision of the user to determine if it is worth incurring substantial cost for the creation of a couple of animations. All of the software mentioned can be found through the usual Internet search engines such as Google (www.google.com).

8.6.3 Animation cartographic design issues

There are certain cartographic design considerations that are necessary to consider when creating individual frames for an animation. First, it is important to indicate to the viewer what time period is being depicted. A clock, or other time representation, is therefore needed. Visual representations of analogue clocks (with two hands) can be tricky to generate on a map and have the added disadvantage that the hand position for antemeridian (am) times is the same for postmeridian (pm) times. It is often better to simply add a line of text onto the map that shows the time. This works if the animation frames relate to hours, days, weeks, months or years. If the actual time period is not that important and you wish to show a relative position from start to end of the animation, a moving bar progressing from left to right can be created. This has the advantage that it is easy to quickly glance at, giving the viewer an immediate impression of the relative time in the animation without demanding so much attention that it distracts from the main information. An acceptable solution is a moving bar, to show relative position within the animation, supported by a small line of text to show the actual time period being portrayed.

The image size is another facet to consider. Large image sizes allow enough detail to be seen, but at a price of possibly huge final file sizes. It is best to consider the final viewing medium. If the aim is to show the animation on a laptop connected to a digital projector, there is little point in creating image sizes that are larger than the screen resolution of the laptop or the output resolution of the projector. With modern screen resolutions giving over a thousand pixels for the horizontal display, this would create an animation file of substantial size.

A final consideration is the overall purpose of the animation. If the aim is to convey many different components of a temporal pattern, there may be too much information for a human to process in the time it takes for an animation to run. The best animations are those that have one or two points to make, but then make them clearly.

8.6.3.1 Animation thematic class ranges

When viewing choropleth maps, it is important to retain the same class boundaries for each image. The viewer will perceive that increased shading in an areal map indicates an increase in the frequency of a variable, and will retain an impression of the previous images in their mind. It is therefore important that the thematic class ranges remain constant throughout the images. This usually requires the animator to find the map with the largest values, and use that set of values as the basis for the class ranges for the whole series.

A final design consideration is continuity. Maps created for an animation should retain the same colour and thickness for roads, the same position, font size and colour for titles and so on. The only things that should change from frame to frame are the time details (moving bar, etc.) and the crime distribution. Any other animated feature acts as a distraction that draws attention away from the crime patterns – the whole point of the animation and the highest level of the visual hierarchy (see Chapter 12). For an example of a crime map animation visit http://www.crimereduction. co.uk/toolkits/fa020405.htm.

8.7 Quantifying change over time

Thus far, this chapter has considered basic quantitative measures of change in individual areas, such as mapping percentage change over time in police districts (Figure 8.4) or looked at ways in which animation could be used to portray an impression of the movement of crime patterns. The following section considers two ways to quantify the degree of change from one area to another. The first method is applicable to point patterns and the second to areal data. Both methods, developed by researchers in the field of crime analysis, are useful in describing real-world problems of crime distribution and pattern change.

8.7.1 Centrographic techniques

Centrography employs three basic spatial analysis techniques to describe point patterns. These three methods are the mean centre, the standard distance and the standard deviational ellipse. These three methods can be calculated using CrimeStat, and are described in detail in Chapter 5. The mean centre is a measure of the average location of a crime pattern. Caution should be employed while using the mean centre and it is best to have both a good look at the distribution of the points and a working knowledge of the area and the data collection processes before employing the mean centre as a description of the data. For example, a bimodal distribution can be obtained if an analyst is asked to plot commercial burglaries in an area where there are two distinct commercial districts. The most likely location of the mean centre will be in the void between the two. In this case, it may be more appropriate to examine each cluster individually (LeBeau, 1987).

With this caveat in mind, it is possible to track the movement of mean centres plotted for different periods of structured time. Figure 8.12a shows a hypothetical study area with the mean centre of a crime distribution moving around the centre of the study area from year to year. In this hypothetical example, the sudden movement in 2002 and 2003 indicating a movement to the bottom right of the area might be worthy of further

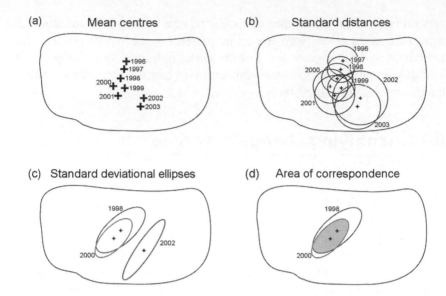

Figure 8.12 Study area indicating the mean centres of offending for eight years (a), the standard distances for each year (b), the standard deviational ellipses for three years (c) and the area of correspondence for two years (d)

examination. LeBeau (1987), drawing on the work of Soot (1975), describes a number of measures that can be used to quantify the change. These include the velocity, acceleration and momentum of the mean centres. Velocity is the measure of the rate of change of the mean centre over a given time period. The distance and direction of change can be calculated either as a measure relative to a fixed location or in relation to the previous measurement. Acceleration is the rate of increase in the velocity of a mean centre from one time period to another. In Figure 8.12a there is clearly a change in direction and an increase in velocity from 2001 to 2002.

Momentum is the product of the number of crime events and the measured velocity. It has some descriptive value in that the inclusion of the number of crime events has the effect of standardising the measure from one time period to another, even if the number of events changes.

The standard distance can be used to determine a rate of change from one time period to another. It produces a circle, centred on the mean centre, that indicates the range of one standard deviation from the mean centre. The standard distance is described more thoroughly in Chapter 5. The size of the circle gives an indication of the 'spread' of the crime distribution. As with the mean centres, it is possible to plot changes in the radius and centre of the standard distances (Figure 8.12b). The movement of the circle across space indicates a change in the general dispersion of points and the mean centre of those locations.

Standard deviational ellipses are an extension of the standard distance. They are able to incorporate an indication of the spatial trend of the points by orientating themselves to the major axis of the point distribution. Standard deviational ellipses are therefore an enhancement of the standard distance because they also provide an indication of the directional component of the main distribution. Figure 8.12c shows the standard deviational ellipses for three years in the example study area. Plotting ellipses over time can give an indication of both the morphological change and the movement of crime hotspots.

8.7.2 The coefficient of areal correspondence

Centrographic techniques can be used as descriptive visualisation tools so that the change in a crime distribution can be plotted over time. One enhancement that provides a quantitative measure of the degree of overlap of an areal centrographic unit from one time to another is the coefficient of areal correspondence, as described by Dent (1999, p. 99). It can be applied to both standard distances and standard deviational ellipses. The equation for the coefficient (C_a) is

$$C_a = \frac{\text{Area covered jointly by both phenomena}}{\text{total area covered by two phenomena}}$$

The range of possible values for the coefficient is from 0.0 (no correspondence) to 1.0 (complete congruence). Figure 8.12d shows a shaded area that represents the area of correspondence between the standard deviational ellipses of the 1998 and the 2000 data. The advantage of the coefficient of correspondence is that the analyst is presented with a figure that can be compared between datasets and between time periods.

8.7.3 Weighted displacement quotients

Centrography and the coefficient of correspondence are appropriate measures for crime patterns that are expressed as individual point clouds. When data are already aggregated to areal units, it is possible to examine any rate of change using weighted displacement quotients (WDQ). This method was formulated by Bowers and Johnson (2003) so that they could examine the impact of crime reduction strategies in England and Wales. The task called for a tool that could both identify and quantify any displacement of offending from the target area to surrounding neighbourhoods, while also indicating if any actual crime reduction had taken place in the strategy area.

The first stage is to determine three areas: a target area that has been the subject of some crime reduction activity (A in Figure 8.13); a displacement area that is used to check that crime has not simply been moved to a similar or neighbouring area (B); and a broader control area that acts as a control on general crime patterns in the region (C). These can be determined in two fashions, either as a concentric nested model (Figure 8.13(1)) or as discrete areas that are established based on a theoretical estimation of the likely locations of possible displacement (Figure 8.13(2)).

Calculating the WDQ first requires calculating a success measure for the target area in comparison with the total study area. If the success measure is negative, then the crime prevention or reduction activity was a success and reduced crime in relation to the control area. If, however, the success measure is positive, then this shows that crime has increased in the treatment area. Further calculation is usually unnecessary. The success measure (SM) can be calculated like this:

$$SM = \frac{A_{t_1}}{C_{t_1}} - \frac{A_{t_0}}{C_{t_0}}$$

where each variable is measured before the initiative (t_0) and afterwards (t_1). For example, A_{t_1} represents the count of the crimes in area A after or during the crime reduction initiative.

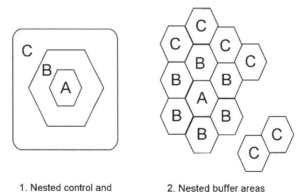

1. Nested control and
buffer areas

2. Nested buffer areas
and remote control areas

Figure 8.13 Hypothetical study areas showing the region that is the target of a crime prevention activity (A), buffer zones (B) and control areas (C). If theoretically sound, it is possible to take a nested approach (1) or to use remote control areas that are not contiguous with the buffer and target areas (2)

Assuming that the success measure is a negative value, indicating that crime has reduced in the target area, the next stage is to calculate the buffer displacement measure (BDM). This checks to see if there has been any displacement or diffusion of benefit to the buffer zone.

$$\mathrm{BDM} = \frac{\mathrm{B}_{t_1}}{\mathrm{C}_{t_1}} - \frac{\mathrm{B}_{t_0}}{\mathrm{C}_{t_0}}$$

If the BDM is positive, then this suggests that some crime displacement has occurred, increasing crime in the buffer zone. If the BDM is negative, this raises the possibility that there has been some diffusion of crime reduction benefit from the crime prevention area (A) into the buffer. This is because there has been a reduction in crime in the buffer zone relative to the overall area, even though there was no specific crime reduction programme operating in the area.

The final stage is to calculate the WDQ which is simply a ratio of the two earlier calculations.

$$\mathrm{WDQ} = \frac{\mathrm{BDM}}{\mathrm{SM}}$$

The WDQ still lacks a simple probability test to determine if any displacement that is observed is significant from a statistical perspective, and not the result of random variation in the crime pattern, though Bowers and Johnson (2003) provide a guide to significance. This said, the WDQ is a useful addition to the arsenal of the crime analyst, as it is easy to calculate and provides an indication of not just displacement, but also diffusion of benefit.

8.7.3.1 WDQ example

To demonstrate with an example, consider a single police district. A crime analyst is asked to evaluate a robbery reduction strategy that local officers conduct in one neighbourhood. The target neighbourhood (A) is surrounded by other neighbourhoods that the analyst thinks may be possible displacement sites. These are therefore designated as buffer zones (B). The whole police district becomes the broader control area (C). The police operation lasts for six months, therefore at the conclusion the analyst collects frequencies for the six months before the police operation (t_0) as well as for the six months of the strategy (t_1). These values are shown in Table 8.1.

Table 8.1 Example values from a hypothetical police operation. Reproduced by permission of Kluwer Academic

Robberies	Area A (target)	Area B (buffer)	Area C (control)
t_0	8	12	230
t_1	5	18	260

$$SM = (5/260) - (8/230) = 0.0192 - 0.0348 = -0.0156$$
$$BDM = (18/260) - (12/230) = 0.0692 - 0.0521 = 0.0171$$

In this case, it can be seen from SM that the robbery reduction programme was a success in the target area, the negative value indicating a relative reduction in robberies compared to the control area. However, there also appears to have been a degree of displacement (indicated by the positive BDM value) that may have been caused by offenders driven by police activity to offend in neighbouring areas. The WDQ value is close to -1.

$$WDQ = \frac{BDM}{SM} = \frac{0.0171}{-0.0156} = -1.0968$$

Bowers and Johnson (Bowers and Johnson, 2003) provide a guide to interpreting the WDQ value, reproduced below (Table 8.2). In this example it appears as though the programme resulted in a displacement effect that was about equal to the direct effects of the target crime reduction programme to area A, indicating there was no net benefit from the programme.

Table 8.2 Interpretation guide for WDQ values

WDQ	Diffusion/displacement	Overall programme effects
$WDQ > 1$	Diffusion greater than direct effects	Positive net effect of the programme
WDQ near 1	Diffusion about equal to direct effects	
$1 > WDQ > 0$	Diffusion, but less than direct effects	
$WDQ = 0$	No displacement or diffusion	
$0 > WDQ > -1$	Displacement, but less than direct effects	
WDQ near -1	Displacement about equal to direct effects	No net benefit to programme
$WDQ < -1$	Displacement greater than direct effects	Programme worse than doing nothing

8.8 Aoristic analysis

Towards the start of this chapter, time span was identified as one of the five temporal characteristics. The time span represents the range of possible times over which an offence could have occurred. It is often necessary to use a time span technique with property offences such as burglary and auto theft when the victim left their house or car for a few hours and returned later to find the crime had happened while they were away. The range of hours (or, in some cases, days) during which the offence happened at some point is referred to as the time span. Aoristic analysis (Ratcliffe and McCullagh, 2000; Ratcliffe, 2000, 2002) is a spatio-temporal tool that calculates the probability that a crime event occurred within a given time period within the time span. It has been designed to assist with the analysis of offences where it is not possible to determine the actual time of offence.

The difficulty that can occur with offences of indeterminate time is best illustrated with an example. Consider three burglaries that occur at residential premises over the course of a weekend. In the first burglary, shown as (a) in Figure 8.14, the owners left their home at midday on Friday only to return on Sunday morning to discover the burglary. In the second burglary, the victims left the home on Saturday morning and returned on Sunday morning (b). In the third offence, the victims were away from the home for a few hours during Saturday (c).

Difficulties begin to arise when a seemingly simple question is asked: How many burglaries happened on Saturday? The problem with offences (a) and (b) is that both of them might have occurred on Saturday, but there is also the possibility that they may not have. Offence (a) may have occurred late on Friday or in the early hours of Sunday morning, while offence (b) could also have happened on Sunday morning. The only offence that definitely happened on Saturday was (c).

So what can be said about the number of offences that occurred on Saturday? It can be said that one definitely occurred on Saturday and

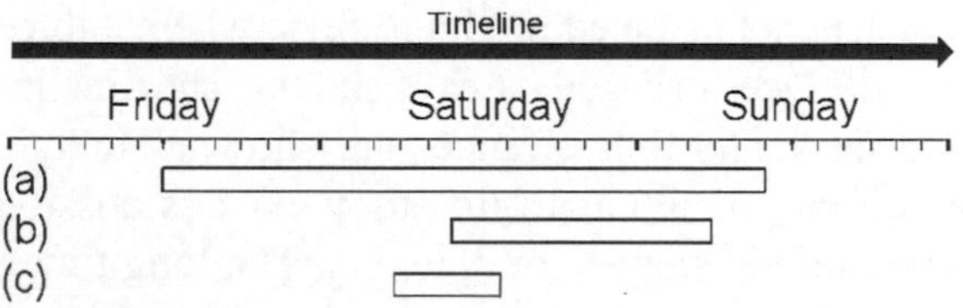

Figure 8.14 Three burglaries (a, b and c) shown along the timeline. Tick marks are indicated at every two hours on the scale

two others may have. Another alternative is to estimate a proportional value of each crime based on the portion of the offence time span that falls within Saturday. Offence (a) has a time span of 46 hours, of which 24 fall within Saturday. If a crime can only occur once and can therefore be given a value of 1, we could allocate 0.52 (24/46=0.52) of the offence probability to Saturday. This can be expressed as either 0.52 or 52%. Offence (b) has a time span of 20 hours, of which 14 fall within the temporal range of Saturday, resulting in a value of 0.7 (70%). Add these values to 1.0, to represent the one crime that definitely occurred on Saturday (c), and we can say that the aoristic value of Saturday burglaries is 2.22.

8.8.1 The spatial dimension

The aoristic value for each offence can be visualised and further analysed spatially. Most of the techniques in this book consider that the value of each offence that is mapped is one. That is, an offence happens once and has a value of 1.0 for the purposes of any calculations. This value can be adjusted by the aoristic weight to permit an analyst to map offences by the probability that the crimes happened within the temporal parameter of the map. In other words, a simple map with a title of Saturday burglaries could show the three crimes as point symbols, coloured or shaded according to their aoristic value. For example, this can be achieved by shading offences with a value of 1.0 as the darkest shading, reflecting the certainty that these offences happened in the time period being examined. Other offences would be shaded according to the probability that they occurred on that day. This approach would increase the accuracy of the map by reflecting the possibility that some offences may have occurred outside the map's temporal limits of 'Saturday'. For completeness, the symbols that had aoristic values of less than 1.0 should appear on other maps, coloured or shaded accordingly. For example, the offence labelled (a) in Figure 8.14 would appear on the Friday, Saturday and Sunday maps, if these were produced.

This concept is illustrated in Figure 8.15, where a three-dimensional representation of offences in a study area is shown. The timeline is expressed as the vertical access, and four offences are shown. Three have start and end times that are far enough apart that they exist as possible offences in both of the two temporal snapshots, t_1 and t_2. The long time span of these offences means that they have relatively low aoristic values. There is one offence that only appears in the t_1 snapshot. This crime, with a relatively short period between the start and end time, has a high aoristic value, signified

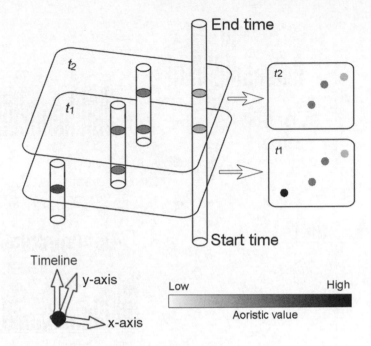

Figure 8.15 Aoristic spatio-temporal values expressed in three dimensions

by the darker colour at the intersection of the spatial location and the temporal dimension.

Case study: Aoristic analysis of vehicle crime in Sydney's Eastern Suburbs

While Australia has noticeably lower rates of violent crime and homicide than the United States, it does share with Europe a property crime problem. Some areas of the country are more affected than others. For example, the neighbourhoods in the Eastern Suburbs of Sydney are particularly vulnerable to vehicle crime. Among some of the oldest parts of the city, properties in this area are highly sought after, being close to the city centre and the Central Business District as well as close to Bondi and Bronte beaches. The competition for space in this area means that many people, though wealthy, have to park their vehicles on the street overnight.

An aoristic analysis of vehicle crime in the area confirms the vulnerability of cars overnight. The beach areas (areas 1 and 4 in Figure 8.16) have the highest probabilities of vehicle crime during the evening and

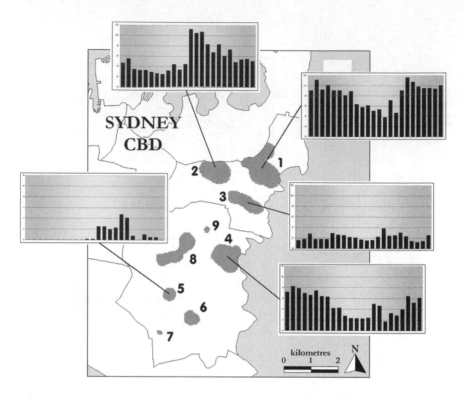

Figure 8.16 Temporal patterns of vehicle crime for each hour of the day for five of nine hotspot areas across the Eastern Suburbs of Sydney, Australia. Source: Ratcliffe (2002). Reproduced by permission of Springer Science and Business Media

early hours of the morning (Ratcliffe, 2002). By comparison, area 2 is a shopping district, and vehicle crime is at its highest probability during the middle of the day when people visit over lunchtime. Area 5 is a local shopping centre, and the aoristic pattern suggests that the most vulnerable times are in the afternoon.

Aoristic analysis shows that areas that are spatially close can have significantly different temporal patterns of offending, driven by the different opportunity structures of the targets of crime.

8.9 Summary

This chapter has outlined a number of techniques that enable the spatio-temporal mapping of crime patterns. The exciting aspect to this area of crime analysis is that new developments are occurring all of the time. The methodological and theoretical aspects of temporal crime analysis are still being developed and there continues to be opportunities for analysts and crime researchers to explore new and innovative ways to map changing crime distributions over time. Aoristic analysis is not a strict statistical technique but a flexible exploratory method that is a simple way for analysts to better understand the spatial and temporal patterns of their crime data. Aoristic values can be added to a hotspot analysis for a truly spatio-temporal analysis. However, as this chapter has shown, there are a variety of different ways to explore spatio-temporal crime patterns.

Further reading

Spatio-temporal analysis of crime is still in its infancy; however, a number of texts can provide the reader with a flavour of the temporal analysis studies being undertaken.

Krimmel, J.T. and Mele, M. (1998). Investigating stolen vehicle dump sites: An interrupted time series quasi experiment. *Policing: An International Journal of Police Strategies & Management*, 21(3), 479–489.

Useful for a crime-related discussion of ARIMA-interrupted time series analysis.

Ratcliffe, J.H. (2002). Aoristic signatures and the temporal analysis of high volume crime patterns. *Journal of Quantitative Criminology*, 18(1), 23–43.

Although there are a number of papers on aoristic analysis, this one provides the clearest description of the process as well as provides a practical application with an examination of vehicle crime in Sydney, Australia.

Polvi, N. and T. Looman, *et al.* (1991). The time course of repeat burglary victimization. *British Journal of Criminology*, 31(4), 411–414.

A paper that describes in detail the importance of time as a factor in understanding and combating repeat victimisation.

References

Barr, R. and Pease, K. (1990). Crime placement, displacement, and deflection. In M. Tonry and N. Morris (eds) *Crime and Justice: An Annual Review of Research*. Volume 12, pp. 277–318. Chicago: University of Chicago Press.
Bowers, K.J. and Johnson, S.D. (2003). Measuring the geographical displacement and diffusion of benefit effects of crime prevention activity. *Journal of Quantitative Criminology*, 19(3), 275–301.
Dent, B.D. (1999). *Cartography: Thematic Map Design*. Boston: WCB/McGraw-Hill.

Dorling, D. and Openshaw, S. (1992). Using computer animation to visualize space-time patterns. *Environment and Planning B: Planning & Design*, 19(6), 639–650.

Hesseling, R. (1994). *Displacement: A Review of the Empirical Literature*, Monsey, NY: Criminal Justice Press.

Langran, G. (1989). A review of temporal database research and its use in GIS applications. *International Journal of Geographical Information Systems*, 3(3), 215–232.

Langran, G. (1992). *Time in Geographic Information Systems*. London: Taylor & Francis.

LeBeau, J.L. (1987). The methods and measures of centrography and the spatial dynamics of rape. *Journal of Quantitative Criminology*, 3(2), 125–141.

Peterson, M.P. (1995). *Interactive and Animated Cartography*. Englewood Cliffs, NJ: Prentice-Hall.

Peuquet, D.J. (1994). It's about time: A conceptual-framework for the representation of temporal dynamics in Geographical Information Systems. *Annals of the Association of American Geographers*, 84(3), 441–461.

Peuquet, D.J. and Niu, D.A. (1995). An event-based spatiotemporal data model (ESTDM) for temporal analysis of geographical data. *International Journal of Geographical Information Systems*, 9(1), 7–24.

Ratcliffe, J.H. (2000). Aoristic analysis: The spatial interpretation of unspecific temporal events. *International Journal of Geographical Information Science*, 14(7), 669–679.

Ratcliffe, J.H. (2001). Policing Urban Burglary. *Trends and Issues in Crime and Criminal Justice* 213, 6.

Ratcliffe, J.H. (2002). Aoristic signatures and the temporal analysis of high volume crime patterns. *Journal of Quantitative Criminology*, 18(1), 23–43.

Ratcliffe, J.H. and McCullagh, M.J. (2000). Aoristic crime analysis. *International Journal of Geographical Information Science*, 12(7), 751–764.

Soot, S. (1975). *Methods and Measures of Centrography: A Critical Survey of Geographic Applications* (Paper 8). Geography Graduate Student Association, Urbana-Champaign, IL.

Tufte, E. (1990). *Envisioning Information*. Cheshire, Conn.: Graphics Press.

Vasiliev, I. (1996). Design issues to be considered when mapping time. In C.H. Wood and C.P. Keller (eds) *Cartographic Design: Theoretical and Practical Perspectives*, pp. 137–146. New York: John Wiley & Sons.

9
Mapping for Operational Police Activities

Learning Objectives

Crime mapping is now routinely practised in police departments across the world. Instead of this being a specialised activity undertaken by academics, police department analysts are able to map crime hotspots and offender behaviour on a regular basis. How is this new information being used to change policing? This chapter looks at ways in which maps are incorporated into the operational thinking of police organisations, and are used to inform police officers wanting to reduce crime.

After reading this chapter you will be able to:

- describe the principles of CompStat;

- understand how CompStat functions in a busy police department;

- describe the four intelligence products of Britain's National Intelligence Model, as well as what they are used for;

- understand the importance of repeat victimisation in crime prevention research; and

- outline the main spatial and temporal components of the hotspot matrix.

GIS and Crime Mapping   Spencer Chainey and Jerry Ratcliffe
© 2005 John Wiley & Sons, Ltd

9.1 Introduction

Earlier chapters in this book have revealed techniques that can make sense of crime data in a spatial sense. In this chapter we explore ways that this spatial information can be used in an operational context. More specifically, we are interested in exploring how maps are used to inform operational police activity. We will look at three particular ways in which operational crime information can influence crime reduction. The first, CompStat, originates from the USA and the New York Police Department, in particular. Two case studies examine the methods of CompStat from a GIS practitioner's viewpoint and from that of a police chief.

Secondly, we look at the operational aspects of the UK National Intelligence Model that are relevant to crime mappers. One of the four products of the model is a problem profile, and one useful guide to problem areas is the degree of repeat victimisation. We provide an overview of the value of the identification of repeat victimisation in the crime prevention effort.

Finally, we explore a matrix that can be used to disaggregate spatial and temporal components of crime hotspots and organise potential crime prevention solutions. All of these tools are ways that crime mappers can guide and direct crime reduction professionals to the most appropriate and needy areas, locations and people. The last is particularly geared to the value that crime mappers can influence crime reduction activities by identifying specific spatial and temporal patterns within the individual events in a crime hotspot.

The concentration on these areas of activity is not to denigrate the value of mapping in other areas of police activity. For example, GIS has been used for security purposes in London to map the individual serial numbers used by search teams that seal public utility access covers and manhole covers in the roads prior to a military or royal event. It has also been used by a variety of police services to map public demonstrations and marches. For example, the Victoria Police in Australia used GIS to provide live 'real-time' mapping of crowd movements at the 2000 meeting of the World Economic Forum in Melbourne (McLean, 2000). These are a variety of operational police uses of GIS, but many are more geared towards command and control functions rather than police activities that primarily have a crime prevention and reduction focus. In this chapter we concentrate on the crime control elements, and we start with CompStat.

9.2 CompStat

CompStat (short for Computer and Comparative Statistics) has been one of the most successful collaborations of GIS and policing strategy. CompStat is an operational management process and is much more than just maps of crime; however, the GIS does form an integral part of the overall strategy. This strategy combines computer technology, operational strategy and managerial accountability to determine the provision of crime reduction policing (Walsh, 2001). The CompStat process sits between the operational and strategic levels of police activity, though the appeal for many law enforcement managers leans towards the operational possibilities. From its early origins with the New York City Police Department, the CompStat idea has been rapidly adopted in many locations, often with a revised name and sometimes with a different format. For example, in the Australian state of New South Wales, the New South Wales Police Service hold an Operations and Crime Review (OCR) meeting for each of the 80 Local Area Commands, where a Local Area Commander can expect to attend an OCR every three months, while in the Philadelphia Police Department CompStat meeting, a District Captain can expect a reappearance every 28 days. Figure 9.1 shows a CompStat meeting of the Philadelphia Police Department in progress. The senior executives of the department are lined up on the left, the mapping personnel are in the central location and the district captains are on the right. As you can see, the map plays a central role in the discussions taking place.

The basic aim of CompStat is to provide detailed intelligence to operational commanders on a regular basis so that they can determine an appropriate crime fighting strategy. The commanders are usually empowered to initiate their chosen policy, but in return for this flexibility are held accountable to the police chief for the success of the strategy. This empowerment can be handed downwards, so it is often the case that operational commanders then hold their own CompStat meetings at a local level and seek the views of Inspectors and Sergeants with regard to crime problems. The computer system to analyse and map crime patterns is at the core of the CompStat process (Walsh, 2001, p. 352). Crime maps form the crux around which police can visualise operational decisions, resource allocation and managerial accountability. The general application of the CompStat has four central principles (Schick, 2004):

1. Timely and accurate intelligence
2. Effective tactics
3. Rapid deployment
4. Relentless follow-up and assessment.

Figure 9.1 Philadelphia Police Department CompStat meeting

Timely and accurate intelligence
This means that police have to be served by an information system that can rapidly convert crime reports, calls from the public and any other relevant information into useful and spatially referenced intelligence within a useful time period. A useful time period is that where the intelligence is actionable. Within a more strategic time frame this could be months or years; however, timely intelligence in a CompStat environment is measured more in days and weeks. The rapid and accurate geocoding of crime-related knowledge is central to getting the information in an analytical format, and as will be seen in the case study to follow from the Philadelphia Police Department, automation goes a long way to improving reliability and speed in this area.

Effective tactics
This is probably the most difficult area for police commanders, yet without the choice of an appropriate and effective tactic, there will be little chance of their actions having an effect on crime reduction. The difficulty is that most police managers were not selected for promotion to their position of leadership on the basis of their ability to reduce crime. In many jurisdictions, promotion is based on an examination that tests knowledge of the law, discipline procedures, departmental administration arrangements, personnel

matters and a knowledge of budgetary and financial obligations. There are no police departments, to our knowledge, that have a significant section of their promotion procedure that requires the candidate to know what works in crime reduction. So effective tactics are a difficult area for police managers to get right first time, and the CompStat process should be adjusted to reflect that commanders are being asked to do something that they have not been trained for and have little previous experience of doing.

Rapid deployment
The criminal environment is a dynamic one and there is little point in trying to get timely and accurate intelligence (principle one) if there is no corresponding prompt marshalling and deployment of all necessary resources once an effective strategy has been selected. The limiting factor is often that resources outside the immediate local command are often not under the control of the local commander, and so must be requested from a central command structure. Central resources often include plain-clothed units and vehicles, canine and mounted patrols, and crime prevention officers. To respond effectively to a local commander's request requires two things: that the central command recognise the viability of moving resources quickly from one place to another, and secondly that the central command are prepared to justify removing resources from one, possibly equally needy, area to another. The latter does require that police commanders at the highest level have a clear understanding of the relative crime problems across all of the areas under their command. Whenever resources are underutilised in one place, they must be moved to somewhere they will be more useful.

Relentless follow-up and assessment
Given the more practical and less introspective nature of the police role, there is probably little worse to a police commander than to assess his or her performance based on their crime reduction ability. In many cases within law enforcement, the mere fact of deciding a course of action and implementing a 'solution' is seen as a measure of success. This was the case in a number of evaluations conducted by the British Home Office:

> One of the most concerning aspects of the information provided by many forces was the way in which success was interpreted. Based on their information, most of the initiatives described by them as successful did not achieve all or most of the initiative's objectives. In many cases, a claim of success seemed to be based merely on the existence of the initiative rather than on what it achieved. (HMIC, 1998, p. 29)

Officers are consciously aware that promotion and indeed tenure are associated with performance at CompStat meetings. It is the one opportunity that police commissioners have to assess the performance of their middle and senior staff and a lot can weigh in the balance. As Laycock rightly notes, 'the measurement of police performance was, and still is, a sensitive matter' (Laycock, 2001, p. 71).

To the four general principles that are the cornerstone of CompStat, McDonald, writing in regard to the originators of CompStat – the New York City Police Department, added a fifth principle at the start of the process: *Specific objectives* (McDonald, 2002). This is a useful addition to consider, as it requires the police leadership to articulate exactly what they want to achieve. McDonald suggests three to five objectives, or to stipulate a definitive time frame for an activity. These are useful in that there are often mixed or confusing messages passed to line officers from management. With clear direction as to where the department wants to go, line officers are more empowered to consider viable strategies on target objectives that are clear.

The mapping system can respond to this direction and clarity in objectives. For example, if the police department has a stated objective of increasing community awareness and community contact, then it may be valuable to map the boundaries of neighbourhood watch areas or to be able to map those schools that receive regular crime prevention advice. If, however, the stated objective of the police service is a zero-tolerance style of law enforcement then an ability to map police quality-of-life interventions such as arrests for public drunkenness may be more useful.

9.2.1 CompStat in practice

The operationalisation of these principles has a decidedly geographic focus. No operational commander will ever tell you that they have enough resources and so a geographic focus is one of the most effective ways to utilise limited resources and people. The whole process usually starts with a CompStat meeting, which can be a stressful process for an operational commander. At these meetings, operational managers and the police chief get together and use maps of crime distribution, projected onto a wall for the whole room to see, as the catalyst for a discussion about the crime situation. These maps can also be enhanced with time series charts showing changes in crime volume over time. If the room is busy, and there can often be over a hundred people in the room for a large police department, the lack of success in previous operations can be clear for all to see.

At this point, the operational commander has to justify their strategies since the last meeting and explain what they will be doing between the present and the next meeting. In the New South Wales OCR, an officer makes notes throughout the meeting, recording the comments of everyone present. When the Local Area Commander returns three months later, the meeting is reminded of the tactics the commander promised to employ at the last meeting.

Crime maps provide a mechanism to provide the required timely and accurate information in an appropriate form. Hotspot surface maps (see Chapter 6) are therefore preferable to point pattern maps in many cases, as three months' worth of data can often swamp a display and can fail to show repeat victimisation in hotspot areas. Point dot maps can also allow a commander to distract the meeting by discussing individual events and small crime patterns. While anecdotal information regarding individual cases may be entertaining, these are not really of relevance at the operational level. This is not to say that individual cases are not important, but a better discussion is to focus on the more strategic issues of why a hotspot exists. For example, 'who is committing a group of offences?', 'are there any general characteristics of the victims' and 'why is the hotspot at that location?' are all better operational questions, and are more likely to generate discussion that examines the root causes of crime and hopefully longer-term crime prevention strategies. These more fundamental issues are more problem-oriented and, in the problem-oriented policing arena (Goldstein, 1990), likely to produce a more permanent crime prevention result.

The mapping component of CompStat is likely to be the easiest part of the process for any police department to initialise. It requires a technical solution and a training environment, but the groundwork has been completed by a number of innovative police departments. The hardest part of the process is most likely to be for police management to instil a culture of innovation and experimentation in the choice of effective tactic employed by police middle management. While we are seeing the development of a reasonable canon of knowledge in regard to 'what works' and 'what does not' in policing (see for example Sherman, *et al.*, 1998), this does not automatically presume that this knowledge filters to operational police commanders. Even though new research is published on a regular basis, 'the incorporation of research findings into a new body of accepted wisdom is a slow and uneven process in which the intellectual persuasiveness of the research is a great deal less relevant than its political appeal' (Weatheritt, 1986, p. 16).

Case study: CompStat mapping in the Philadelphia Police Department

Rachel Weeden Manager, Crime Analysis and Mapping Unit, Philadelphia Police Department

Since 1998, the Philadelphia Police Department has fully integrated GIS-enabled crime mapping into their CompStat process. The ability to display crime and other related incidents in a highly interactive and detailed mapping environment allows CompStat to move beyond a basic question-and-answer session, and instead elevates it to a process whereby those in charge can assess information for themselves. It is difficult to imagine what CompStat would be like without the strong emphasis on mapping. This success can be attributed to several characteristics:

- Currentness of information
- Variety of base layers
- Flexible interactive environment
- Customizations and advancements.

For GIS to be the hub of CompStat, the data behind the maps must be up to date. In Philadelphia, CompStat examines crime for a 28-day period which ends only a few days before the meeting is held. As result of this, the Department's Mapping Unit has only a short time to prepare the necessary data and maps. The success of the Unit's daily download routines is attributed in part to recognizing the importance of developing a thorough and versatile address scrubbing process that requires very little interaction from the staff. For this, a Visual Basic project with MapObjects components carefully examines each incident address before assigning x and y coordinates. With this process in place, the Unit rarely sees a geocoding success rate below 98%.

While accurate and timely incident data is the foundation of a successful CompStat project, GIS staff must also focus on mapping crime in context – this means creating a map that provides a very detailed and complete representation of the City. Over the years, the constant attendance of GIS staff at CompStat means they are able to determine what base layers would be relevant to crime analysis, and then work quickly to create the relevant map-based files. Examples include polygons depicting the deployment of the Narcotics Task Force, and points representing nightclubs, resulting from discussions about violence after closing time. Police personnel may not verbalize the need for these layers, so the Unit anticipates ways in which additional data can advance CompStat; this has resulted in the creation of over 60 data layers, and has become an ongoing process in terms of map files updates, attribute accuracy and metadata. Although this can take time, any time invested is worth the benefit of a more accurate and informative proximity analysis.

Having a variety of base layer data is beneficial to the CompStat process, but equally important is the way in which the crime data itself is viewed. For the current 28-day period, Police will focus on crime categories that might be problematic at that specific time. For example, one captain might be asked to discuss handgun robberies and shootings in their area, while another might be called on regarding stolen vehicles and narcotics activity in a different part of the City. The GIS project must not only be flexible enough to allow for these specific lines of questioning, but should also provide other pertinent information about the incident itself, aside from simply showing its location.

To meet these requirements, the Unit creates GIS layers for each crime and symbolizes these based on key attributes, such as if an arrest was made, if the incident was domestic-related, and even at what time of day it was reported. Therefore, for any one layer, the map uses numerous symbols to depict not only the broad categorization of the crime, but also nuances that may influence Police response to that pattern.

Showing which incidents resulted in arrests is one way for police to gauge if a problem has been resolved, at least in the short term. Another way of measuring police response is to display routine activity in conjunction with serious incidents. For example, the Unit is regularly called on to display points representing car or pedestrian investigations, truancy stops, and curfew violations. Using a visual overlay of police activity with reported crime gives commanders the ability to asses the logistics of their strategies and then make appropriate adjustments.

Beyond the ability to display simple points depicting incidents and activity, the Unit also has the advanced functionality to link related incidents to each other. The most useful application is the ability to display points symbolizing stolen cars, and then draw a link to another symbol representing the recovery location of the vehicle (Figure 9.2). Additionally, this process of linking incidents based on common fields can be used to track firearms used in various crimes, or to show the geographical relationship between incidents and the offender's residence.

During a CompStat meeting, it is important for GIS staff to be able to navigate the software efficiently in response to developing conversations, in real time. For this purpose, GIS staff are constantly tailoring and automating the project not only to meet the needs of CompStat, but

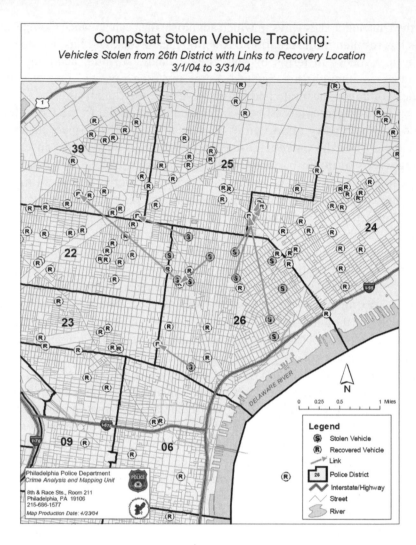

Figure 9.2 Example of a stolen vehicle CompStat report from the Philadelphia Police Department

also to make the project easier to operate. The Unit relies on customized toolbars with underlying ArcObjects programming to make this possible. If a captain is relaying information about shared characteristics of robbery victims, for example, GIS staff can label incidents with victim age – but rather than navigating through a maze of menus, they can instead click on a pre-programmed button that accomplishes this. Anticipating the GIS tasks used often at CompStat, and then simplifying them with ArcObjects, allows the staff to keep up with dynamic conversations, and also to interact with the project in ways that clarify the topic(s) at hand.

Using GIS as a dynamic illustrator for detail-oriented discussions is what elevates the CompStat process. The power of visualization when analyzing crime speaks directly to the tenets of CompStat – development of effective tactics, rapid deployment, and focused assessment and follow-up.

9.2.2 The place of CompStat in a crime reduction framework

A simple model can be used to explore the conceptual framework of intelligence-driven policing, such as the 3i Model (Ratcliffe, 2004b, p. 8). This model, shown in Figure 9.3, recognises that the role of the intelligence or crime analysis unit is to interpret the dynamics of the criminal environment. This is an active process of seeking out the information necessary for effective crime analysis (hence the arrow running from the unit to the criminal environment). Unless the intelligence unit is the decision-making body in a particular law enforcement environment, it is more common to use the available intelligence to influence a decision-maker, often a local commander. This is a part of the process that is often unclear to crime analysts. It is sometimes the case that analysts do not convey their intelligence to the best person or group at a police station. Often, the intelligence and crime maps are printed and adorn an office wall without ever making it on to the radar of a significant decision-maker. This is less of an issue with CompStat as the decision-makers should all be in the room during the CompStat meeting.

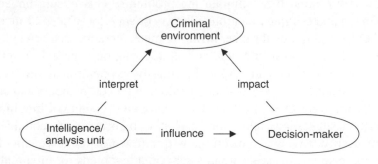

Figure 9.3 A model of intelligence-led crime reduction. Source: Adapted from Ratcliffe (2004b)

The final stage of the process requires that managers use the intelligence that they are given so that the law enforcement system can actually have some sort of an impact on the criminal environment. As said, this requires some knowledge of effective tactics on the part of the decision-maker and this is not an assumption that can be made in current law enforcement. Although it would seem that experience in the police service is sufficient to know how to reduce crime, this focus on crime reduction is still relatively new. Until the last few years the focus in most police departments has been on reactive policing and crime investigation. This is still the case in much of the world. The next case study is from Lincoln, Nebraska (US) where the police chief has been running CompStat meetings for many years. Tom Casady's experiences are valuable for all police managers.

Case study: CompStat from a management perspective

Tom Casady, Chief of Police, Lincoln, Nebraska

At the Lincoln Police Department, the chief of police hosts an internal geographic crime analysis meeting that we call ACUDAT, an acronym for Analyzing Crime Using Data About Trends. These meetings are loosely based on the CompStat model, but differ from many agencies' similar processes in their focus. Most CompStat-style meetings focus on holding commanders accountable for the crime in their geographic area of responsibility. Such meetings are typically attended by senior managers and headquarters staff. In the typical CompStat meeting, an area commander reports about the crime trends in his or her specific region, and what is being done to ameliorate the problems. This occurs under the questioning and critique of a group of peers or superior officers. Lincoln's ACUDAT meetings, on the other hand, are much more focused on exploring crime trends together as a group, and strategizing on approaches to these trends. These meetings are targeted primarily to operational staff—field supervisors and detectives, who attend along with commanders and crime analysts. We want these meetings to be an open exchange of intelligence information, suggestions, and comments—not a barbequing of a commander. We invariably leave these meetings with a few cases that we can clear, a few crime series that we were unaware of, a few plans for interventions that we have brainstormed about, and a positive feeling that something was actually accomplished.

Our meetings tend to be short—rarely longer than 90 minutes—and crime maps are only one of several information sources brought to bear. The meetings are highly interactive, both in terms of the GIS projects themselves and the discussions which emerge. We always use live GIS projects—never static images. It is common for us to run new queries on-the-fly, and to link to other data, such as our on-line police reports, records management system, digital photographs, and a variety of web sites such as the State sex offender registry, Department of Motor Vehicles records, and State criminal history repository.

One of our goals in these meetings is to expose the audience to new or little-known information resources. We always try to run a new query, link to a new on-line data source, or demonstrate an updated application. At a recent meeting, a discussion took place about a suspect in a crime series and the type of automobile he was believed to be using. During the discussion, an analyst went to a public-domain automotive website, and within seconds was displaying a photo of the make and model we were discussing. She noted that with a right mouse click, the photo could be printed, saved and dropped into a document, or emailed to a colleague. As valuable as this was to the current discussion, the real value was to remind everyone at the meeting that getting a representative photo of a motor vehicle is quick and easy. Sometimes these brief 'stealth training sessions' happen naturally, but we often plan such short detours in advance, when the Chief and the crime analysis unit staff get together for an hour-long planning session in the afternoon prior to the evening ACUDAT meeting.

From my perspective as chief, there have been other valuable outcomes of ACUDAT. For the first time in my career, the commanding officers of the police department are primarily talking about our core business—enhancing the quality of life and safety of our community. Our management staff meetings have been consumed in the past with the minutia of large bureaucracies—budget reports, personnel issues, and so forth. Now, rather than discussing what we are going to do about lack of evidence storage space and the vacancy in the Training Unit, we are actually discussing crime and disorder, and what we are doing about it. It has been a refreshing change.

Another valuable outcome has been more personal. I am convinced that the officers, sergeants, and detectives who attend these meetings are actually a bit surprised to learn that the brass actually know a thing or two about the city, the crime, the criminals, and the issues officers confront on the street. In fact, due to the ACUDAT meetings, they now know quite a lot.

9.2.3 How often should CompStat run?

CompStat has been described as a new paradigm in policing (Walsh, 2001), yet the 'rules' of an effective CompStat process are not yet in place. There is no real clarity on the best way to run CompStat. Some police services have their local commanders at the meeting every three months (for example, New South Wales Police Service in Australia), while others have a 28-day return period (Philadelphia) and others (such as Los Angeles, see Schick, 2004) an even shorter period, meeting every week.

The reason this is worth understanding better is that longer return periods allow local commanders time to initiate and run an operation that has time to succeed or fail. Indeed, if a problem-oriented policing solution is sought, then it is clear from case studies that the solution to a complicated crime problem can be many months (Scott, 2000). Shorter time periods would seem to run the risk of running into potential problems.

First, the short time between crime reporting periods can leave the police department chasing individual crime events and failing to see the overall crime problems emerging. It is analogous to not seeing the wood for the trees. The police end up chasing the individual crime events from the previous week instead of focusing on the longer-term problem areas. Of course, if a spree of homicides suddenly started occurring over the space of a few days, one would hope that the police might notice this trend prior to a CompStat meeting, but occasional vehicle thefts in a low crime area may not be as easy to tackle as a long-term problem in one particular neighbourhood. Secondly, the lack of significant crime trends from week to week can tempt the police executive to micro-manage their local commanders. Finally, and most worryingly, is that it may lead local command staff to seek short-term solutions to crime problems. The immediacy of CompStat, with its focus on timely intelligence and rapid response, can lead to short-term solutions for long-term problems, and an inevitable return to the particular crime nuisance in a few weeks. This ends up resembling the reactive policing model that CompStat was trying to avoid in the first place.

Of course, there are negative consequences of a long return period to a CompStat meeting. The local commander may think that they are off the hook for some time and return to their station and relax, only to exert some crime reduction effort just prior to the next meeting. It is also recognised that there are examples of negative consequences from well-meaning attempts to reduce crime (Grabosky, 1995, 1996). If a police commander instigates a policy that actually exacerbates a problem, then a long gap between CompStat meetings can allow the negative policy to run longer than is healthy. Negative crime reduction policies are easier to correct by the police executive if they are made aware of them at more frequent CompStat meetings. The balance of a long or short return period for a CompStat meeting is certainly an issue for any police department considering CompStat. CompStat would certainly appear to be here to stay, but there are still some methodological issues to be ironed out before it can be widely used across police departments in any fashion of uniformity.

9.3 Intelligence products in the UK

In the UK, CompStat is less developed than in the US, partly due to the growth of the National Intelligence Model. The inability to control a growth in crime in the 1980s, and the developing recognition of the importance of repeat victimisation as fuelling crime problems were but two of a number of driving forces that contributed to the development in Great Britain of intelligence-led policing, and in turn the NIM. The recognition that a small number of offenders are responsible for a much larger contribution to the crime problem was highlighted in a 1993 Audit Commission report (Audit Commission, 1993). The NIM evolved from experience across a range of different police services and their attempts to formulate a single doctrine for crime management. By May 2000, the National Criminal Intelligence Service (NCIS) were distributing copies of the NIM across British police services (Flood, 2004). The NIM is a CD-based business model for the management of crime and intelligence-related information. The homepage from the CD is shown in Figure 9.4.

The core of the NIM is the tasking and coordination meetings that work infrequently at the strategic level, and regularly at the tactical level. In a number of forces, the Tactical Tasking and Coordination meeting happens every two weeks and is tasked with oversight of the Control Strategy for that area (NCIS, 2000). As Flood summarises, the NIM 'describes four intelligence products: strategic assessments, tactical assessments, problem profiles and target profiles that individually or collectively support the setting of strategy, the application and monitoring of a menu of tactical options and the conduct of operations that deliver against the "tactical menu"' (2004, p. 48). The advantages for the crime mapper in this situation is that there are only four products that decision-makers expect under the NIM.

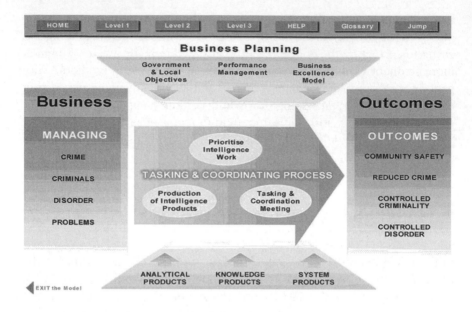

Figure 9.4 Homepage of the National Intelligence Model CD[1]

9.3.1 NIM's four intelligence products

As explained in the NIM, there are four basic intelligence products. The purpose of this limitation on different outputs from an intelligence unit is to achieve uniformity across regions and areas. This means that one police area can pass their intelligence on to another area and the receiving department will be able to recognise the basic format of the intelligence. The four products have different aims and purposes.

1. *Strategic assessments*: These aim to identify long-term issues and predict growing criminal areas. Their purpose is to establish priorities, resource allocations, inform decision-makers and set the control strategy. To many analysts, strategic assessments are the most difficult to create, because they ask for a level of prediction and foresight which few analysts are trained to provide. Indeed, it may not be possible to provide accurate assessments of the future criminal environment beyond the immediate future. The strategic assessment might include an assessment of the relevant political, economic, sociological, technological,

---

[1] Reproduced by permission of the National Criminal Intelligence Service (NCIS), London and the Association of Chief Police Officers of England and Wales (ACPO).

environmental and legal dimensions of the future criminal environment (Heldon, 2004, p. 111).

2. *Tactical assessments*: These aim to identify short-term preventable issues and monitor progress with the control strategy. Their purpose is to assist operational management and reallocate resources if the situation changes. These products are designed to act as an update to ongoing police activities and to be shorter and more operationally relevant. They will usually include an update on current police operations in the area.

3. *Problem profiles*: These aim to identify established and emerging crime/incident series and hotspots. Their purpose is to assist investigative needs, hotspot management, targeting, crime reduction and prevention. In this instance, problems are not necessarily geographical in nature, but can include problems of a new emerging type of criminal behaviour, such as a new modus operandi for an old type of crime. It is, however, the spatial criminal behaviour which is most likely to be of interest to the crime mapper.

4. *Target profiles*: These aim to provide a detailed picture of offenders and associates. Their purpose is to assist decision-makers with target selection and to guide investigations. It is anticipated that many target profiles will naturally flow from earlier problem profiles (Flood, 2004). When a problem is noted in one area, it may be valuable for the local crime management team to get a picture of who may be committing and are involved in the local criminality.

Crime mappers can expect to provide some mapping support to all four intelligence products. Strategic assessments may include an analysis of future trends in population demographics which can be ascertained from census data, or from long-term hotspot maps showing chronic problem areas. Tactical assessments can benefit from maps that show allocation of police and crime prevention resources at a local level, and the recent crime events in the area. Target profiles can show the areas of residence of known offenders, and the addresses of individuals recently released from prison. All of these intelligence products can be enhanced by the addition of carefully considered maps, but probably none more so than the problem profile.

Problem profiles are at the core of many crime mapping activities around the world. Although they are rarely termed as such, many crime analysts outside the UK would recognise the aim of a problem profile, and the value that a map would provide. Emerging patterns of problems can often be noticed due to the physical proximity of the crime events. In other words, a problem can become a problem when someone notices a large number of offences happening in the same area. The very act of mapping crime events with a kernel density programme may identify a hotspot of which decision-makers were unaware.

This recognition of the value in targeting resources to concentration areas of offending is not new. In the US considerable research effort was put into understanding the dynamics of streetcorner drug markets. Place-based anti-drug strategies were tested in places like Oakland, California. Although traditional law enforcement activities made a dent in the criminality, the real long-term gains were achieved through non-traditional law enforcement, such as by employing building code violations against drug dealers (Green, 1995). These approaches were successful by first identifying the high drug-crime problem locations, and having mapping tools that can identify high repeat victimisation sites can be a quick first step to the identification of a problem area.

Repeat victimisation studies are well advanced in the UK. The value that an understanding of the dynamics of repeat victimisation can bring to a crime analyst is discussed in the next section.

9.4 Repeat victimisation

Crime and disorder problems are concentrated among a relatively few number of offenders, victims and places. The value of identifying homes, places and people that are the victim of repeated criminal events as a way to target crime prevention activities is well known in the UK (Farrell and Pease, 1993; Ellingworth *et al.*, 1995; Spelman, 1995; Laycock, 2001). Repeated studies have shown that a small number of victims are the target of a considerable amount of the crime that is reported to the police. If it is possible to identify the characteristics of crime targets that are more likely to be revictimised then it may be possible to prevent repeat incidents at the most vulnerable locations and for the most vulnerable people.

Numerous studies have noticed that there is a definite repeat time course for burglary, where the risk of a repeat event is greatest immediately after the event (Polvi *et al.*, 1991). The risk of another burglary then decreases as time from the first event increases. Eventually, the level of risk returns to the background risk for burglary after a few months (Ratcliffe and McCullagh, 1998). This study, as with many others, shows that there is a near-exponential flow to the repeat time course of events. This includes non-residential burglary (Bowers *et al.*, 1998) and other crime types. Over a third of targeted schools were the victims of repeat incidents within a week (Burquest *et al.*, 1992). Combining census data with repeat victimisation information has shown that deprived areas are more likely to suffer repeat victimisation than affluent areas, even taking into consideration differences in crime rates (Ratcliffe and McCullagh, 2001).

Strangely, given the strong empirical evidence for the value of identifying repeat victimisation, there has been less enthusiasm in the US. Spelman, in his study of fast-food restaurant crime in Texas, argued that it may be necessary to consider the longer-term criminality of an area and identify a long-term problem-oriented solution (Spelman, 1995), but either way the value of being able to identify repeat victimisation seems pretty clear. Even if a short-term solution is ignored in favour of a longer problem-oriented resolution of the crime victimisation at a site, being able to clarify the amount of victimisation at a site is an important step. Unfortunately, that is sometimes not an easy task for the crime mapper.

It has long been recognised that police databases are not set up to identify repeat victimisation easily: 'The problem of identifying repeats in police records is immense' (Anderson *et al.*, 1995, p. 10). The problem is that police databases are generally more concerned with providing statistical returns and management information to law enforcement and organisations outside the police domain (Ericson and Haggerty, 1997). Counting offences has become the most important application of recorded crime databases and record management systems, because the crime rate is often the feature that managers and executives are concerned with and to which they are accountable. As long as the location is accurate to within the right police district, the management statistics will be relatively accurate. For a more micro-level application, such as the identification of repeat victimisation, official crime records 'uniformly fail to highlight the extent to which crime victimization is concentrated on particular individuals and households' (Ellingworth *et al.*, 1995, p. 360).

9.4.1 Mapping repeats

Once identical crime locations have been geocoded to the same place, the task for the crime mapper is relatively simple. In MapInfo for example, it is possible to create two fields that contain the x and y coordinates of each point. A simple query can then be written to output x and y coordinates along with the number of incidents, as shown in Figure 9.5. The resulting table can be mapped and the size of the symbol at each point varied to reflect the number of repeat incidents at each site. This has been done in Figure 9.6 where the image on the right shows the number of events at

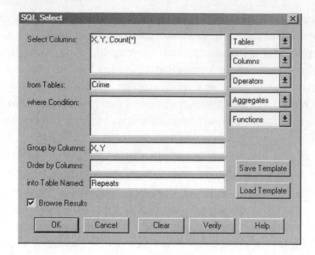

Figure 9.5 MapInfo query screen that will output only repeat locations as *x* and *y* coordinates, along with the number of incidents at each site

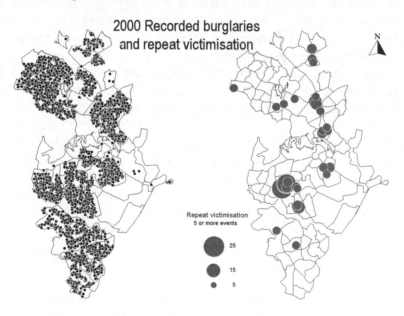

Figure 9.6 Burglaries in Canberra, Australia. Map on the right shows the sites with five or more burglaries in 2000

each site that has been the victim of five or more events during 2000. As has been done in Figure 9.6, it is always worth checking that the GIS has drawn the largest symbols behind the smaller circles. The most targeted places are non-residential, such as schools and shops.

Therefore GIS can be an excellent way to identify repeat victimisation (Ratcliffe and McCullagh, 1998). The challenge for crime mappers is therefore not to identify the repeat locations once they have been geocoded to the same site. A GIS can easily complete this task. The real challenge is to ensure that the same location is geocoded in the same way so that it retains the same x and y coordinate every time the location is in the event database. This will require the crime mapper to carefully ensure that events are recorded in the database in the same way. For example, it is easy for locations such as schools (which can cover a city block) to be recorded in a database as a variety of street addresses. As the case study from Rachel Weeden (Philadelphia Police Department) indicates, setting up systems to assist with geocoding can help with increased productivity. However, it is also a way to achieve greater continuity across a database. One method that can help is to set up an alias table that identifies key locations, such as commercial buildings and schools, and adds a single x and y coordinate to the database. This is done before the individual street address entered by the police officer is able to add an x and y coordinate that may differ from others relating to the school or commercial site.

Once repeat victimisation sites have been mapped, these locations can be studied to better understand the phenomenon of repeat victimisation, or the locations can become the site for targeted crime prevention activity or police preventative patrols.

9.5 The hotspot matrix

One of the difficulties that occurs in any police service is the lack of any real 'institutional memory'. This can be significant because police services with strong and charismatic leadership can often fail to pass on lessons that were learned at considerable cost when leaders are replaced (Dupont, 2003). Organisational or institutional memory is a valuable tool in the development of operational strategies, as unfortunately evidenced by the flexibility and adaptability of Columbian drug cartels (Kenney, 1999). While there is a recognition that organisational memory is a selective capacity, and one that is fairly subjective in interpretation (Mulcahy, 2000), it is generally agreed that there is a need to pass on positive knowledge throughout an organisation. This section concerns itself with crime reduction, and the difficulty with crime reduction tactics is that lessons learned are usually not passed on to other policing areas, regions and departments. Lessons are learned locally and often forgotten locally.

The hotspot matrix is an attempt to find a way to communicate successful ideas from one operational area to another (Ratcliffe, 2004a). It tries to categorise crime hotspots into a 3×3 matrix of generic temporal and spatial characteristics, so that crime reduction ideas can flow from police agency to police agency. The idea works on the principle that it may be possible to broadly categorise crime hotspots into three spatial and three temporal types. Once a solution to a temporal pattern has been discovered somewhere, it can be passed on to other police and crime reduction agencies who have a similar type of hotspot problem. The value of this for crime mappers is that it provides a framework with which to conduct spatial analysis of hotspots (see Chapters 5 and 6) as well as a framework for temporal studies (see Chapter 8). The matrix components are described as follows:

9.5.1 Hotspot categories

9.5.1.1 Spatial categories

The important point to consider with all of the following categories is that they relate to the dispersal of crime events *within* the crime hotspot. In other words, it does not significantly matter how you determine a crime hotspot (we review a number of different methods in Chapter 6). The important feature is the distribution of points *within* the crime hotspot area.

Dispersed
This is a type of crime hotspot where the crime events are distributed around the whole hotspot area. An example might be where the location of stolen vehicles indicates that cars are stolen from a variety of different places within a housing project, or across a number of car parks surrounding a large shopping mall. This type of spatial hotspot is characterised by a lack of clustering at any one significant location.

Clustered
When there is clustering at a location, the hotspot is referred to as a clustered hotspot. This does not negate the possibility of crime events at other locations within the hotspot, but there is clear evidence of crime surrounding a particular feature. An example might be where a particular bar is a crime generator for late-night alcohol-fuelled assaults. Often these assaults will not occur actually at the bar (perhaps due to the presence of bouncers and security). These will cluster in the vicinity of the bar in the surrounding street, and there may also be some assaults in other parts of the hotspot.

Hotpoint

A hotpoint is a crime hotspot that is caused by the repeat victimisation of a single location. An example could be the generation of a crime hotspot due to repeat burglaries at a school, or a pattern of continued robberies at a corner store late at night. There may be an occasional crime event elsewhere in the hotspot, but the vast majority of crime events in a hotpoint hotspot will be at one central location. This type of crime hotspot is often identified visually with kernel density software as a symmetrical shape, such as a circle.

Visually, when the pattern of crime events is examined within the hotspot, these three spatial types appear as in Figure 9.7, from Ratcliffe (2004a, p. 11) which shows the dispersed (a), clustered (b) and hotpoint (c) spatial patterns.

9.5.1.2 Temporal categories

As well as spatial hotspot types, there are three general temporal patterns to crime hotspots.

Diffused

A diffused temporal pattern occurs in a hotspot when there is no discernable pattern to the time that the crime events occur. In other words, if the crime hotspot is one in which vehicle thefts occur in a public car park, there is no particular time or part of the day when these events occur. A car could as easily be stolen at 1 am as at 3 pm. This does not negate the possibility of a temporal histogram (see Chapter 8) showing some rises and drops across the hours of the day or night; however, in a diffused temporal pattern these changes are not significant enough to be useful from a crime prevention perspective. In other words, the differences from hour to hour have little value operationally.

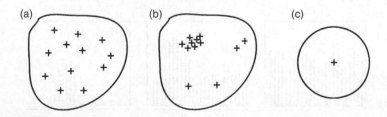

Figure 9.7 Three different types of spatial hotspot: dispersed (a), clustered (b) and hotpoint (c). Source: Ratcliffe (2004a, p. 11)

Focused

With a focused temporal pattern, there is a time or a block of time, where criminal activity is significantly more focused than at others. This could be, for example, during the late evening around bar-closing times, for a crime hotspot of aggravated assault. There may be other assaults taking place throughout the rest of the day in the hotspot, but there is a noticeable increase in criminal activity during key blocks of time. Alternatively, it could be a block of a few hours during the day when residential burglaries occur (Bottoms and Wiles, 2002). The operational value of a focused crime series is that it is significant enough to allow police resources to be targeted more effectively to the best deployment times.

Acute

The temporal equivalent of a hotpoint hotspot, the acute temporal pattern has a significant clustering of crime events into one short period of time, such as a block of three or four hours. The acute temporal pattern differs from the focused one in that there is almost a negation of the possibility of criminal activity happening at other times outside the acute time periods. Although the acute pattern is not common, an example could be shoplifting at a store that is only open for a few hours in each day.

If a temporal histogram is constructed as shown in Chapter 8, then the patterns of temporal hotspot events can be generalised as shown in Figure 9.8 (from Ratcliffe, 2004a, p. 12) which shows the diffused (A), focused (B) and acute (C) temporal patterns.

When the spatial and temporal components are combined, they form a 3×3 matrix that can be populated with crime reduction ideas. Because the hotspot matrix establishes a common language of crime problems, it then becomes possible to share successful crime reduction and prevention ideas between practitioners. A hotspot matrix (adapted, as all matrices are, from Ratcliffe, 2004a) is shown in Figure 9.9, along with some possible ideas to combat a local burglary problem (the target of the matrix is usually shown in the top left box).

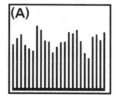 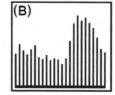 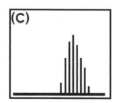

Figure 9.8 Three different types of spatial hotspot: diffused (A), focused (B), and acute (C). Source: Ratcliffe (2004a, p. 12)

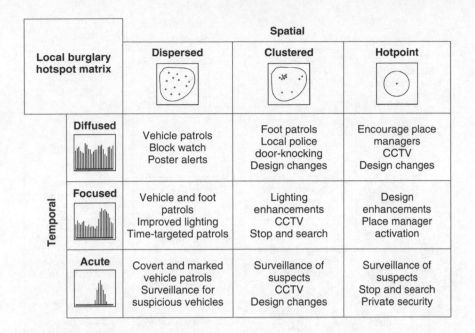

Figure 9.9 An example of a hotspot matrix for a local burglary problem

Case study: A street crime hotspot matrix

Professor Nick Tilley, University College London and Nottingham Trent University, UK

Renowned crime researcher, Professor Nick Tilley, has been recently working on a project for the British Home Office. Tilley and his colleagues have been working on ways to combat street crime across the UK, and in doing so have been exploring practical solutions to the problem of street crime in British cities. Nick Tilley and his colleagues have used the hotspot matrix as a way to draw together guidance on dealing effectively with street crime through problem-solving.

Tilley's work comes from a problem-oriented perspective and aims to have practical crime reduction value at the point of the criminal event, and this focus is reflected in the hotspot matrix that he put together to illustrate some ideas to prevent street crime within different types of crime hotspot. Tilley's hotspot matrix for street crime is shown in Figure 9.10, where HVP indicates High Visibility Patrols and pulse patrols are occasional visits to a location.

Street crime hotspot matrix		Spatial		
		Dispersed	**Clustered**	**Hotpoint**
Temporal	**Diffused**	Pulse patrols Warning notices Design changes Mobilisation of wardens	CCTV Place-targeted HVP Design changes	Design changes CCTV Lever place managers Access control for likely offenders
	Focused	Lighting upgrades Pulse- and time-targeted HVP	Lighting upgrades CCTV Time- and place-targeted HVP Stop and search	Lighting upgrades CCTV Place managers' mobilisation and sensitisation
	Acute	Time-targeted, pulsed HVP Revolving crackdowns Covert timed suspect-focused patrol	Time- and place-targeted HVP Alert key place managers Cluster-targeted lighting/CCTV	Covert patrol Lighting upgrades Stop and search Mobilisation of relevant place managers

Figure 9.10 Hotspot matrix for street crime, as suggested by Tilley (personal communication)

9.5.2 Using the hotspot matrix

The key to using the hotspot matrix is not to get hung up on exact definitions of each type of spatial and temporal pattern. It was not designed as a definitive categorisation of crime hotspots, but as a broad categorisation to enable different crime reduction practitioners to have a common frame of reference so that a dialogue could take place on crime prevention activities (Ratcliffe, 2004a). In fact there may be some value in identifying crime hotspots that lie on the boundaries between these different classifications and in deploying strategies from neighbouring boxes on the hotspot matrix.

The aim of the matrix is as a starting point for thinking and communicating ideas about how to solve different types of crime problem. A matrix can be created for any type of criminal activity that has a hotspot pattern. Vehicle crime, drug activity, burglary, robberies or assaults can all have a hotspot matrix created for them. The matrix can be used to extend the thinking of analysts tasked with solving crime problems, and as such it is closely linked to problem-oriented policing and the aims of the NIM.

> The secret of both medium and long term success lies in securing synergy between the areas of activity so that lessons learned and intelligence gleaned from 'crime series' investigations, 'hot spot' management and target operations are absorbed and used to inform new tactical choices or consolidate the gains through well-informed preventive work. (Flood, 2004, p. 50)

9.6 Summary

When crime analysis actually influences crime reduction tactics is when crime analysts earn their salary. It is at this point that crime mappers have a chance to influence decision-makers. In the CompStat process, the maps are central to informing a room full of police officers. Crime patterns are monitored and their change from meeting to meeting are plotted. The key to CompStat is the accountability of the police officer in charge of each area, and the maps are an essential component to achieving that accountability.

The NIM focuses more on the formalisation of intelligence products and the mechanisms by which they inform the decision-making process of the organisation. Repeat victimisation analysis is one way to identify a problem area for further analysis with a problem profile. In regard to burglary, it has been shown that the highest risk for burglary at a home is in the first couple of weeks after the initial event. Identifying vulnerable locations or places that have been victimised more than once can be a good start to targeting crime prevention. The hotspot matrix is one way to categorise crime prevention activities in line with the different types of crime patterns within a hotspot, and to share successful measures.

This chapter has addressed the main challenge facing people working in the criminal justice system today. The challenge for crime reduction practitioners is to convert criminal intelligence and crime mapping analysis into effective crime prevention on the streets.

Further reading

Willis, J.J., Mastrofski, S.D., Weisburd, D. and Greenspan, R. (2004). *CompStat and Organizational Change in the Lowell Police Department: Challenges and Opportunities*, p. 96.

Can be downloaded from www.policefoundation.org. An in-depth examination of the experiences of one US police department as they implement CompStat.

Cope, N. (2004). Intelligence led policing or policing led intelligence?: Integrating volume crime analysis into policing. *British Journal of Criminology*, 44(2), 188–203.

Nina Cope's article explores the meaning of intelligence-led policing from within police organisations and from the perspective of analysts at the forefront of this new policing paradigm.

Flood, B. (2004). Strategic aspects of the UK National Intelligence Model. In J.H. Ratcliffe (ed.) *Strategic Thinking in Criminal Intelligence*, First edition. Sydney: Federation Press.

One of the few publicly available articles about the National Intelligence Model written by one of its primary architects.

Ratcliffe, J. (2004). The hotspot matrix: A framework for the spatio-temporal targeting of crime reduction. *Police Practice and Research*, 5(1), 5–23.

The original article that first refers to the hotspot matrix.

Hirschfield, A. and Bowers, K. (eds) (2001). *Mapping and Analysing Crime Data*. London: Taylor & Francis.

A recent book that has a number of chapters that refer to repeat victimisation. Also contains a number of other chapters that are useful to the reader as references for other areas of crime mapping.

References

Anderson, D., Chenery, S. and Pease, K. (1995). Biting back: Tackling repeat burglary and car crime. *Police Research Group: Crime Detection and Prevention Series*, Paper 58, 1–57.

Audit Commission (1993). *Helping With Enquiries: Tackling Crime Effectively*. London: HMSO.

Bottoms, A.E. and Wiles, P. (2002). Environmental criminology. In M. Maguire, R. Morgan and R. Reiner (eds) *The Oxford Handbook of Criminology*, Third edition. London: Oxford University Press.

Bowers, K.J., Hirschfield, A. and Johnson, S.D. (1998). Victimization revisited. *British Journal of Criminology*, 38(3), 429–452.

Burquest, R., Farrell, G. and Pease, K. (1992). Lessons from schools. *Policing*, 8, 148–155.

Dupont, B. (2003). Preserving Institutional Memory in Australian Police Services. *Trends and Issues in Crime and Criminal Justice,* 245, 6.

Ellingworth, D., Farrell, G. and Pease, K. (1995). A victim is a victim is a victim? *British Journal of Criminology*, 35(3), 360–365.

Ericson, R.V. and Haggerty, K.D. (1997). *Policing the Risk Society*. Oxford: Clarendon Press.

Farrell, G. and Pease, K. (1993). Once bitten, twice bitten: Repeat victimisation and its implications for crime prevention. *Police Research Group: Crime Prevention Unit Series,* Paper 46, 32.

Flood, B. (2004). Strategic aspects of the UK National Intelligence Model. In J.H. Ratcliffe (ed.) *Strategic Thinking in Criminal Intelligence*, First edition. Sydney: Federation Press.

Goldstein, H. (1990). *Problem-Oriented Policing*. New York: McGraw-Hill.

Grabosky, P. (1995). Counterproductive crime prevention. *Crime prevention conference 1994*, Centre for Crime Policy and Public Safety, Griffith University. Brisbane: Griffith University.

Grabosky, P. (1996). *Unintended Consequences of Crime Prevention*. Volume 5. Monsey, NY: Criminal Justice Press.

Green, L. (1995). Cleaning up drug hot spots in Oakland, California: The displacement and diffusion effects. *Justice Quarterly*, 12(4), 737–754.

Heldon, C.E. (2004). Exploratory analysis tools. In J.H. Ratcliffe (ed.) *Strategic Thinking in Criminal Intelligence*. Sydney: Federation Press.

HMIC (1998). *Beating Crime*. London: Her Majesty's Inspectorate of Constabulary.

Kenney, M. (1999). When criminals out-smart the state: Understanding the learning capacity of Colombian Drug Trafficking Organizations. *Transnational Organized Crime*, 5(1), 97–119.

Laycock, G. (2001). Hypothesis-based research: The repeat victimization story. *Criminal Justice*, 1(1), 59–82.

McDonald, P.P. (2002). *Managing Police Operations: Implementing the New York Crime Control Model – CompStat*. Belmont, CA: Wadsworth.

McLean, G. (2000). *Victoria Police and its Use of GIS for Intelligence Led Policing*, Crime mapping: Adding value to crime prevention and control conference, Adelaide, Australia, 21–22 September 2000.

Mulcahy, A. (2000). Policing history: The official discourse and organizational memory of the royal ulster constabulary. *British Journal of Criminology*, 40(1), 68–87.

NCIS (2000). *The National Intelligence Model*. London: National Criminal Intelligence Service.

Polvi, N., Looman, T., Humphries, C. and Pease, K. (1991). The time course of repeat burglary victimization. *British Journal of Criminology*, 31(4), 411–414.

Ratcliffe, J. (2004a). The hotspot matrix: A framework for the spatio-temporal targeting of crime reduction. *Police Practice and Research*, 5(1), 5–23.

Ratcliffe, J.H. (2004b). *Strategic Thinking in Criminal Intelligence*. Sydney: Federation Press.

Ratcliffe, J.H. and McCullagh, M.J. (1998). Identifying repeat victimisation with GIS. *British Journal of Criminology*, 38(4), 651–662.

Ratcliffe, J.H. and McCullagh, M.J. (2001). Crime, repeat victimisation and GIS. In K. Bowers and A. Hirschfield (eds) *Mapping and Analysing Crime Data.* London: Taylor & Francis.

Schick, W. (2004). CompStat in the Los Angeles Police Department. *Police Chief,* 71(1), 17–23.

Scott, M.S. (2000). *Problem-Oriented Policing: Reflections on the First 20 Years.* Washington, DC: COPS Office.

Sherman, L.W., Gottfredson, D., MacKenzie, D., Eck, J., Reuter, P. and Bushway, S. (1998). *Preventing Crime: What Works, What Doesn't, What's Promising.* Washington, DC: National Institute of Justice.

Spelman, W. (1995). Once bitten, then what: Cross-sectional and time-course explanations of repeat victimization. *British Journal of Criminology,* 35(3), 366–383.

Walsh, W.F. (2001). Compstat: An analysis of an emerging police managerial paradigm. *Policing: An International Journal of Police Strategies & Management,* 24(3), 347–362.

Weatheritt, M. (1986). *Innovations in policing.* Dover: Croom Helm.

10
Tactical and Investigative Crime Mapping Applications

Learning Objectives

In this chapter we illustrate how crime mapping can support the tactical and investigative requirements of law enforcement and crime reduction. In particular, we explore ways to catch offenders or gather facts that can be useful in targeting where to implement diversion schemes and crime prevention strategies, or focus the control of behaviour. Four main areas are covered:

1. Understanding offenders – a number of analytical techniques can be used to understand the offending patterns of offenders. This chapter presents a range of analyses methods that can be used to explore offending patterns, with particular emphasis on geographical analysis. Importantly, we affirm the need to consider the theory and supporting research that helps to understand offenders' spatial behaviour.

2. Journey to crime – exploring the routes and spatial movements that offenders take to commit crimes is an important aspect of geographical crime analysis. In this chapter we will link certain of the key theoretical concepts discussed in Chapter 4 with what is known about the distances and journeys offenders take to commit crime.

GIS and Crime Mapping   Spencer Chainey and Jerry Ratcliffe
© 2005 John Wiley & Sons, Ltd

3. Geographic profiling – we present the main concepts of geographic profiling and how this important tool has been used to help serial crime investigations.

4. Using maps as evidence – crime mapping is increasingly being used as a way to develop and present evidence that supports an investigation or counter offender alibis. We describe the key requirements to consider when producing maps for prosecution evidence, drawing on examples that describe when crime mapping has been used with great success.

We also conclude the chapter by explaining the concept of 'self-selection', how this relates to geography and how it can assist in tactical strategies for targeting offenders.

10.1 Introduction

Tactical and investigative geographical crime analysis uses mapping and supporting techniques to:

- Understand how offenders behave and move in space, and how this behaviour is linked to their network of other individuals, places or illicit markets that support them in their activities. In these terms the analytical role is one that helps to catch criminals more quickly and more effectively.
- Understand what drives individuals into becoming offenders, particularly in terms of the environmental factors that surround an individual – the push factors to committing crime. Analysis of environmental factors that contribute to why an individual decides to offend can be used to identify the crime generators that can be controlled or modified to help prevent others from being pushed towards a criminal life style.
- Understand what motivates them to the spaces where they commit crime – the pull factors to committing crime. Motivational factors could be personal in terms of the goods they CRAVED (Clarke, 1999, and defined in Chapter 4) or relate to the opportunities presented in the place to which they are drawn.
- Gather and present evidence to support a prosecution and conviction.

Catching criminals is hard work. Detection levels from London's Metropolitan Police from April 2003 to March 2004 show that of the 1 060 930 crimes committed in this year, only 15% were solved (equivalent to 162 981 offence clear-ups). The success in catching criminals ranged between 96.6% for murders to 1.6% for pickpocketing (Table 10.1) and over the years has not drastically improved (Table 10.2). If those working in policing and crime reduction want to improve their understanding of offenders and how to catch criminals, then crime mapping can play a significant role. This chapter will demonstrate, through several examples, the way in which crime mapping is supporting investigative analysis, improving detection, developing evidence and preventing crime.

Surprisingly, tactical applications of crime mapping are still in their infancy, even though theoretical aspects of the geography of crime and the behaviour patterns of offenders in space have been examined for some time – from the Chicago School of the 1930s (Shaw and McKay, 1931) to more modern references to 'The Urban Criminal' (Baldwin and Bottoms, 1976), juvenile delinquency (Herbert, 1976) and environmental criminology (Brantingham and Brantingham, 1981). In some ways the study of the geography of crime and evidence on offenders' spatial behaviour has been held back by technology, difficulties in recording crime records and offender details, and the development of analytical techniques for persistent criminal behaviour. In many cases the data available on offenders are still

Table 10.1 Crime detection levels in London for a sample of crime types (April 2003 to March 2004). Source: Metropolitan Police http://www.met.police.uk/crimestatistics/index.htm

	Murder	Rape	Street robbery of the person	Burglary in a residential dwelling	Theft of a motor vehicle	Theft from a motor vehicle	Pick pocketing	Theft snatches
Total offences	204	2571	37 476	67 996	55 158	103 899	27 238	18 979
Total clear-ups	197	842	4414	7017	4150	2173	428	751
Percentage of crimes detected	96.6	32.7	11.8	10.3	7.5	2.1	1.6	4.0

Table 10.2 Changes in detection levels in London between 2000 and 2004. Source: Metropolitan Police http://www.met.police.uk/crimestatistics/index.htm

	2000/01	2001/02	2002/03	2003/04
Total offences	994 233	1 057 360	1 080 741	1 060 930
Total clear-ups	148 995	148 827	156 554	162 981
Percentage of crimes detected	15.0	14.1	14.5	15.4

incomplete or requires data to be gathered together from a number of agencies. Although an often illusive picture, there is value in generating as full a picture as possible that helps to identify those active in crime, and understand their offending behaviour, offending history, motivations, treatments and their vulnerability to re-offend. For example, sourcing data from a number of local law enforcement agencies, utility and construction companies was essential for solving a spate of burglaries and thefts of equipment and stocks from residential construction sites in Overland Park, Kansas. Pooling data and intelligence together resulted in revealing and confirming the names of suspects, and in addition provided information on who their associates were and their current place of employment (Wernicke, 2000).

Understanding how offenders move around space – between their home and the scenes of their crimes – what influences this movement, what attracts them to certain areas and what motivates them to commit crime are all analytical requirements in tactical and investigative crime mapping. It is possible to explore different crime types and identify the typical profile of the offender who commits this type of crime, explore their personal networks and associations with other offenders and even their associations with the victims of their crimes. It is also possible to map where offenders live to explore if certain areas have high densities of offenders. Using the information that is gathered on offenders who are caught can be vital in helping understand why offenders commit crime, what their motivations are and what attracts them to certain areas and to certain victims. Using this data could then assist in tactical and investigative analysis on other crimes, and used to design a suitable policing response to an area that may have an impact on preventing crime. More important than a tactical response, it can help understand what opportunities the offenders were exploiting so that these opportunities can be denied to future offenders.

Several of the environmental concepts that influence crime are explored in greater detail in the next chapter on strategic analysis, but understanding the environment is also useful for helping understand how offenders behave and how behaviour can be controlled for front-line tactical and investigative requirements. The next section explores offending in detail.

10.2 Understanding offenders

In Chapter 4 we presented many of the key theories that explain why crime happens and how offenders behave. Any analysis of offenders requires the analyst to bear these theoretical principles in mind as they are vital to understanding a crime pattern. This knowledge will often enhance an analyst's ability to interpret any statistically descriptive results. These theoretical concepts help in the interpretation of a number of stages of the SARA problem-solving methodology (see Chapter 3 for a description of the SARA model). An understanding of offender behaviour in space can help police determine, at the scanning stage, which problems are more likely to be part of a pattern than a collection of random events. This understanding also helps in the analysis stage, by focusing the thinking of the analyst on other relevant data. The response should be designed with the thorough understanding of aggregate criminal spatial behaviour, and the assessment of any response should reflect likely criminal behaviour that might be a reaction to the crime prevention activity.

10.2.1 Who commits crime?

Profiling crimes that have been committed based on those that have committed them offers a useful insight to criminal behaviour patterns. In this sense the profile is not focused on a particular individual but instead is focused on a particular crime type or group of crimes that identifies patterns in the behaviour or types of offenders that commit these crimes. These profiles often need to be explored specifically to find common facts about the conditions which lead to crime (Poyner, 1986; Poyner and Webb, 1991; Clarke and Eck, 2003). For example, in one study Poyner and Webb (1991) dissected residential burglaries into two categories – one where offenders had targeted electronic goods and the other where offenders had targeted cash and jewellery. Those that committed cash and jewellery burglaries often did so on foot and targeted properties close to the busy city centre. Burglars that targeted electronic goods such as TVs and VCRs were quite different in their approach, committing crimes in suburban areas that needed a car to transport the stolen items.

An *offending* profile identifies the general characteristics of those that are likely to commit these crimes, and even helps pick out particular individuals who are known to the police and who could become subject to further targeting. In contrast, an investigation of a particular crime (an *offender* profile) would require a good understanding of the type of person that would be a key suspect and requires analysis that: connects the offender to the crime; identifies their motivational factors; and supplies evidence that was non-prejudicial and that clearly supported the case for a conviction.

An offending profile would seek to identify the following:

- comparative age differences in offending;
- comparative gender differences in offending;
- comparative ethnicity differences in offending;
- methods of entry and method of attack;
- their association with other crimes;
- their association with their victims;
- popularity of types of property stolen from an incident;
- hotspots of crime – not necessarily in terms of where certain crimes concentrate but in terms of where certain groups of offenders commit their crimes;
- geographic offender densities (i.e. hotspots of where offenders live); and
- motivational factors for committing crime.

Age, gender and ethnicity profiles would seek to identify if a certain demographic group were more prevalent in committing certain crimes than others. For example, the age profile could group offenders into age categories (e.g. 10–14 years of age, 15–19 years of age, etc.) to produce a count for each group, the gender profile would split the differences between male and female offenders, and the ethnicity profile would seek to group offenders by their ethnic description (e.g. White, Asian, African-American). With these types of profiles it is important to consider comparisons to the population in the area to see if a high number of offenders in one group is not just a reflection of the large number of people who are represented in this group and that live in the local area. Age, gender and ethnicity profiles can reveal if a demographic group is disproportionately over-represented or under-represented as offenders. Rose (2004) has also discussed extending this profile of offenders by viewing offenders as customers, and using geo-demographic data, which are more commonly used in retail marketing, as a means to help extend the profile of offenders, target tactics and direct investigations to areas.

Analysing how a house or car is broken into and the property that is stolen could also reveal useful patterns that help identify particular suspects or the preventative measures that could reduce further offending. For example, patterns of residential burglaries in Liverpool, England, revealed that the alleys that ran along the back of residents' gardens acted as a useful entry point for any burglar. By closing off the end of the alleys using security gates (for which only the residents had the keys) helped to reduce this local burglary problem (Young, 1999). A case study on alley-gating is provided in Chapter 2.

Offenders do not necessarily restrict themselves to just one crime type, but are often prolific in a select few. This tends to happen because certain offenders are more comfortable with committing certain types of crimes rather than others, and after successfully committing a crime, the experience teaches the offender several important lessons that they can add to their specialism when they offend again (Clarke and Eck, 2003). However, even if an offender is prolific in one or two particular areas of criminal activity, analysis of their offending over different crime types could provide additional information that can be used for tactical targeting. In several cases, the act of committing one type of crime may require another to be committed to help them carry out the primary intended criminal act. For example, a series of violent armed robberies that targeted small businesses in Seattle in November 1998 led detectives in the Seattle Police Department (Washington State, USA) to believe these offences were being committed by the same suspects. It was also thought that the suspects used stolen vehicles to help in the act of their robberies. The vehicles tended to be minivans or pickup trucks and were usually stolen shortly before the robberies. Both the theft of vehicles and the armed robberies were being committed within close geographical proximity of each other. Police officers were fully briefed on these offending tactics, so when information was received that a Nissan Pathfinder had been stolen in the area that was under surveillance and that it had been spotted containing two people that matched the possible suspects, police officers moved in and attempted to stop the vehicle. After a chase, the suspects were apprehended, arrested and were found guilty of committing all the violent robberies that had affected the area (Robbin, 2000).

In Chapters 5 and 6 we presented a number of techniques that can aid the process for identifying geographical patterns of crime. Diagnosing these hotspots, based on the activity patterns and the profile of offending in these areas, can provide an important additional dimension to understanding the crime pattern over and above the spatial pattern. For example, Reno (1998) demonstrated that after identifying a recent rise in burglaries in Shreveport, Los Angeles, the simple step of mapping these burglary offences by daytime offences and night-time offences helped in the tactical plan to address this crime problem. The map-based analysis revealed that daytime burglaries were most prevalent and appeared to be linked to offenders of school age, and a possible school truancy problem. The tactical response focused not just on arresting burglary suspects but also correcting the truancy problem. The impact was a 67% reduction in burglary that only saw a slight increase when the school holidays started.

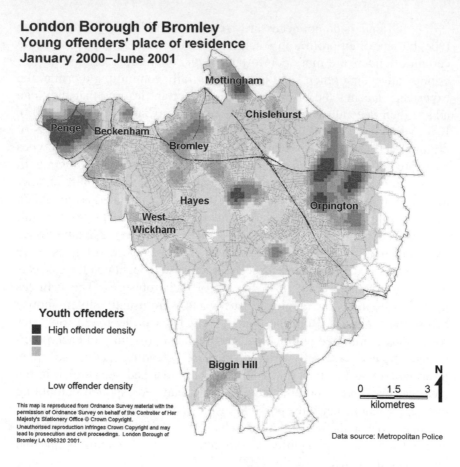

Figure 10.1 Young offenders identified by their place of residence from the London Borough of Bromley's 2001 Crime and Disorder Audit

Mapping does not only have a role to play in identifying where crime happens, but also by mapping where offenders live can reveal if certain areas act as offender hotspots (Figure 10.1). These offender hotspots could be further profiled by age, gender and ethnicity and could be used as a means to target particular individuals that are at a higher risk of committing crime, such as truant school children in the example above.

10.2.2 Understanding the reasons why offenders commit crime

In almost all cases, an offender needs to have some type of motivation to commit a crime, whether it be an offence that is opportunistic or planned.

The theoretical concepts presented in Chapter 4 provide a number of the cues that help understand why and how offenders behave. Some data can reveal trends in the motivational factors of offenders, the products they CRAVED (Clarke, 1999, and see Chapter 4) and their association or relationship with their victims. This information could be subsequently used to help reduce and divert other current offenders, protect those that are vulnerable to being victims or divert those at risk of offending in the future. Often, police data contain very little information on the reasons why an offender committed a certain type or series of crimes. These data are often best sourced through the probation and correctional organisations that may record what these motivational factors were – data recorded as part of the pre-sentence process. Other non-law enforcement data can also help. For example, the 1998/99 British Youth Lifestyles Survey (Flood-Page *et al.*, 2000) identified the most predictive risk factors of 12 to 17-year-olds becoming persistent or serious offenders. If a young person showed evidence of any of the following (listed in order of the most serious risk factors first), then they were vulnerable to turning to crime if they had not done so already:

1. Drug user (has used drugs in the last year)
2. Disaffected from school
3. Hanging around in public places
4. Delinquent friends or acquaintances
5. Poor parental supervision
6. Persistent truant (at least once a month).

Similar research in the UK by the Joseph Rowntree Foundation (2002), Crime Concern (2002) and Flood-Page *et al.* (2000) for the Home Office has revealed additional characteristics of offending, including the age when offending begins and the impact that family life (e.g. evidence of parental criminality), schooling (e.g. low educational achievement), peers (e.g. peer involvement in problem behaviour), early adulthood (e.g. lack of skills or qualifications) and community (e.g. high percentage of children in the community) have on the increased likelihood of becoming involved in offending. This type of survey data offers a useful foundation for exploring offending profiles, which when conceptualised with demographic and socio-economic data from a census or geo-demographic statistics offers a rich information source for exploring correlations with those areas where large numbers of offenders appear to reside (Ashby, 2004).

Research and analysis of risk factors, offending characteristics and offending profiles, when combined with a spatial analysis, are useful for understanding the nature of crime that is committed by individuals. They are also useful in understanding how certain preventative measures can be designed to reduce the risks of future offending, control behaviour and divert those vulnerable to offending into more constructive activities.

10.3 The journey to crime

Offenders are restricted in how they move around space and, as explained in Chapter 4, they tend to follow basic rules in regard to their movement. The behavioural axiom of the least effort principle suggests that people will usually exert the minimum effort possible to complete their tasks, whether those tasks are shopping, performing recreational activities, visiting friends, travelling to work or journeying to crime. This therefore implies that there is a decay in the frequency of activity as distance increases. In other words, in the routine of performing daily activities there are more tasks completed with short trips, and tasks that require longer distances to be made are completed with lower frequency. In essence, this theory means that offenders tend to travel short distances on average to commit their crimes. This distance decay can be represented as a function as shown in Figure 10.2. The routine activities that people take (be they offenders or potential victims) that we introduced in Chapter 4 also need to be considered when understanding the routes and movements that offenders take to commit crime. Appreciating the likely distance that an offender may travel to commit their crime is an important component in tactical and investigative analysis.

Figure 10.2 shows distance decay as a simple linear relationship. This works as a basic theoretical concept; however, the reality of distance

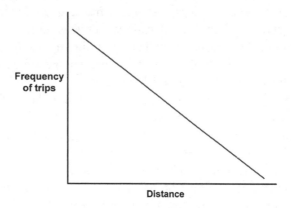

Figure 10.2 The distance decay function – the routine of performing activity results in more frequent shorter trips to complete tasks than longer trips

decay is often more complex. Real distance decay functions can vary in shape and linear orientation (e.g. the decay could be exponential) in response to different types of movement behaviour (Longley *et al.*, 2001). For example, shopping trips can be divided into two general categories.

1. Convenience shopping is characterised by a high frequency of short trips because people will tend to purchase items such as milk or a newspaper from the closest possible source.
2. Comparison shopping occurs when buyers are seeking more expensive items such as electrical appliances, furniture or cars. Theoretically, any longer trips in search of these more expensive goods are seen to be worthwhile because the financial savings from better deals or the ability to review a greater range of products will off-set the longer distance travelled (Harries, 1999).

These differences in distance decay could also be applied to crime where opportunity-based criminal behaviour may be more frequent over shorter distances, compared to organised and planned criminal behaviour that may require travelling further distances to complete criminal acts.

Frisbie *et al.* (1977) performed one of the pioneering analyses of journey-to-crime distances in Minneapolis and showed that:

- more than 50% of residential burglary offenders travelled less than 0.5 miles from their homes to their crimes;
- Commercial burglars tended to travel slightly further, committing 50% of incidents within 0.8 miles; and
- Stranger-to-stranger assaults and commercial robbers tended to travel the longest distances, committing 50% of their crimes within 1.2 miles of their home location.

A British study by Wiles and Costello (2000) in the City of Sheffield drew similar findings, noting that:

- offenders of burglaries to residential properties on average travelled 1.88 miles from their home to their crimes;
- shoplifters travelled an average of 2.51 miles from their home to their crimes;
- theft-from-vehicle offenders travelled on average 1.97 miles to their crimes;
- those that committed theft of vehicles on average travelled 2.36 miles to their crimes;
- young offenders tended to travel shorter distances than older offenders; and
- although distances travelled to crime had increased slightly over a 30-year period, journeys to crime are still frequently short.

For a comprehensive summary of journey-to-crime studies see Rossmo (2000, pp. 105–110).

10.3.1 Distance measurement techniques

The methodology for measuring the distance between a home location and a crime is usually the Euclidean approach. This refers to the straight-line distance between two points, the offenders home address and the crime location (Figure 10.3a). It can be calculated in a GIS using standard SQL functions or using Pythagoras Theorem in a spreadsheet software package. Readers with a broader background in GIS may also be aware of distance measurements using the great circle approach, one that takes into consideration the curvature of the Earth. For crime mapping purposes, there is little value in this because of the generally small distances involved in criminal acts. Three other approaches can also be applied to measuring these distances (other measures do exist but these are the most common; see Webb, 2002 for a comprehensive description of distance measurement functions).

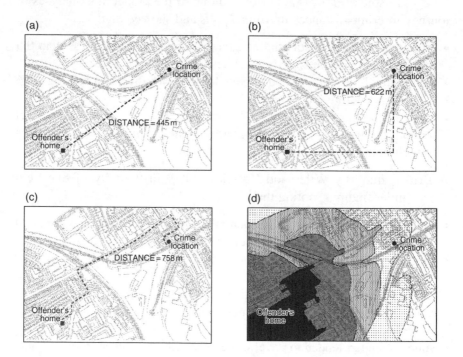

Figure 10.3 Methods for measuring journey distances to crime: (a) Euclidean, (b) Manhattan, (c) Street route and (d) Distance as time. Each method produces a different distance result

1. *Manhattan distance* calculates the shortest distance between two points, with distance being constrained to horizontal or vertical directions (Figure 10.3b). This distance can be calculated in some GIS or in a spreadsheet package such as Microsoft Excel.
2. *Street route distance* calculates the shortest path distance following the street network from the offender's place of residence to the crime location (Figure 10.3c). The calculation of this route requires specialist street routing data and quite often specialist software.
3. *Journey time distances* are a measurement of the time it takes to travel a distance (Figure 10.3d), recognising that travel time can be different on roads with different speeds, and that the mode of travel influences travel time. This type of journey distance measurement has been commonly applied to store location analysis and other forms of accessibility modelling (Birkin *et al.*, 1996; Liu and Zhu, 2004) and is usually represented in geographical form as isochrones – lines connecting points of equal travel time.

The Euclidean and the Manhattan routes do not take into account any of the physical barriers that may obstruct the journey to crime, such as railway lines (as shown in Figure 10.3), rivers, lakes, buildings or open spaces that are not accessible. This often leads many to assume that the street route tends to be the more accurate (and longer) of the three Cartesian distances between the offender's residence and the crime location, assuming that the offender uses the street network to travel to the crime. The offender may of course take a short cut across an area of parkland or along a railway siding (if travelling on foot), yet not enough is known about these movements to robustly model in a GIS. In a study in Croydon in South London (Chainey *et al.*, 2001), Euclidean, Manhattan and street-route distance methods were used to measure journey-to-crime distances. These results were consistent with other research confirming that most journeys to crime are short in distance. When compared by distance measure method the Euclidean method consistently produced the shorter distance, whereas the street route and Manhattan measures were very similar. Contrary to expectations, the street route measure was not the longer distance, being consistently marginally shorter than Manhattan distance in all crime types for which journey-to-crime measurements were calculated (Table 10.3). While many strive to use the street route method as the more accurate means for measuring journey-to-crime distances, its requirement for specialist street routing data and software often prevents its use. However, due to the consistency in distance differences between the Manhattan distance and the street route distance (where the

Table 10.3 Journey-to-crime distances for a range of crime types using Euclidean, Manhattan and Street routing distance methods. Source: Chainey et al. (2001)

Crime	Number of offences	Mean Euclidean distance (km)	Mean Manhattan distance (km)	Mean Street route distance (km)	Ratio in distance between Euclidean and Manhattan	Ratio in distance between Euclidean and Street route	Ratio in distance between Manhattan and Street route
Robbery of the person	1175	3.088	3.854	3.831	0.80	0.81	1.01
Burglary to a residential dwelling	185	2.897	3.716	3.691	0.78	0.78	1.01
Theft from vehicle	280	3.950	4.964	4.892	0.80	0.81	1.01
Theft of vehicle	309	3.441	4.371	4.267	0.79	0.81	1.02
Criminal damage	495	2.391	3.006	2.960	0.80	0.81	1.02
Shoplifting	1332	3.108	3.872	3.784	0.80	0.82	1.02

ratios from the Croydon study ranged between 1.01 to 1.02 across six crime types), the Manhattan distance may be an acceptable proxy for the more complicated calculations. The Manhattan distance is easier to calculate than the street route method, is comparable for most crime mapping purposes and provides greater accuracy than the Euclidean distance which consistently underestimates the actual journey-to-crime distance by approximately 20%.

10.3.1.1 Journey-to-crime considerations

The most common anchor point from which the offender travels, when analysing offender journey distances to their crimes, is their home address. Other anchor points, such as their place of work, frequented premises (e.g. bars and nightclubs) and, particularly for analysing journey patterns of young people, school location are all additional locations from which to consider the point the offender journey starts. For younger offenders, the strong links between offending, truancy and exclusions of young people may de-emphasise the importance of the school location in studying young offenders' journey-to-crime distances. However, their networks with friends that may also truant or be excluded may be important anchor points of young offenders. For adult offenders, Wiles and Costello (2000) note that they often lack a stable workplace or cannot afford to take part in formal social activities, therefore the only other usual anchors may include a family or friend's house (this additional consideration explains our inclusion of a friend's house in Figure 4.3). Data for these other anchor points are, though, often difficult to include, so journey-to-crime analysis is often restricted to a starting point of the residential addresses. This does lead us to suspect that actual journey-to-crime distances for many offenders may be even shorter than estimated, if all possible anchor points are considered.

Certain other factors will also influence an offender's perception of distance. These may include the availability of transportation, barriers that obstruct their mobility (e.g. bridges, railway lines, motorways or highways) and familiarity with a specific region. For example, the 'tower-block rapist' in Birmingham, England, did not have a car (Canter, 1994). This meant that he needed to take into account the distance to walk home from a crime scene. Ted Bundy, a serial rapist and killer who is thought to have committed 36 crimes in the United States did have a car which meant that his mobility was less impaired and saw him travel across the USA to commit his crimes (Winn and Merrill, 1979; Michaud and Aynesworth, 1989).

Case study: The journey to crime and the 'self-containment index'

This case study is an adaptation from Clarke and Eck, 2003. Reproduced with permission of the Jill Dando Institute of Crime Science, University College London

In a study in the West Midlands of England, analysis of over 250000 offender journeys by crime analyst Andy Brumwell revealed:

- 50% of journeys to crime were less than one mile in distance. This ranged between 0.5 miles for 50% of arsonists compared to only 13% of shoplifters committing their offences within a mile of home.
- Females tended to travel further than males, although this was likely influenced by the high proportion of activity in shoplifting.
- When offenders worked with co-offenders their journey distances tended to be longer.
- And, young people tended to not travel as far as older people to commit their crimes; however, this could have partly been influenced by the types of crime they committed.

Based on this journey-to-crime analysis Brumwell has developed the 'self-containment index' that compares offenders who live in an area with the crimes they commit in an area. An index value of 100 indicates that offenders who live in the area commit all the area's crime. An index of zero would indicate that none of the offenders who live in the area commit any of the area's crime. An index of this type is useful in helping to design the type of tactical and strategic response to target into the area. For example, if an alley-gating scheme was introduced into an area that had a high index score for residential burglary, the scheme may result in being less successful than hoped because many of the offenders who commit the local burglaries would themselves have access within the gated scheme.

10.4 Geographic profiling

Where the victim does not know the offender, the ensuing criminal investigation can generate huge volumes of information, particularly when the crime or crimes are serious in nature. For example, in the English Yorkshire Ripper investigation, officers and detectives accumulated 268000 names, visited 27000 houses and recorded 5400000 vehicle registration numbers, all of which were thought to have some connection with the investigation

(Nicholson, 1979). Any tool that can filter the more meaningful information from such data mountains is usually very welcome! Behavioural and psychological profiling act as tools to assist investigations overcome the problems of information overload. Much emphasis is placed on the physical or behavioural clues that can be gleaned from victim interviews or discovered at a crime scene, yet one of the most overlooked pieces of information is the location where the crime occurred.

Geographic profiling is an investigative methodology that uses the locations of a connected series of crimes to determine the most probable area where an offender lives (Rossmo, 2000). The birth of geographic profiling emerged from a particular crime investigation that initiated the idea that an offender may target victims specific to their geographical locations. From the late 1960s until the mid-1980s the Zodiac Killer operated in and around San Francisco, killing as many as 37 people. He was never caught despite his frequent communications with the press and police, including an interesting map he sent of Mount Diablo. Examination of this map revealed all the murders to lie along a line. This showed that the victims appeared to have been chosen not because of who they were but because they just happened to be in the wrong place at the wrong time. After theoretical developments and research by Brantingham and Brantingham (1981), Canter and colleagues (Canter and Larkin, 1993; Canter and Gregory, 1994; Canter, 2003) and Rossmo (2000), geographical profiling was developed to help police locate serial killers, rapists and arsonists, and today is also being applied to many other crime types such as robbery and burglary where an unidentified person is known to have carried out crimes at a series of geographic points (Chainey, 2002).

Geographic profiles are comprised from both quantitative and qualitative components. The quantitative aspects of the profile use a series of scientific geographic routines and quantitative measures to interpret the point pattern created by the offender's crime site locations. These crime locations can include not only the offence location, but also the sites of victim abduction, assault and release (or body dump), if these locations are known to police. The qualitative elements of geographic profiling aim to understand and reconstruct the offender's behaviour and movements across the areas they are familiar with and feel confident. Importantly, the qualitative element also aims to ensure that the crime incidents that are reported are part of a linked series. A geographic profile can also be helped by bringing in other geographic layers of information. For example, demographic, geo-demographic (Rossmo *et al.*, 2004), land use, deprivation data, licensed premises, school locations and the locations of crime hotspots can be added to help inform the geographic profile and support investigation efforts.

The resultant geographic profile can be used in conjunction with various investigative strategies to increase the efficiency and cost-effectiveness of the investigation. These could include:

- Suspect prioritisation based on the cross-referencing of the geographic profile with suspect lists;
- Analysis against known offender registries, parole, youth offending services or probationary information records;
- Comparison with DNA databanks on a spatial basis;
- Patrol saturation and operational stakeouts; and
- Neighbourhood canvasses, including targeted postal mail-outs based on the prioritisation of target areas.

Before describing in further detail the processes involved in creating a geographic profile it is worth reviewing the key theoretical concepts that led to its development and application.

10.4.1 The theory behind geographic profiling

The underlying principles that led to the development of geographic profiling draw from a number of theoretical and analytical concepts. Although these concepts and approaches have already been presented in this book (Chapter 4), it is worthwhile to briefly review and describe them in the context of serial offending. The theoretical principles behind geographic profiling include: Rational Choice Theory, Routine Activity Theory, Mental Maps and Journey-to-crime analysis.

- *Rational Choice Theory* draws on the concept that individuals act as rational human beings in the actions they perform. It is a theory that is used to help answer the question 'Why do people engage in deviant and/or criminal acts?' The focus of Rational Choice Theory is on the individual's personal choice and the concept that individuals act rationally about their personal choices (Brantingham and Brantingham, 1981).
- *Routine Activity Theory* tries to help explain why crimes occur. Routine Activity Theory argues that three things must converge at the same time and in the same space, in order that a crime has the potential to occur:

  - availability of a suitable target
  - lack of a suitable guardian to prevent the crime from happening
  - presence of a likely and motivated offender (Felson and Clarke, 1998).

  In other words, for a crime to occur a likely offender must find a suitable target in the absence of a capable guardian.

- The *Mental Map* is a cognitive image of one's surroundings developed through their experiences, travel routes, reference points and centres of activity (Brantingham and Brantingham, 1981). Places where people feel safe are usually located within their mental maps and the same is true for offenders. The more an offender lives or offends in an area, the more confidence they gain in that area, and the more their crime area tends to expand. This suggests that the acts of an offender are likely to be close to where the offender lives or relates to an area for which they have extensive knowledge (e.g. the area in which they grew up, they previously lived, where they work or where a relative lives). As offenders grow bolder, their maps may change and they may then increase their range of criminal activity, or become more confident by committing crime closer to home.

    The mental map may also differ depending on the type of offender an individual may be. Popular definitions for offenders include:

    - Hunter (searches for a specific victim in the offender's home territory)
    - Poacher (travels away from home for hunting)
    - Troller (opportunistic encounters while occupied in other activities)
    - Trapper (creates a situation to draw a victim to them) (Rossmo, 2000).

- *Journey to crime* – The distances that offenders travel to their crime play an important role in geographic profiling. The concept of distance decay suggests that people will usually exert the minimum effort possible to complete tasks of any kind. This suggests that it is more likely that offenders live close to the sites of their crimes than far away. However, in the case of geographic profiling it is also suggested that a buffer zone exists around the offender's immediate anchor point within which the frequency levels are slightly lower (Brantingham and Brantingham, 1981). As distance from the anchor point increases, frequency in activity begins to increase until reaching a modal trip distance, before following a distance decay effect (Figure 10.4). Applying this concept of buffered distance decay in geographic profiling involves overlapping distance decay functions that are centred on each crime location to produce a resulting probability surface that indicates the relative probabilities of offender residence across different areas.

Case study: Geographic profile for Operation Lynx

This case study is based on material in Chainey, 2002

Operation Lynx was the largest police manhunt in Britain since the Yorkshire Ripper case. The operation focused on a series of unsolved rapes in the

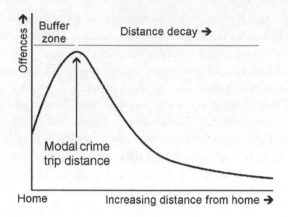

Figure 10.4 Distance decay function with buffer

Leeds area of Northern England. The police investigation had managed to collect DNA from one of the crime sites, and a partial fingerprint. Analysis of the DNA revealed it was not on the national database, and the print fragment was too small to search using the Automated Fingerprint Identification System. The investigative analysis suggested that the offender was more likely to have robbery or fraud in his background than sex crimes.

Police efforts were focusing on suspects in the Greater Leeds area, but where the population was of several million people. A geographic profile was used to assist the investigation and help narrow down the police search for suspects. This identified a much smaller area to focus efforts. The investigation then turned towards identifying those police stations that fell into the geographic profile's high score areas as a means of prioritising the search for useful intelligence that may help identify key suspects. Fingerprints from all robbery and fraud offenders were hand compared against the partial print. A match occurred at the second police station on the list, and DNA confirmed the suspect as the rape offender.

10.4.2 Geographic profiling in practice

Several computer systems have been developed to assist in the analysis of geographic profiling. The Criminal Geographic Targeting model built into the Rigel software (Rossmo, 2000) incorporates mathematical models of known offending movement patterns and hunting behaviour, journey-to-crime distances and includes a method to calculate the relationship between sets of crime locations (e.g. contact, assault, release sites) and

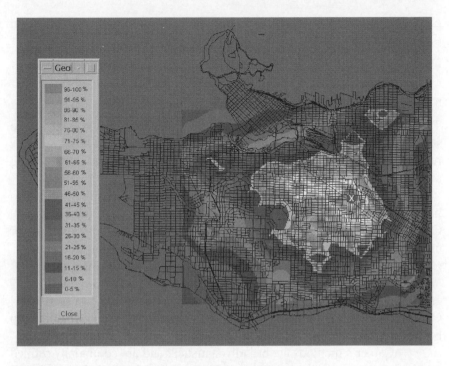

Figure 10.5 An example of a jeopardy surface produced as a result of a geographic profile. The area of highest probability of where an offender lives is shown by the dark red and orange areas. The white cross on a blue circular background positioned at this location is where the offender actually lived. Source: Rossmo (2000) (see Plate 3)

offender residence. The software process assigns scores to the various points on a map that represent the offender's hunting area. These scores are based on the crime site locations and certainty in the linkages between crimes. The map output of a Rigel geographic profile, called a 'jeopardy surface', identifies the most probable area of offender residence or anchor point (Figure 10.5). Rigel allows for the adoption of existing data used to create the profile (as new intelligence becomes available) and for the addition of new information from new crimes that enter the series.

Calculating the probabilities is a four-stage process. First, the offender's hunting area is defined, based on the extent of the crime sites linked to the series. This is essentially the minimum-bounding rectangle for the crime series. Next, the distance from every crime on the map to every other crime on the map is measured. These measurements are calculated using the Manhattan distance method. The third stage of the process is to use these distances as the independent variable in a function that produces a value that:

- if the point lies outside of a buffer zone (shown in Figure 10.4), uses a distance decay function to generate a smaller value for greater distances; or
- if the point lies inside the buffer zone, becomes larger for greater distances.

The value is calculated using this function for each crime location in the series. The final stage is for these multiple values to be added together, resulting in a single score for each map point. The greater the resulting score, the higher the probability of offender residence.

An alternative methodology for geographically profiling offenders has been developed by David Canter from the University of Liverpool. Canter is a professor of psychology and has advised on a number of major crime investigations on the likely psychological profile of the offender. Canter uses the principles of environmental psychology to make deductions about an unknown offender's behaviour and extends this to include inferences about where that offender may be located (Canter and Gregory, 1994; Canter, 2003). Environmental psychology examines the interrelationship between environments and human behaviour.

Canter's methods are broadly statistical and are continually refined as his database of offenders grows. By studying these statistics, Canter has defined broad offender groups, and the crimes and behaviour of the unknown offender are compared to this control group. The outcome is a list of characteristics that are likely to be possessed by the unknown offender by virtue of their similarity to the known offender group.

As a psychologist, Canter's methodology is not confined to understanding the geography of an unknown offender, but encompasses the areas of psychological profiling, as well as that of geographic profiling. In terms of analysing the geography of offenders, Canter has developed a model of offender behaviour known as the 'circle theory' which incorporates two models of offender behaviour – the 'marauder' and the 'commuter'. The marauder model assumes that an offender will 'strike out' from their home base to commit crimes locally whereas the commuter model assumes that an offender will travel a distance from their home base before engaging in a criminal act. The circle theory concept is demonstrated in Figure 4.4.

Canter's theoretical models are implemented within a software system called Dragnet (Canter and Larkin 1993; Canter and Gregory, 1994; Canter, 2003). Dragnet produces a type of map output similar to that of Rigel, seeking to identify the area in which to begin the search for an unknown offender, derived from the location of the offender's crimes.

The Rigel and Dragnet methods for geographic profiling differ quite considerably but both have been used around the world on numerous high-profile cases. At present, little published research exists that independently assesses the relative accuracy of geographic profiling methods. Geographic profiling is often quoted as helping an investigation to become focused to a small area of the community, thus reducing the level of resources that may be required for an investigation. It also helps with prioritising suspects and developing strategies for linkage analysis of information. However, it is also argued that geographic profiling relies too strongly on analysis of a single manifestation of behaviour, cannot distinguish between two similar offenders operating in the same area and relies on the subjective interpretation of data and so is prone to the interpreter's experience and skill in linking together a series and interpreting results (Rich and Shively, 2004).

Paulsen (2004) has begun work to investigate the use of the Rigel and Dragnet geographic profiling methods in comparison to simple centrographic statistics and other forms of spatial modelling. This work is still in its infancy, but is an excellent example of the types of critical review that need to be conducted more frequently on all crime mapping methods, not just geographic profiling. Both Dragnet and Rigel have been used successfully to assist a large number of serial investigations, and as reviews of the type conducted by Paulsen begin to report their findings it can only help in developing the robustness of crime mapping tools that are used in policing and crime reduction.

10.5 Using maps as evidence

As crime mapping has developed to support crime analysis, map outputs from spatial analysis are increasingly being called upon to support the presentation of evidence in court. Prosecutors have begun to realise that maps can offer a clear, powerful, unbiased and persuasive device to help them in the presentation of their case material. In particular, maps can be useful for helping to describe a complex sequence of events that take place over space and time, corroborate other evidence (such as eyewitness statements) and counter any alibis that a suspect is using (Schmitz *et al.*, 2000). In all cases, the map as a courtroom exhibit needs to be designed so that the information it portrays is factual, impartial and not construed as being prejudicial against the defendant.

Maps produced for the courtroom are likely to go through a number of iterations prior to being considered an ideal exhibit that a prosecutor can use to help deliver their case. Indeed, producing a map as an exhibit should be a process that includes consultation with not only the prosecutor but also the detective and other officers who are working on the case. This collaboration will ensure that the necessary adjustments and corrections can be made to the maps so that it can become a convincing and strong piece of evidence. In Chapter 12 we list items to consider in map design in greater detail. Features important for the court include not only that maps should be clear, cartographically accurate and well designed, but also that they are of a good size for all to see. Indeed, most maps produced for the courtroom should use a large-size plotter to generate the output and be mounted on boards.

Maps used in the courtroom frequently prove the adage that 'a picture paints a thousand words'. Maps can often act as an important framework around which a prosecutor's case unfolds, helping the members of the jury and the judge to better understand what can often be a complex and technical investigation of evidentiary material.

Case study: Using maps as evidence in a murder trial in Florida

This case study is an adaptation from Moland, 1998. Reproduced with permission from the US Police Executive Research Forum

The discovery of a partially dismembered and beheaded male body near the Florida Pier in St Petersburg launched a lengthy and complex investigation to identify and convict the 19 year old victim's killers. The victim had been killed as a result of mistakes he made in a drug deal, and the two primary suspects that the police identified were the drug dealers whose profits were hit hard from the bungled deal.

The investigation disclosed that the young man paid for his mistake with his life, and how the two suspects plotted to meet him, take him against his will to a secluded area, and how they tortured him and carried out a execution-style killing using a Samurai sword. Immediately after the murder the suspects dumped his body into Tampa Bay and his head in a river near the crime scene. The Samurai sword was thrown into a different waterway. With very little in the way of eyewitness statements to use as evidence, it was the type of case for which it is often difficult to achieve a successful conviction. The forensic evidence was minimal, with the main challenge of the investigation being to prove the connection of the suspects to the victim and his murder.

A method that one of the investigators was familiar with was the ability to track movements from cellular telephone calls. When one of the suspects was arrested, he had with him his cellular phone. By reviewing the suspect's cellular phone account and the phone's memory bank the investigator believed it would be possible to determine the movements of the suspects before, during and after the time of the murder, and that the trail left by the phone would help to establish connections between the suspects, the victim and the places they visited during this time period. This map trail would also serve to prove or dispute the alibis that the suspects had recorded in their statements to the police.

Producing map information that would be robust and admissible in court would require careful checking to ensure that the information was accurate and non-prejudicial. This included the requirement to verify that each of the cellular telephone antennae were working correctly. The other major challenge was to as clearly as possible describe to the jury the concept of the cellular phone map trail and how the cellular phone antennae could locate an individual using a cellular phone as they moved about the area.

To get the maps exactly right required the crime mapper to work closely with the investigating detectives and the prosecutors. The mapping outputs went through a number of iterations before the prosecutors felt that what they had was an exhibit that would help the presentation of their case, when the evidence that required explaining was complex, technical, but a core part of their strategy to achieving the convictions.

The final map products that were presented in court allowed the prosecutors to use visual cues to unfold their case against the suspects, and helped to improve the jurors' understanding of all the important case material. The maps clearly described the suspects movements, where and when they were and the directions they moved as they made calls on the cellular phone, verifying, and corroborating with eyewitness testimonies the suspects movements and connections to all the key sites that were involved in the murder. The map products provided an easy to interpret and digestible platform from which to present the main investigative and evidentiary material, as well as overcoming the difficulty in explaining the technical evidence provided from the cellular phone records. Indeed, the prosecutors, the members of the jury and judge all commented on how the map products improved their understanding of the case, which resulted in convictions of both of the suspects.

10.6 Detecting offenders through their self-selection

For all the crimes that Al Capone was alleged to have committed, it is well known that he was eventually convicted on tax evasion charges. Arresting and putting Capone behind bars exploited a weak spot in Capone's arsenal. Even with many more advanced tools at the disposal to police today, catching and convicting an offender is difficult to achieve. However, in a similar way to exploiting Capone's lack of proper accounting processes and completion of his tax returns, new ways are being found to help identify, arrest and convict active criminals known to the police.

A problem with operations that target particular offenders is the constant need to update and be accurate with profiles that describe the behaviour, actions and locations of those that are frequently committing crime. In addition, excessive targeting could degenerate into harassment, which could result in additional problems if those targeted are not current offenders or if the policing activity also impacts the families of the known offenders (Chenery *et al.*, 1999). Targeting offenders through 'sting' operations also requires offenders to present themselves through their criminal activity in order for the police to then respond.

A subtle way of offender targeting is when offenders 'self-select'. Self-selection is based on the hypothesis that 'people who are the most committed criminals are also the most versatile, and will not willingly be bound by law or convention of any kind' (Chenery *et al.*, 1999, p. 1). In other words, those that commit crimes break other rules, rules that non-criminals may also break, but rules that a high proportion of active criminals break quite often. Identifying these rule breakers may also reveal those that commit more serious criminal activities. For example, Kelling and Coles (1996) identified that a substantial minority of Squeegee merchants in New York City also had outstanding warrants for other criminal offences. Their unsolicited activity in cleaning car windscreens self-selected them as worthy of targeting, based on the value of exploring if they had warrants out for more serious crimes. Another example comes from the work of Gerry Rose (Rose, 2000), who identified that those that repeatedly commit serious traffic offences are likely to commit mainstream offences as well, finding that 50% of dangerous drivers had a previous conviction and 30% had a conviction for car theft. It is also useful to remember that the action that led to the arrest of Peter Sutcliffe, the Yorkshire Ripper, was not as a result of a sting operation, but as a result from being picked up for having suspicious car registration plates.

Offenders that self-select are seen as an ideal of offender targeting – 'actions disproportionately undertaken by prolific offenders provide means of inducing self-selection' (Chenery *et al.*, 1999, p. 1). Geography plays an important role in self-selection, and how the activities of individuals in space can be exploited for targeting offenders. This includes not only targeting *activity locations* where the actions of individuals lead to self-selection, but also through targeting their actions performed at *non-active locations*. These two concepts are explained below.

Activity locations are those where an offender self-selects by travelling to that location and performing an action. Squeegee merchants are an example of this but it also extends to targeting offenders that self-select through:

- *Illegal parking* – many people that park in 'no parking' areas are not active criminals, yet a significantly high proportion that park in these areas are, and are versatile criminals. This self selection could warrant gathering information that selects them for offender targeting. The case study below reveals how current criminals were prone to self-selection by illegally parking in disabled parking bays (Chenery *et al.*, 1999).
- *Driving in bus lanes or car pool lanes* – the logic of rule breaking and self-selection of active criminals could also apply to illegal use of car pool lanes or bus lanes.
- *Fare evasion on public transport* – a significant minority of those who evade paying for a train ticket or a bus fare are also believed to commit other crimes (Newton, 2004). Targeting offenders that self-select through this action could be performed by focusing on locations such as train stations or particular bus routes where there are high rates of fare evasion.
- *Repeat victimisation* – those who repeatedly commit crime against the same target have stronger criminal tendencies than those who commit one-off offences against a target (Ashton *et al.*, 1998; Gill and Pease, 1998; Everson, 1999). These repeat offenders therefore self-select by returning to the same location or target.

The notion of self-selection through non-active locations concerns offenders volunteering themselves by revealing their residential address through an action they have not performed at this address. This concept suggests that offenders could be targeted through their self-selection in the following ways:

- *Failing to pay vehicle road tax* – in the United Kingdom, and similar in many countries around the world, a vehicle licence or registration tax is required to be paid to use a car on a public highway – the 'road tax'.

Driving registration databases contain the home address of car owners. If the car's road tax is unpaid but is spotted on the public highway, the action can reveal the residential address of the car owner. The logic from this is that a significant minority of those that fail to pay their road tax may also be involved in other criminal activity, thus road tax evasion could offer a source of self-selection.

- *Failure to complete a census return form* – the failure to complete a census form is a criminal act in many countries around the world. Logic suggests that if an offender wants to conceal their identity, then an action they will not complete is the completion of their personal details on a census form. If census returns could be matched against a definitive residential address file then what could be revealed are those addresses that did not generate a census form. These addresses could be targeted for further investigative action on the grounds of the failure to complete a form. This information could potentially self-select offenders of other unsolved crimes, or even as one way to begin the arduous task of identifying terrorist cells.

- *Failure to pay for a television license* – in the United Kingdom, anyone that owns a television must also, by law, possess a TV License. The UK's TV Licensing department claim to know all those addresses that do not hold a TV License but who own a TV. Indeed the UK's TV Licensing department release a League of Shame each year of those cities and towns that head its lists (e.g. In 2004, London topped the table with 37 298 evaders). By using their database of over 28 million addresses (a database which records whether a property in the UK is licensed or unlicensed) they can dispatch enquiry officers to only visit properties where there is no record of a TV Licence. The department use detection equipment including detector vans and handheld scanners to conduct surveillance on suspect properties that are not on their list. If the TV Licensing department know where their evaders live then it is possible that a significant minority of current offenders that are of interest to the police may also self-select by volunteering their residential address through their actions of failing to pay for a TV License.

Some caution, however, is levied at expecting all ideas of self-selection to return significant results. Recent work on penalty notices served to the owners of vehicles that committed an offence in the West Midlands of England found that this approach did not appear to be an efficient predictor of future, more serious criminality. This research did, though, offer some indication that the issue of multiple vehicle penalty notices may produce a better self-selection hit rate (Wellsmith and Guille, 2005).

Case study: Self-selection of offenders through illegal parking in disabled parking bays

This case study is an adaptation from Chenery et al., 1999

Chenery *et al.* (1999) embarked on an initiative in Huddersfield, England to assess the scope for self-selection of current offenders through their actions of illegally parking in disabled parking bays. Their hypothesis being that 'such parking will disproportionately be a practice of active offenders' (Chenery *et al.*, 1999, p. 1). In the UK, disabled parking spaces are reserved for vehicles bearing an orange badge on the windscreen indicating that the driver or passenger has a disability. Illegally parking in one of these spaces is regarded as a particular act of selfishness, as well as criminal.

Traffic Wardens were used to collect certain items of data over a 5 month period on any vehicle that they found was parked illegally in a disabled parking bay. This data included:

- the status of the road tax license;
- the condition of the tyres; and
- the registration mark.

For every illegal vehicle checked, the traffic wardens were required to also collect this same information from the nearest legally parked vehicle. This data could then be used as the control group to compare findings between the two types of parking venues. They also collected data on how difficult it was for any car that was parked illegally to park legally in a nearby space.

The registration details for both groups were then checked with the Police National Computer (PNC) and the local police forces' computer-based intelligence systems. This information was categorised into five groups:

1. Immediate police interest: The vehicle in question was a stolen vehicle, the registered keeper was wanted and/or the vehicle 'did not exist' (i.e. it was a cloned vehicle)
2. The registered keeper had a criminal record
3. The car had a history of traffic violations
4. The car was known or suspected to have been previously used in the commission of crime
5. The car was currently 'illegal': The vehicle road tax license was absent, photocopied or had expired, and/or the tyres were defective (Chenery *et al.*, 1999).

Table 10.4 Illegal parking in disabled bays by each category ($n = 178$). Source: Chenery *et al.* (1999)

Category	Illegally parked (%)	Legally parked	Significance level
1. Immediate police interest	21	2	0.001
2. Criminal record of keeper	33	3	0.001
3. Vehicle's history of traffic violations	49	11	0.001
4. Past use in crime	18	0	0.001
5. Current vehicle illegality	11	1	0.005

The primary group of interest was category 1 as this would self-select offenders for targeting. In addition, category 5 would also enable a police action due to law breaking. Categories 1 and 5 together would indicate the number of cases where a police action could be taken. The results (Table 10.4) revealed that:

- 'One in five of those illegally parked in a disabled parking space would occasion immediate police interest, contrasted with 2% of legally parked vehicles
- One in three keepers [of cars] illegally parked in a disabled space have a criminal record, contrasted with 2% of legally parked vehicles
- Half of those vehicles illegally parked in a disabled space had a history of traffic violations, contrasted with 11% of legally parked cars
- One in five of those vehicles illegally parked in a disabled space were known or suspected to have been previously used in crime. None of the legally parked cars were
- One in ten of those vehicles illegally parked in a disabled space were currently in an illegal condition, compared to 1% of the legally parked cars
- One in four of those vehicles illegally parked in a disabled space were [categorised] as to require or justify police action [categories 1 and 5 combined], as contrasted to 2% of the illegally parked cars' (Chenery *et al.*, 1999, p. 3) .

Following the study, traffic wardens routinely made checks on the local police force's Offender Information System and if anything of interest was found, daily reports would be completed and returned to the police intelligence office. This process revealed that one in three vehicles parked illegally were connected to other offences, ranging from unpaid parking tickets, drugs, assault, vehicle crime, theft and burglary.

Self-selection can act as an effective means for identifying, targeting and detecting offenders for their criminal activities through the legitimate means of operating police actions based on their other rule or law breaking actions. Self-selection does, though, rely on the need for accurate data collection and quite often partnership collaboration, but for the additional effort that is placed on refining and adapting to new ways of working, it can offer significant returns in detection and crime reduction. Geographical analysis and enquiry play an important role in self-selection. Identifying the types of activities, when self-selection is successful, would then operationally result in targeting those locations where these activities take place. Similarly, by active criminals not performing an activity from their home, they can potentially self-select themselves by revealing their residential address.

10.7 Summary

This chapter has presented many of the ways in which crime mapping can play a role to support tactical and investigative analysis. These forms of analysis mainly focus on identifying, understanding and detecting criminals, both in terms of exploring their behaviour patterns and movement around space, their profile and motivational factors, and the influence that their surrounding environment has on their actions. The analytical methods presented are not an exclusive list for tactical and investigative analysis, but instead identify and demonstrate the general role that geography and crime mapping can play in these forms of policing and crime reduction. Offending profiles of crime types offer rich opportunities to understand who are likely to commit certain types of crime, their behaviours and what motivates them. Understanding the routes and spatial movements that offenders take to commit their crimes helps to refine the targeting of a tactical response to deal with crime problems, and geographic profiling uses the clues of the locations where crimes happen to support investigations on serial cases. Maps in the court room can provide reliable evidence and excellent tools for supporting the prosecution of offenders. The final application of self-selection has demonstrated pioneering ways that offenders can (give due reason) be targeted and caught more effectively and demonstrates how the geography of crime acts as an important component that can be exploited to improve policing responses and reduce crime.

Further reading

Osbourne, D. and Wernicke, S. (2003). *Introduction to Crime Analysis: Basic Resources for Criminal Justice Practice*. New York: Haworth Press.

This book describes the wider (non-geographic) range of tactical and investigative analysis techniques.

Clarke, R.V. and Eck, J. (2003). *Become a Problem Solving Crime Analyst*. London: The Jill Dando Institute of Crime Science. www.jdi.ucl.ac.uk.

The starting point of a tactical and investigative analysis should not begin by loading data into a statistical software package or mapping crime data. Instead, these types of analysis should follow a problem-solving path that begins with the scanning process. This manual acts as an excellent reference to help establish where to start and how to progress with an analysis project.

Wiles, P. and Costello, A. (2000). The road to nowhere: The evidence for travelling criminals. Home Office Research Study 207, Research, Development and Statistics Directorate, Home Office. www.homeoffice.gov.uk/rds/hors2000.html.

This Home Office research paper is a good starting point for exploring the concepts of, and distances, that offenders take to commit their crimes.

Rossmo, K. (2000). *Geographic Profiling*. CRC Press, Boca Raton, Florida.
Canter, D. (2003). *Mapping Murder: The Secrets of Geographical Profiling*. London: Virgin Books.

These texts are the most accessible established references for Rossmo's and Canter's approaches to geographic profiling.

Ratcliffe, J. (2004). Jerry's top ten crime mapping tips v2.1 http://jratcliffe.net/papers/index.htm#tips

In Chapter 12 we describe the steps and functions to perform to design effective maps. This reference offers a useful addition to help crime mappers prepare their material for presentation.

References

Ashby, D.I. (2004). Crime mapping and the neighbourhood: A new approach to the profiling of crime, fear of crime and policing performance at a local level. The 2nd UK Crime Mapping Conference, 9–10 March 2004, London, England. www.jdi.ucl.ac.uk/news_events/conferences/index.php.
Ashton, J., Senior, B., Brown, I. and Pease, K. (1998). Repeat victimisation: Offender accounts. *International Journal of Risk, Security and Crime Prevention*, 3, 269–280.

Baldwin, J. and Bottoms, A. (1976). *The Urban Criminal: A Study of Sheffield*, London: Tavistock.

Birkin, M., Clarke, G.P., Clarke, M. and Wilson, A.G. (1996). *Intelligent GIS*. New York: Wiley.

Brantingham, P.J. and Brantingham, P.L. (eds) (1981). *Environmental Criminology*. London: Sage.

Canter, D. (1994). *Criminal Shadows*. London: HarperCollins.

Canter, D. (2003). *Mapping Murder: The Secrets of Geographical Profiling*. London: Virgin Books.

Canter, D. and Gregory, A. (1994). Identifying the residential location of rapists. *Journal of the Forensic Science Society*, 34(3), 169–175.

Canter, D. and Larkin, P. (1993). The environmental range of serial rapists. *Journal of Environmental Psychology*, 13, 63–69.

Chainey, S.P. (2002). Geographic Profiling of Serial Offenders. Criminal Justice Management, September 2002.

Chainey, S.P., Holland, F., Patrick and Austin, B. (2001). Insights into the analysis and patterns of offender journeys to crime in the London Borough of Croydon. Infotech Enterprises Europe consultancy report for the Croydon Crime and Disorder Reduction Partnership (Unpublished).

Chenery, S., Henshaw, C. and Pease, K. (1999). Illegal parking in disabled bays: A means of offender targeting. Home Office PRCU Briefing Note 1/99. London: Home Office.

Clarke, R.V. (1999). Hot products: Understanding, anticipating and reducing demand for stolen goods. *Police Research Series*, Paper 112. London: Home Office. http://www.homeoffice.gov.uk/rds/prgpdfs/fprs112.pdf.

Clarke, R.V. and Eck, J. (2003). *Become a Problem Solving Crime Analyst*. London: Jill Dando Institute of Crime Science. www.jdi.ucl.ac.uk/publications/adhoc_publications/index.php.

Crime Concern (2002). Keeping young people safe and out of trouble. www.crimereduction.gov.uk/youth22.pdf.

Everson, S. (1999). Repeat victimisation and criminal careers. PhD Thesis, University of Huddersfield.

Felson, M. and Clarke, R.V. (1998). Opportunity makes the thief: Crime Detection and Prevention Series, Paper 98. *Police Research Group*. London: Home Office. www.homeoffice.gov.uk/rds/prgpdfs/fprs98.pdf.

Flood-Page, C., Campbell, S., Harrington, V. and Miller, J. (2000). Youth Crime – findings from the 1998/99 Youth Lifestyles Survey. Home Office Research Study 209, Research, Development and Statistics Directorate, Home Office. www.homeoffice.gov.uk/rds/pdfs/hors209.pdf.

Frisbie, D.W., Fishbine, G., Hintz, R., Joelson, M. and Nutter, J.B. (1977). Crime in Minneapolis: Proposals for prevention. St Paul, MN: Community Crime Prevention Project, Governor's Commission on Crime Prevention and Control.

Gill, M. and Pease, K. (1998). Repeat robbers: Are they different? In M. Gill (ed.) *Crime at Work: Increasing the Risk for Offenders*, Volume 2. Leicester: Perpetuity Press.

Harries, K. (1999). *Mapping Crime: Principles and Practice*. Washington, DC: United States Department of Justice.

Herbert, D.T. (1976). The study of delinquency areas: A social geographical approach, Transactions, Institute of British Geographers, NS1: 472–492.

Joseph Rowntree Foundation (2002). A national survey of problem behaviour and associated risk and protective factors among young people. www.jrf.org.uk/knowledge/findings/socialpolicy/432.asp.

Kelling, G.L. and Coles, C.M. (1996). *Fixing Broken Windows: Restoring Order and Reducing Crime in our Communities*. New York: Free Press.

Longley, P., Goodchild, M., Maguire, D. and Rhind, D. (2001). *Geographic Information Systems and Science*. Chichester: John Wiley & Sons.

Liu, S. and Zhu, X. (2004). An integrated GIS approach to accessibility analysis. *Transactions in GIS*, 8(1), 45–62.

Michaud, S.G. and Aynesworth, H. (1989). *Ted Bundy: Conversations with a Killer*. New York: Penguin Books.

Moland, R.S. (1998). Graphical display of murder trial evidence. In N. LaVigne and J. Wartell (eds) *Crime Mapping Case Studies: Successes in the Field*, Volume 1. Police Executive Research Forum, Washington, DC.

Newton, A. (2004). Exploring the link between 'bus related' crime and other crimes. The 2nd UK Crime Mapping Conference, 9–10 March 2004, London, England. www.jdi.ucl.ac.uk/news_events/conferences/index.php.

Nicholson, M. (1979). *The Yorkshire Ripper: The Authoritative Study of the Most Vicious Series of Murders this Century*. London: Star.

Paulsen, D. (2004). Geographic profiling hype or hope? Preliminary results into the accuracy of geographic profiling software. The 2nd UK Crime Mapping Conference, 9–10 March 2004, London, England. www.jdi.ucl.ac.uk/news_events/conferences/index.php.

Poyner, B. (1986). A model for action. In G. Laycock and K. Heal (eds) *Situational Crime Prevention*. London: HMSO.

Poyner, B. and Webb, B. (1991). *Crime Free Housing*. Butterworth – Architecture, Oxford.

Reno, S. (1998). Using crime mapping to address residential burglary. In N. LaVigne and J. Wartell (eds) *Crime Mapping Case Studies: Successes in the Field*, Volume 1. Police Executive Research Forum, Washington, DC.

Robbin, C.A. (2000). Apprehending violent robbers through a crime series analysis. In N. LaVigne and J. Wartell (eds) *Crime Mapping Case Studies: Successes in the Field*, Volume 2. Police Executive Research Forum, Washington, DC.

Rich, T. and Shively, M. (2004). A methodology for evaluating geographic profiling software. A report prepared for the United States National Institute of Justice, Cambridge, MA: Abt Associates. http://www.ojp.usdoj.gov/nij/maps/gp.pdf.

Rose, G. (2000). *The Criminal Histories of Serious Traffic Offenders*. Home Office Research Study 206. London: Home Office.

Rose, S. (2004). Criminals as customers: Applying the principles of customer relationship management and the use of geodemographics to policing. The 2nd UK Crime Mapping Conference, 9–10 March 2004, London, England. www. jdi.ucl.ac.uk/news_events/conferences/index.php.

Rossmo, K. (2000). *Geographic Profiling*. Boca Raton, Florida: CRC Press.

Rossmo, D.K., Davies, A. and Patrick, M. (2004). *Exploring the Geo-demographic and Distance Relationships Between Stranger Rapists and Their Offences* (Special Interest Series: Paper 16). London: Research, Development and Statistics Directorate, Home Office.

Schmitz, P., Cooper, A., Davidson, A. and Roussow, K. (2000). Breaking alibis through cell phone mapping. In N. LaVigne and J. Wartell (eds) *Crime Mapping Case Studies: Successes in the Field*, Volume 2. Police Executive Research Forum, Washington, DC.

Shaw, C.R. and McKay, H.D. (1931). *Social Factors in Juvenile Delinquency*. Washington: US Government Printing Office.

Webb, A.R. (2002). *Statistical Pattern Recognition*. Chichester: John Wiley & Sons.

Wellsmith, M. and Guille, H. (2005). Fixed penalty notices as a means of offender selection. *International Journal of Police Science and Management*, 7(1).

Wernicke, S. (2000). Reducing construction site crime. In LaVigne, N. and Wartell, J. (eds) *Crime Mapping Case Studies: Successes in the Field*, Volume 2. Police Executive Research Forum, Washington, DC.

Wiles, P. and Costello, A. (2000). The road to nowhere: The evidence for travelling criminals. Home Office Research Study 207, Research, Development and Statistics Directorate, Home Office. www.homeoffice.gov.uk/rds/hors2000.html.

Winn, S. and Merrill, D. (1979). *Ted Bundy: The Killer Next Door*. New York: Bantam Books.

Young, C.A. (1999). The Smithdown Road pilot 'Alleygating' project. Evaluated on behalf of the Safer Merseyside Partnership (Unpublished).

11
Policing the Causes of Crime

Learning Objectives

In this chapter we develop the ideas of crime control beyond the immediate level that many police officers are familiar with, and explore longer-term strategic concepts of crime reduction. Tactical arrests and the targeting of offenders can help temper problems in the short term, but longer-term gains in crime reduction need a more comprehensive understanding of the crime problem. We start by reviewing three current policing styles: community policing, intelligence-led policing and problem-oriented policing, as well as look at the long-term concepts of community partnerships.

The chapter then explores some of the underlying drivers of crime, through theoretical research areas such as collective efficacy, social disorganisation and the rational choice perspective. Many of the underlying causes of crime function at different spatial levels. For example, neighbourhood studies of collective efficacy rely on an accurate measurement and spatial understanding of the idea of 'neighbourhood' or 'community'. We conclude the chapter by looking at the different spatial levels available to researchers, and how these different levels can fit together.

GIS and Crime Mapping   Spencer Chainey and Jerry Ratcliffe
© 2005 John Wiley & Sons, Ltd

11.1 Introduction – the level of strategic crime control

Most people understand the term 'spatial'. It is especially relevant to policing and crime reduction because policing has an inherently spatial component, and crime reduction policies are most successful when targeted and tailored to particular crime environments or areas. 'Strategic' is also a term that many people think they can define, but it is often easiest to define strategic within the crime reduction or policing sphere in relation to other types of crime control.

Whereas the tactical level can be viewed as the front-line level, where investigations and other operational areas tend to have a case-specific and offender focus, and the operational level is the decision-making level for area commanders and crime reduction practitioners seeking to deploy resources effectively (Ratcliffe, 2004, pp. 4–5), strategic crime control aims for a longer-term view.

With a strategic crime control perspective, 'strategic' can often mean different things to different people. For example, a small US police department with only half-a-dozen sworn officers can view a long-term objective as crime reduction in a neighbourhood over a number of months, where it is equally possible that a large police service with tens of thousands of officers can have a strategic aim of reducing region-wide crime levels over a number of years. Where both will often find a degree of commonality is in the recognition that a spatial focus can help to concentrate crime reduction activity (either police-centred or not) in the most appropriate areas.

This is because crime clusters in some spatial areas for a reason. An understanding of the root causes of crime that underpin clustering in certain areas and neighbourhoods can help crime reduction practitioners both focus on the right areas and understand the mechanism (Pawson and Tilley, 1997) that is causing the crime. With this understanding, effective crime reduction policies can be formulated.

This chapter re-examines the main theoretical structures that can help explain the spatial distribution of crime, some of which were introduced in Chapter 4, with the aim of understanding the spatial studies that have tested these theories. The chapter then goes on to examine the different types of geographical scale that have been employed to study the link between theoretical understanding of crime and its spatial extent. We start, however, with an overview of the current paradigms in policing and whether these can be effectively employed in a spatial sense to prevent crime.

11.2 Policing for crime reduction

Observers of policing contend that law enforcement has, in many places, abandoned the professional era of policing, where the emphasis was on fast police response, the rule of law and an emphasis on professionalisation, and entered the community policing era, with an emphasis on community contact and problem-solving (Walker and Katz, 2001). The enthusiasm for community policing in many Western countries is certainly strong, though whether this current epoch is one of community policing or more simply a post-professional period is debatable.

Many police services were beginning from the 1970s to recognise that the existing ways of policing were not working. Officers were being called to the same address again and again, and faster police cars and more effective command and control systems were not reducing crime. Detectives were overloaded with cases and clear-up rates were not improving or keeping pace with the increasing crime rate. There really was only one model of policing, a reactive one, and it was not stemming the rise in crime. Although there was a significant push in the US towards community policing, in other parts of the world there was more enthusiasm for different models. This section explores a number of these current policing styles and the place of crime mapping within them.

11.2.1 Community policing

Community policing seeks a more meaningful relationship with the community and a greater integration of police and community decision-making. However, community policing also suffers from a problem of 'buzzword adoption', where many police agencies claim to be practising community policing with little knowledge of the fundamental philosophy of the paradigm. In general, community policing has its origins in the lack of police interaction with the community that led to the 'crisis of legitimacy' of the police during the time of the urban race riots in the 1960s (Sherman *et al.*, 1998). Recognising that the police placed too much emphasis on the efficient rule of law and were too closely tied to their police cars, police researchers advocated an increase in the level and quality of dealings the police had with the community.

Definitions of community policing are difficult to nail down. According to the United States Community Policing Consortium, a group comprising the International Association of Chiefs of Police (IACP), the National Sheriffs' Association (NSA), the Police Executive Research

Forum (PERF), the Police Foundation and the National Organization of Black Law Enforcement Executives (NOBLE), community policing is 'a collaboration between the police and the community that identifies and solves community problems. With the police no longer the sole guardians of law and order, all members of the community become active allies in the effort to enhance the safety and quality of neighborhoods' (CPC, 1994, Preface).

Community policing has been around long enough for a number of significant studies of the paradigm to be published. Unfortunately some police tactics, which are commonly attributed to community policing, do not appear to work as tactics for long-term crime reduction. Two of the most popular ones, Neighbourhood Watch and the Drug Abuse Resistance Education (DARE) programme, have both been studied in some depth. A thorough study funded by the US National Institute of Justice NIJ which reviewed the literature surrounding both of these strategies concluded that there was no evidence that either were effective as crime prevention strategies (Sherman *et al.*, 1998). Indeed in the case of Neighbourhood Watch there was some suggestion that it actually increased fear of crime in local residents (Skogan, 1990), and longitudinal studies of the DARE programme have shown no long-term effect (Rosenbaum *et al.*, 1994; Sigler and Talley, 1995; Clayton *et al.*, 1996). One study actually found that drug abuse increased in some suburban children that had been through the DARE programme (Rosenbaum and Hanson, 1998).

Crime mapping does have a role in a more community oriented policing style. Even though community policing is difficult to define, a general move towards greater integration with the community can result in a need for an increased level of communication about crime problems. One particular area that has grown in recent years is in the provision of crime information to the community through Internet maps. The number of agencies that provide this service is still very small; however, static and interactive crime mapping sites are beginning to appear on the Internet.

At the time of writing, the NIJ's MAPS programme had a list of known police Internet crime mapping sites at www.ojp.usdoj.gov/nij/maps/weblinks.html. While on first examination these mapping sites may appear to be an invaluable tool to the community, Ratcliffe (2002a) raises some ethical doubts about the value of Internet crime maps, including: concerns regarding inaccurate or incomplete mapping of crime; differences between different hotspot methods (see also Chapter 6); and the potential for spatial labelling and its implications for house prices and social cohesion.

11.2.2 Intelligence-led policing

Intelligence-led policing originated from the realisation in many developed world countries during the late 1980s and early 1990s that the traditional mode of policing was not combating rocketing crime rates. In the UK, the problem was compounded: as crime rates rose, there was no matching increase in police resources. The rise in crime occurred at a time of significant financial constraint on police services in the UK, and the lack of effective tactics to combat the seemingly inextricable rise in offending was starting to sap public support for the police.

The first stage in the development of intelligence-led policing was the 1993 publication, by the UK Audit Commission, of a report titled 'Helping With Enquiries – Tackling Crime Effectively' (Audit Commission, 1993). Not only did this report appear to offer the first readable guide to cost-effective law enforcement, but it also did so in a style that appealed to many street police officers. The emphasis on crime fighting and targeting the criminal had more appeal to police on the streets than more 'soft' options, such as community policing.

Shortly after the Audit Commission report, Her Majesty's Inspectorate of Constabulary weighed in with a document supporting the use of intelligence in policing (HMIC, 1997), and the Home Office published a number of case studies that documented the move various British police forces had made towards intelligence-led policing (Maguire and John, 1995). These reports and publications indicated a significant change in the direction of policing in the UK (Ratcliffe, 2002b). Once adopted by the NCIS, and incorporated as a central philosophy of the NIM (Christopher, 2004), the central role of intelligence-led policing in the fabric of British policing in the new millennium was cemented.

Intelligence-led policing has a specific 'strategic, future-oriented and targeted approach to crime control' (Maguire, 2000, p. 316). The targeting of criminal organisations places a centrality on the analysis role to support decision-making, and crime mapping has a place in the analysis role. Furthermore, the tactical tasking and coordination aspect of the NIM focuses police activity on four areas: targeting offenders; management of crime and disorder hotspots; investigation of linked crimes and incidents, and the application of a range of preventative measures (NCIS, 2000). Although the central role of intelligence gathering is on surveillance and the use of informants to gather information in regard to organised criminal groups, there has been a call for a harm reduction approach that makes more use of an understanding of environmental criminology (Sheptycki and Ratcliffe, 2004). In this, it would be clear that crime mapping, with its inherent spatial understanding, would play a significant role. Indeed each of the NIM activity areas (mentioned above) can be enhanced with crime mapping applications.

11.2.3 Problem-oriented policing

The definitive origins of problem-oriented policing are difficult to identify; however, the formal beginnings of what is currently referred to as 'problem-oriented policing' lie in the collaboration of Herman Goldstein and the Madison (Wisconsin) Police Department in the early 1980s (Scott, 2000). Since then, Goldstein's problem-oriented policing model has been adopted by numerous police services across the world. In particular, the model received a considerable boost in the US with the formation of the COPS office, also known as the Office of Community Oriented Policing Services (www.cops.usdoj.gov). The role of the COPS office was to provide funding to police agencies in the US in support of community policing. The COPS office define community policing as 'a policing philosophy that promotes and supports organizational strategies to address the causes and reduce the fear of crime and social disorder through *problem-solving* tactics and police-community partnerships' (COPS website, emphasis added). The need for problem-solving tactics, built into the COPS office definition, was timely for problem-oriented policing which was still emerging as a policing style. This relationship helped advance the cause of problem-oriented policing as the two were perceived to be linked in the minds of many. However, proponents of problem-oriented policing do not necessarily agree and see considerable differences between community policing and problem-oriented policing (Clarke and Eck, 2003; Tilley, 2003). We agree: while community policing is often ill-defined and vague in terms of the mechanism it is trying to affect, problem-oriented policing has a stronger pedigree.

Problem-oriented policing has been described as 'an approach/method/process within the police agency in which formal criminal justice theory, research methods, and comprehensive data collection and analysis procedures are used in a systematic way to conduct in-depth examination of, develop informed responses to, and evaluate crime and disorder problems' (Boba, 2003, p. 2). This would suggest that crime mapping has a significant place as a tool that can assist with spatial research methods, data collection and analysis procedures.

Problem-oriented policing has been applied in the US to a range of problems, including drugs, vehicle crime, prostitution, graffiti, false alarms, gangs and college-related crime (Simpson and Scott, 1999). In the UK, a number of Home Office publications have lauded the use of problem-oriented policing (Leigh *et al*., 1996) though also finding that local police can resist or grow disillusioned with problem-oriented policing if there is insufficient support for the concepts. Implementation issues still exist (Leigh *et al*., 1998).

As the collected wisdom of a number of senior analysts, practitioners and academics noted, 'problem analysis *is not* merely creating maps to see where crimes have occurred. It is not merely conducting statistical analysis on secondary data to compare levels of incidents over time. It is not merely identifying trends and patterns in the frequency or magnitude of crime or supporting the police function, but it is examining the underlying conditions of both the simple and complex problems police are tasked to solve' (Boba, 2003, p. 3, emphasis in original). This challenges the crime mapper to explore beyond the pin map in order to determine some understanding of the root causes of the crime problem so that a possible solution can be communicated to the decision-makers in the organisation. As such, it is important to understand the underlying drivers that are the structural causes that underpin the crime issue (Heldon, 2004). If therefore, the root causes of crime are the subject of examination, this will require the crime mapper to explore beyond the crime pattern and to understand the possible social, economic and opportunistic influences that may be behind the crime pattern.

11.2.4 Choice of policing style?

The choice of policing approach adopted locally is probably beyond the control of most crime analysts; however, it is worth bearing in mind that problem-oriented policing and intelligence-led policing have different aims and approaches. Problem-oriented policing is interested in patterns of events and whatever necessary solutions can be found to the 'problem' that is causing a crime pattern, while intelligence-led policing is more focused on individuals and groups of offenders and the law enforcement solutions that can be brought to bear against them (Tilley, 2003). In reality, it may be necessary that both approaches are in the arsenal of the police in order that the right method can be tailored to each problem or criminal group as they are encountered.

From a crime mapping perspective, both intelligence-led policing and problem-oriented policing place a centrality on the role of the analyst. With both, quality analysis is the driving force in the proactive targeting of police resources (in the case of intelligence-led policing) and crime prevention resources (in the case of problem-oriented policing). If both of these strategies are to be central to the policing in the Western world, then crime mapping will be a significant player in the analysis of future crime problems. Either way, understanding the underlying drivers is certainly a vital part of problem-oriented policing, and would be helpful to the prioritisation phase of intelligence-led policing (Sheptycki and Ratcliffe, 2004).

11.2.5 Crime and disorder reduction partnerships

Popular in UK – mainly as a result of legislation – Crime and Disorder Reduction Partnerships (CDRP) are not a specific policing initiative, though there is no doubt that police are central players in partnerships. In Chapter 7 the concept of these partnerships was presented, particularly in terms of the rich non-police data that can be shared to support crime reduction decision-making. The role that these partnerships can play in supporting strategic crime control can be summarised into four actions:

1. Auditing and strategy setting
2. Continual auditing
3. Strategic analysis
4. Monitoring and assessment.

The measure, analysis and interpretation of crime from an audit provide the means by which partners can decide on: what they should do to reduce it; the targets that they should set for crime reduction; the priorities; and how they will act to improve information gathering. In England and Wales this is a formal process that requires each of the CDRPs to produce, at least every three years, a strategy for crime and disorder reduction. In other countries this process is not as formal, but draws on the similar concept of analysing and measuring the problem before deciding what is going to be done, how it will be done and what the strategy will aim to do.

If a strategy determines targets and priorities for the future, a continual analysis function is required that feeds the management information that decision-makers employ. It will show if the reductions in crime are on target, what needs to be done if they are not, what are the problems that persist or have emerged, and improve on the information gaps that were missing that prevented certain types of analysis at the previous auditing stage. In a crime mapping context, this continual auditing process can use maps and other spatial analysis outputs as a rich media to help describe how the partnership is performing and the problems that remain.

A real opportunity that these types of partnerships can bring to supporting strategic crime control is the ability to provide comprehensive problem-solving-based analysis of crime issues. Often, those working as analysts in the policing community find it difficult to find the time (or convince their bosses of the need) to spend more than a few hours on an analysis project. Police analysts are often overwhelmed with operational and tactical analysis requests, and a strategic analysis project that may take several weeks to complete may be difficult because of these competing pressures.

The local partnership may also employ an analyst who, with access to both police and non-police data and useful links into partnership contacts, is well placed to perform analysis projects that investigate the causes of crime, and generate analysis products that fit with the partnership ethos of working across a number of different agencies. For example, an analysis of a residential burglary problem may reveal that the highest rates are in areas of poor terraced housing and are related to the ease of access to the back of terraced properties. The analysis may also find that a contributing factor is the poor quality physical security on these premises (e.g. poor door locks, lack of property alarms, weak doors and door frames and lack of window locks). Often, local residents cannot afford to fund any security improvements themselves; so rather than these areas persisting as police high call-volume areas, the partners can work together to seek support and additional funding from local sponsors and their regional government. This support can provide a crime prevention initiative that builds security gates at the entrances to alleys, replaces the front doors of all the properties with strengthened doors and frames, and improves the door and window locks. The strategic analysis in this example would have helped to identify the problem, diagnose the problem, provide evidence of the problem (to support their bid for the government grant and to convince funding from sponsors), help determine the response that would be most effective, and provide a basis from which the scheme could be monitored to assess its success after the security improvements were introduced (i.e. linking to the SARA problem-solving methodology (Eck and Spelman, 1987) defined in Chapter 3).

Monitoring and assessment are important roles that the partnership can play in strategic crime control by reviewing what works, what does not, what can be applied elsewhere and what needs to be improved – that is, performing the last 'A' of SARA, something which is missed by many working in policing and crime reduction because of demands on their time. They are too often either asked to spend more time on the 'S', 'first-A' and 'R' of SARA, or they forget about assessment all together, in anticipation of the next challenge that the day presents.

Case study: Supporting strategic crime analysis in London, England

London is predominantly covered by one police service, but is divided into 33 local authority areas that also constitute the 33 Crime and Disorder Reduction Partnerships (CDRP). Each of the 33 CDRPs employ an analyst in addition to the analytical duties that are performed by the police-based intelligence analysts. The CDRP analyses tend to be oriented to auditing processes and supporting the understanding of crime problems that can be solved through strategic initiatives.

In a review conducted by the Government Office for London (GOL) of the role and tasks being performed by the CDRP analysts, a number of issues were identified:

- a lack of direction and support in developing strategic information;
- improvements on data used for crime analysis; and
- the need for support in strategic analysis and problem-solving methodologies.

In addition GOL recognised the need for pan-London analysis outputs, particularly analysis on cross-border problems or problems that were evident across the whole or parts of London. These were not being identified because of the tendency for CDRPs to focus only on data for their own discrete area. In an effort to help meet these needs, a team of five Crime Information Analysts have been recruited to provide direction, support and guidance to help improve the level of consistency, content and quality of information that is used to support crime reduction in London. Their activities include a function that is focused towards improving raw data that is collected and processed by the CDRP partners in London, from which they can identify areas for improvement and gaps in data that need to be filled. Secondly they provide an information and intelligence provision role, performing analysis to understand cross-border, sub-regional and pan-London issues.

Strategic analysis of the team is developed through their Analysis Work Programme. This programme is structured around three categories of analysis:

1. *Crime-based* – exploring and understanding patterns and trends in offences, offending and victim targeting.
2. *Information-based* – identifying the core information requirements and necessary strategic information improvements to ensure better levels of consistency, content and quality of information to support crime and disorder reduction activity.
3. *Problem-solving-based* – to provide the lead and break new ground in problem-solving analysis, and improving the development of techniques and methodologies that will support the borough analysts in understanding their crime and disorder problems.

11.2.6 Policing models and crime reduction

The second half of this chapter concerns itself with the underlying causes of crime. In the end, the various models of policing that the first part of the chapter explored are of little value in crime reduction if they do not have any impact on criminal behaviour. The way that offenders behave directly impacts the chances that a policing strategy will be successful in preventing or detecting crime. The next section explores the role of criminal behaviour in the local context.

11.3 Analysing the underlying drivers of crime

Researchers have utilised crime mapping technologies to go beyond exploratory maps of crime and have been engaged in seeking out explanations for the root causes of crime in different places. Within these studies, the local neighbourhood has become a significant place of criminological study. Chapter 4 introduced a number of theoretical explanations for varying rates of crime in different locations, and the following section documents the different variables and scales that have been used to test some of these theories. These studies provide considerable opportunity to extend crime prevention and crime reduction beyond the tactical and operational limit of policing. These studies provide an opportunity to understand the underlying drivers that fuel crime in local areas by providing long-term strains on the social fabric of local neighbourhoods.

11.3.1 Routine activities and rational choice

As described in Chapter 4, routine activities theory tells us that the necessary ingredients for a crime are the coming together in time and space of a motivated offender, a suitable target and the absence of a capable guardian (Cohen and Felson, 1979). Rational choice theory tells us that offenders make a purposeful decision to offend primarily based on the environmental cues from the prospective target. This is the second decision in a two-part process, the first of which is to become involved in crime generally (the criminal involvement decision) and the second of which is the criminal event decision. The two theories are linked in that they are less interested in the residential situation of offenders and more interested in the opportunities that are presented to offenders when they interact with potential targets. GIS can be used to map the availability of opportunities and the incidence of crime in these areas.

Researchers have attempted to use various measures of routine activity as independent variables in order to test the validity of routine activity theory. It is hypothesised that increased levels of crime will be clustered in the vicinity of particular locations because they draw offenders and victims to the area. Such variables include hotels and motels (Rice and Smith, 2002), schools (LaGrange, 1999), shopping malls (Nelson *et al.*, 2001) and bars and subway stations (Ouimet, 2000).

Features such as the number of shopping malls and the number of bars are useful independent variables of routine activity. The number of offenders living in the area is not an appropriate choice of dependent variable because offenders can be drawn into an area from outside, attracted by the bars and shopping malls. Researchers have generally operationalised the dependent variable as a crime count of some description. Dependent variables have included counts of robbery, homicide, vandalism, vehicle theft, burglary and sexual assault.

Groff and LaVigne generated a raster surface map of burglary opportunity for a neighbourhood in Charlotte, North Carolina (Groff and LaVigne, 2001). By coding a fine mesh of grid cells (25 feet resolution) with variables believed to represent increased risk of burglary due to increased opportunity. They were able to test this predictive raster surface against actual burglary rates. Their independent variables included housing tenure, vacant housing, proximity to bus stops, existence of street lighting, proximity to major thoroughfares and a measure that indicated if the grid cell was within a buffer of between 500 and 1500 feet of the home address of a convicted burglar (Groff and LaVigne, 2001, Table 1). While their model had modest success in predicting high burglary areas, it was particularly good at predicting low burglary areas, an unexpected feature of the study and one that may merit further investigation.

All of these studies are highly spatial in focus, which is ideal for problem-oriented policing strategies. The offenders are often local, with local connections and areas of operation. This therefore provides opportunities for intelligence-led policing operations.

11.3.2 Social disorganisation

The theory of social disorganisation grew from the pioneering studies of the Chicago School of Sociology and, in particular, Shaw and Mackay (Shaw and McKay, 1942), whose work was influenced by Burgess's zonal model of urban development shown in Chapter 4 (Burgess, 1925). Shaw and Mackay were drawn to the zone in transition (Figure 4.1), recognising that the constant movement of people in and out of the area, the melting pot of different racial groups and the range of economic factors that influenced the lives of people in this zone also influenced the delinquency rate. Parts of the city that suffered physical deterioration and economic deprivation were correlated with 'population instability and cultural fragmentation' (Bottoms and Wiles, 2002). These negative influences affected the delinquency rate in a process termed 'social disorganisation'. Social disorganisation theory posits the idea that increased levels of delinquency, especially juvenile delinquency, exist because of the lack of a local social fabric where the structure and culture of the community are strong enough to provide a concerted influence over local residents. Given that adolescents are a cause of much local strife in a community, this group are often the local residents that are in need of influence and direction. Social disorganisation suggests that if there is a high degree of cultural heterogeneity and a high turnover of residents, the community are unlikely to be able to agree to a common standard for behaviour in the street, and that few residents are likely to know the young people on the street or their families. With no clear rules as to acceptable behaviour and few sanctions available (you cannot tell a child to stop misbehaving or you will tell his parents, if you and the child both know that you do not know his parents) to curb adolescent exuberance, juvenile delinquency increases. Shaw and Mackay identified three distinct negative influences that related to social disorganisation: poverty, residential mobility and ethnic heterogeneity (Shaw and McKay, 1942).

The difficulty with this theory is that it can be tricky to construct variables which are a direct measure of 'social disorganisation'. It is unlikely that a household survey that asked the respondents to rate the level of social disorganisation in their neighbourhood from one to ten would be very successful, because few people would have a clear and common notion of social disorganisation. We have to rely on indirect measures that provide an indication of possible precursors of social disorganisation. Traditional meas-ures have included poverty, residential mobility and ethnic heterogeneity, and studies since Shaw and Mackay have either attempted to replicate these variables or add to them in some way. Studies that have examined social disorganisation have employed (among others) the following independent precursor variables as an attempt to model social disorganisation: single parent families with children, percentage of immi-grants, percentage of blacks, heterogeneity (measured in various ways), percentage of residents who have moved in the last five years, Socio-Economic Status (SES), population density and family disruption. As said, the problem with measuring social disorganisation is that these measures have to stand as an indirect 'proxy' for the real variable under examination. In response, some researchers have tried to measure the reverse of social disorganisation, social or collective efficacy.

Measuring and identifying social disorganisation is one thing, but having a policing impact on this issue in order to improve the crime situation is quite another. Policing strategies are often a little limited when faced with these more ingrained causes of crime, and it may be better to seek a more situational approach to crime control or align police tactics with other local initiatives. This type of approach is being adopted in the UK, where the police's role in helping to tackle social disorganisation sees it working with local strategic partners (e.g. local government regener-ation programmes) (Chainey, 2004).

11.3.3 Collective efficacy

In many respects, collective efficacy is the opposite of social disorgan-isation. It can be defined as the 'social cohesion among neighbours combined with their willingness to intervene on behalf of the common good' (Sampson *et al.*, 1997, p. 918). Collective efficacy can be found in areas where neighbours co-operate on issues of mutual interest, share some areas of agreement with the people who live around them and, possibly most important of all, are prepared to intervene if local youths are behaving in a manner unacceptable to local norms. This requires enough implicit or explicit communication between neighbours in order to define and agree the standard for local normative behaviour. It is therefore argued that areas high in collective efficacy are well suited to resisting crime at the local level by being able to influence local young people and exercise some control over a group in their peak offending years.

Collective efficacy has been directly measured using community-based surveys which have attempted to measure neighbourhood social and institutional processes. Collective efficacy is related to the notion of 'social capital', a feature that some researchers have operationalised as the number of interactions that take place with neighbours. Social capital is a measure of the skills and social position that a person possesses that provide them with the power to effect a positive social change on their local environment. Robert Sampson and his colleagues have lead the research in this area, actively seeking to measure collective efficacy. Their survey of over 8000 Chicagoans asked if the respondents felt it was likely if neighbours could be counted on to intervene if children were spray-painting a local building, if a fight broke out or if children were skipping school. They were also asked if they lived in a close-knit community where neighbours could be trusted (Sampson *et al.*, 1997). Their study also included census measures of race, poverty, immigration, the labour market, home ownership and residential stability (among others). The study concluded that collective efficacy could be reliably measured and could act to control the level of violent crime.

For both collective efficacy and social disorganisation, the dependent variable can be either a measure of the number of offenders in an area or the frequency of criminal activity. This is because the negative effects of social disorganisation (and conversely the positive effects of collective efficacy) are posited to work on both offenders and offending. Social disorganisation reduces the ability of an area to resist crime, so counts of property or personal crime are appropriate measures. However, these theories also deal with the concept of exerting control over local youths, so arrest records (particularly of juveniles) can be an acceptable dependent variable.

11.3.4 A measure of heterogeneity

The measure of heterogeneity in a population is commonly used in crime causal studies; however, there is rarely consensus as to the method of calculation. Some studies simply calculate the percentage of the population in an area that are not white; however, in some parts of the US this is not ideal. Areas with equal numbers of black, Hispanic and white residents have a different degree of heterogeneity than areas with only black and white residents.

One solution is to use a proportional measure that is flexible enough to consider a variable number of racial groups. If the number of residents of a racial or ethnic group can be expressed as a fraction of the total population, then the equation (from Blau, 1977) is as follows:

$$h = 1 - \Sigma p_i^2$$

where h is the heterogeneity measure and p is the proportion of each racial group (i) expressed as a fraction. The following table demonstrate the calculation with an example.

Racial/ethnic group	Population	Population Rate (p_i)	Rate^2
White	100	(100/400)=0.25	$(0.25)^2 = 0.0625$
Black	100	(100/400)=0.25	$(0.25)^2 = 0.0625$
Hispanic	100	(100/400)=0.25	$(0.25)^2 = 0.0625$
Asian	100	(100/400)=0.25	$(0.25)^2 = 0.0625$
Total	400		0.25

With the distribution of the population as shown, $h = 1 - (0.0625 + 0.0625 + 0.0625 + 0.0625) = 0.75$. In the next table, we show the effect of a less balanced population distribution, which is more realistic. With a less balanced population distribution, h becomes 0.41 ($1 - 0.59$).

Racial/ethnic group	Population	Population Rate (p_i)	Rate^2
White	300	(300/400)=0.75	$(0.75)^2 = 0.5625$
Black	60	(60/400)=0.15	$(0.15)^2 = 0.0225$
Hispanic	30	(30/400)=0.075	$(0.075)^2 = 0.005625$
Asian	10	(10/400)=0.025	$(0.025)^2 = 0.000625$
Total	400		0.59125

With four different ethnic groups, maximum population heterogeneity is 0.75 (as calculated from the first table) where each group is represented in the general population equally. Maximum heterogeneity is 0.66 for three groups, 0.75 for four groups, 0.80 for five groups and continues to approach 1.0 as the number of racial/ethnic groups is increased. As *h* approaches zero, heterogeneity decreases until at zero there is no heterogeneity and the population of an area is dominated by one racial group. The advantage with this calculation is that it can be calculated for small areas such as census tracts and is directly comparable with other areas as long as the same number of racial/ethnic groups are counted.

Figure 11.1 shows the heterogeneity measure for block groups in Philadelphia for residents (who expressed one race/ethnicity from White, Black, Hispanic and Asian at the 2000 population census). Because there are four groups considered, the maximum theoretical population heterogeneity value is 0.75. It can be seen that there are distinct clusters of heterogeneity in Philadelphia, a correlate of social disorganisation according to Shaw and Mackay.

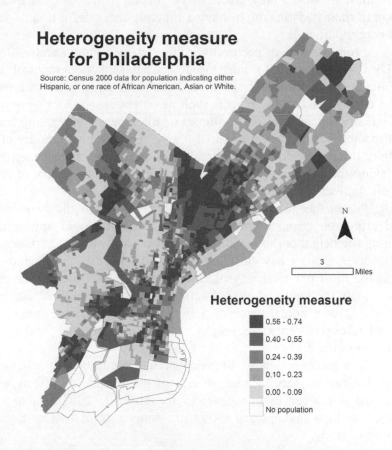

Figure 11.1 Heterogeneity rate calculated for Philadelphia block groups

11.4 The geography of neighbourhood studies

The difficulty with concepts such as collective efficacy or social disorganisation is that they are often presumed to exert a strong influence at the neighbourhood level. But how big is a neighbourhood? Where are the boundaries and do all residents on a neighbourhood agree in the name and scale of the neighbourhood in which they live? Local councils and city authorities often designate parts of a city as neighbourhoods, but this can often be more for administrative purposes in the provision of services – the neighbourhood influences on crime may recognise other, less definite boundaries. It is also possible that different people view their 'neighbourhood' differently. For example, a child might say the boundary is a few streets away, as this is the limit of where they are allowed to play and roam, whereas a person in their twenties might have a much broader definition of their own neighbourhood.

From a mapping perspective, this also raises the spectre of the MAUP (Openshaw, 1984), introduced in Chapter 6. Geographical data are often aggregated in order to present the results of a study in a more useful context, and spatial objects such as output areas, block groups or police district boundaries are examples of the type of aggregating zones used to show results of some spatial phenomena. These zones are often arbitrary in nature, and different areal units can be just as meaningful in displaying the same base level data. In other words, the choice of areal unit often affects the study and the outcome.

Given the growing enthusiasm for studies of the neighbourhood effect on crime (Sampson *et al.*, 2002), it would appear that defining the neighbourhood is therefore of considerable importance. In reality, most studies have used census areal units in some fashion as this is the most practical way of converting data recorded in national censuses to the study of neighbourhoods (though some researchers have employed school districts or police divisions). The question exists as to what level of census geography best approximates the local notion of neighbourhood.

In a recent review of 40 major studies, Sampson and colleagues found that the majority of areal studies of neighbourhood effects were conducted at the census tract level (Sampson *et al.*, 2002). Table 11.1 summarises their findings and shows the dominance of census tracts in this type of study.

Table 11.1 Summary of 40 studies on neighbourhood effects of crime published from 1996 to 2001 ($N=41$ because one study examined block groups and census tracts). Source: Sampson *et al.* (2002)

Areal unit	Number of studies
Block groups, enumeration districts or face blocks	7
Census tract	18
Neighbourhoods, tract clusters or neighbourhood clusters	8
Police boundaries	1
Postal or ZIP districts	5
Political boundaries	2

11.4.1 Census geography

In the UK, census output areas and their larger counterparts, the census super output area and census ward, are common areal units for neighbourhood studies. Output areas are designed to maintain a degree of social homogeneity in the area they cover and, where possible, are a standard size across the country (Martin, 1997, 1998). They are slightly smaller than the previously used enumeration districts, with every output area having a minimum of 40 households and 100 residents to ensure data confidentiality. The average size is 125 households.

In the US, although the census geography starts at the block level the next aggregation units of the block group and then the census tract are more popular for research purposes. There are over 8 million blocks in the 2000 census, though the individual block is infrequently used for social science research. More common as a research tool is the block group or the census tract. The average number of residents in a US census tract is about 4000.

As a point of comparison, consider the City of Philadelphia on the east coast of the US. With a population of about 1.3 million (the greater metropolitan area has three times the population), the US Census Bureau decided that for the 2000 census the City of Philadelphia consists of 381 census tracts, which are further sub-divided into 1816 block groups, which in turn are divided into 17 199 blocks. The census tracts and block groups are visible in Figure 11.2. Each of the smaller aggregation units can be nested easily into a larger unit, so there are no instances where a block group is split between two census tracts. Each block is contained within only one block group (along with a number of other blocks) and

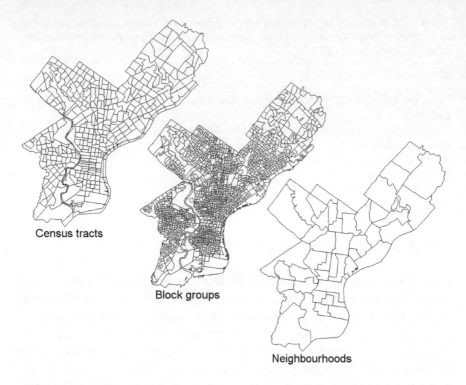

Figure 11.2 Boundaries of census tracts, block groups and neighbourhoods in Philadelphia

each block group belongs to only one census tract. The average number of housing units in a Philadelphia block group is 365 (835 residents), while the average census tract in the city has 1737 housing units (with an average population of 3983). If Philadelphia is a good example, the US block group is approximately the same size as three UK output areas.

11.4.2 Neighbourhoods

When measuring the collective behaviour of groups of people, it is important to consider their collective living arrangements. Few people would be able to say what census tract they lived in; however, they would undoubtedly be able to name their neighbourhood. Some researchers have tried to create neighbourhood boundaries to represent these areas. Collective efficacy measurement can include involvement in neighbourhood organisations and community groups and these do not exist at the census tract level, so there is some merit in attempting to construct boundaries that are meaningful in terms of the size and shape of a neighbourhood.

Ouimet (2000) divided Montreal into 84 neighbourhoods based on municipal boundaries, while Sampson and colleagues combined 847 Chicago census tracts into 343 'neighbourhood clusters' (Sampson *et al.*, 1997). Figure 11.2 shows census block groups and census tracts in Philadelphia; however, the last image shows neighbourhoods, as defined by the Philadelphia Health Management Corporation. The advantage of neighbourhoods is that they are readily identifiable to residents of the city by name. Asking people if they live in Chestnut Hill or Center City is more meaningful than if they live in block group 421 010 004 001. The problem is that some concepts of neighbourhood can be nested. For example, one of the authors lives in Philadelphia. To people outside the city explaining that the author lived in Center City would probably be sufficient in explaining the general neighbourhood; however, if speaking to a Philadelphian, stating Society Hill would be more useful as this smaller sub-division of the city centre provides greater location precision.

The difficulty with neighbourhoods is that they are sometimes an emotional barrier as much as a physical one, and the barriers between neighbourhoods can be variable for different people as well as fuzzy in their definition. Furthermore, as time passes, the distinction between two neighbourhoods can change for the same individual.

11.4.3 Census tracts or neighbourhoods?

Maps and any sort of spatial analysis that employs aggregate data will inevitably suffer from a loss of detail from the original data source. When using census boundaries, there is no avoiding this as it is not possible to access the original data. The level of aggregation that census agencies allow is designed to protect the identity and details of individual people. As the size of the areal unit becomes smaller, the population within that unit is likely to become more homogenous resulting in a better study. Unfortunately, as the size of the unit gets smaller, the computed rates for variables and population become less reliable.

Ouimet studied the impact of choosing census tracts or neighbourhoods on a study of opportunity theory and social disorganisation in Montreal (Canadian census tracts are located in urban areas and had an average population of around 4000 at the 2001 Census of Canada). His study of 495 census tracts and 84 neighbourhoods found that the explanatory level of the study improved at the neighbourhood level (a larger areal unit) but at the price of reducing the statistical power of the findings. He concluded that census tracts were small enough to provide enough variation such that statistical tests were applicable while also providing areas large enough that some approximation of a neighbourhood is possible. He concluded that the 'results show that decisions regarding both the aggregation level and the type of dependent variable are critical. Most important, those decisions are likely to have an impact on the results obtained and, consequently, on the theory that will be supported by the data' (Ouimet, 2000, p. 148).

11.4.4 Multilevel studies

Researchers from the Chicago School envisioned a neighbourhood as an organic structure that developed naturally due to the competition for land use by businesses as well as the competition for affordable housing by the population. In this way various drivers (ecological and cultural) dictated which groups occupied certain spaces (Sampson *et al.*, 2002). The organic structure of neighbourhoods also includes them being a part of hierarchical community structure, each neighbourhood being a sub-division of a larger community areal unit. As an example of a hierarchical structure, consider that political decisions that favour one part of a city will influence developments in a number of neighbourhoods, while a religious organisation may only have influence on the social cohesion of a smaller locale. The picture of collective efficacy or social disorganisation is therefore complicated by the possibility that different characteristics that influence social processes act at a variety of geographical scales.

Analytical methods that explicitly test for significant variables at different spatial units of aggregation may have value in explaining the relationships between crime and community relationships. One such technique is Hierarchical Linear Modelling (HLM).

Hierarchical Linear Modelling is a statistical technique that recognises that there are varying levels of relationships in many datasets. For example, a group of offenders may be related by a range of ties, including family, neighbourhood or school ties. It is also possible that some offenders that are members of a group at one level are members of different groups at another level. The same group of convicted offenders may have attended the same school and live in the same neighbourhood, but may visit a different probation office, which brings them into contact with offenders from different neighbourhoods and school groups. HLM is a statistical technique which takes into account such dependencies, arranged in relevant hierarchies by the user (for example, it tests the correlations between offenders within schools, correlations between schools within city districts, and correlations between city districts within the city). In essence, multilevel modelling techniques such as HLM attempt to model the hierarchical relationships that are found in the real world.

Hierarchical Linear Modelling is an advanced statistical technique and there is not the space in this book to explain the process of HLM modelling (whole volumes are dedicated to its application). For researchers interested in understanding more complex relationships between crime and socio-economic factors, it may be a technique worth investigating. The further reading list of this chapter provides details of a worthwhile introductionary text by Kreft and de Leeuw.

Case study: Street corner geography for street corner problems?

In 2001, in response to an increase in violence and drug dealing, the Philadelphia Police Department commenced Operation Safe Streets. The mapping section of the police department worked with local officers to identify the most prominent 300 corners of the city that were used for street-level drug dealing. Police officers were posted to continuous static patrols at each of the city's most dangerous drug corners.

In 2004, a review was conducted of these street corner locations. Drug dealing is a dynamic activity that is able to respond to policing activity, and it was strongly suspected that drug dealers were responding to police behaviour. Crime data were gathered from weapon offences, drug dealing, drug possession and anonymous calls to the city's narcotics tips hotline. These locations were geocoded, but what was the best way to define a suitable geography for street corners?

The problem is one that is ideally suited to GIS. The ability to move beyond administrative boundaries that are artificial and not necessarily responsive to urban geography is a significant strength of GIS. In this case, it was decided that the geography that was centred on each individual street corner was an appropriate mechanism.

A programe was written to geocode every junction in the study area, by finding each individual street name and then writing a single line of text that placed each street name with each other street name, separated by a junction symbol (&&). The text file of thousands of lines of possible junctions was geocoded, such that only street junctions that exist were geocoded. From these individual geocoded street corners, Thiessen polygons were generated that created a new geography, one that reflected the street corner geography of the city.

Generated from a set of points, Thiessen polygons are a particular kind of geometric shape where the boundaries define the area that is closest to each point relative to all other points. In other words, the internal area of a Thiessen polygon is closer to the central point than to any other point. An example of this can be seen in Figure 11.3, which shows a detail of the city.

In Figure 11.4, each street corner polygon is thematically shaded to show the number of drug offences, weapon offences and narcotics tips that were closer to each street corner than any other corner. Chapter 12 advises that areal units should not be thematically shaded with raw count

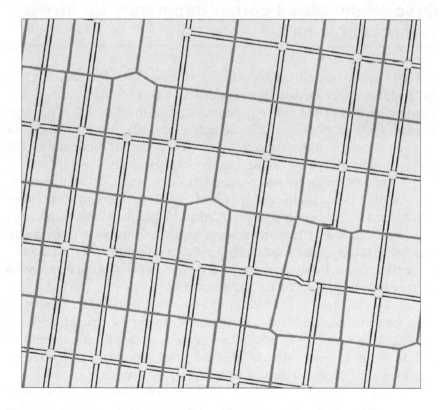

Figure 11.3 Detail of the street network in one part of Philadelphia, with street corners identified as white circles. Thiessen polygons for each street corner are shown as dark grey outlined polygons

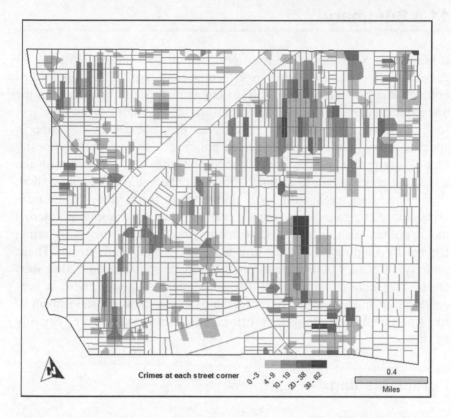

Figure 11.4 Theissen polygons shaded to indicate the number of offences that were closest to each corner

data, as larger areas tend to dominate the display, however in this particular case, raw counts at each street corner do help to prioritise police activity. What is noticeable is that a number of the high crime corners are on the edges of the mapped area, which is one police district. These could be explained perhaps by displacement from existing police activity, or by drug dealers taking advantage of any border confusion between police districts.

Although the long-term strategic crime prevention benefit of police standing on-street corners is not known, there is probably a benefit in terms of public reassurance which may help to instil greater confidence in the police. The street corner geography is a way to use the most appropriate geography to explore street corner policing and street corner crime.

11.5 Summary

Long-term crime reduction is often difficult to achieve. Policing in particular has had difficulty in effecting long-term reduction as law enforcement has rarely been able to either understand or influence the underlying causes of crime. Problem-oriented policing tries to effect long-term crime reduction by identifying the root causes of crime in specific areas, while intelligence-led policing tries to reduce crime by identifying problem areas and the repeat offenders associated with the areas, and removing these criminals from circulation through incarceration. Police can also work with local partnerships to try and recruit non-crime-specific organisations into the war on crime. Organisations external to the police, such as health departments, can have an impact on crime figures by helping to address the underlying causes of crime. These underlying causes include features that can be spatially modelled such as those associated with social disorganisation. Although there is general support in the UK where partnerships are flourishing, the impact of local partnerships on long-term crime reduction is yet to be effectively evaluated.

Further reading

http://www.cops.usdoj.gov

The US Department of Justice's Office of Community Oriented Policing Services has a range of readable and downloadable problem-oriented policing guides to a range of problem areas, including disorderly youths in public places, assaults in the vicinity of bars and graffiti.

http://www.popcenter.org

The Center for Problem-Oriented Policing is designed to advance the concept and practice of problem-oriented policing by making available readily and accessible information about ways in which the police and others working in crime reduction can more effectively address specific crime and disorder problems.

http://www.crimereduction.gov.uk

The Crime Reduction website at the UK Home Office has a wealth of information that is geared towards crime reduction practitioners and especially those in the 376 Crime and Disorder Reduction Partnerships in England and Wales.

Tilley, N. (2003). Community policing, problem-oriented policing and intelligence-led policing. In T. Newburn (ed.) *Handbook of Policing* (pp. 311–339). Cullompton, Devon: Willan.

Nick Tilley's chapter in the Handbook of policing provides an overview of the different approaches, strategies and underlying crime control philosophies of community policing, problem-oriented policing and intelligence-led policing.

Kreft, I.G.G. and de Leeuw, J. (1998). *Introducing Multilevel Modeling*. Thousand Oaks, CA: Sage.

One of the less mathematical and more understandable introductions to a growing area of spatial investigation.

Taylor, R.B. (2001). *Breaking Away From Broken Windows*. Boulder, CO: Westview.

A thoroughly researched and detailed investigation of crime in neighbourhoods and the long-term mechanisms to improve neighbourhood life. Focuses on Baltimore neighbourhoods, but is applicable in a broader context.

References

Audit Commission (1993). *Helping With Enquiries: Tackling Crime Effectively*. London: HMSO.

Blau, P.M. (1977). *Inequality and Heterogeneity*. New York: Free Press.

Boba, R. (2003). *Problem Analysis in Policing* (p. 55). Washington, DC: Police Foundation.

Bottoms, A.E. and Paul, W. (2002). Environmental criminology. In M. Maguire, R. Morgan and R. Reiner (eds) *The Oxford Handbook of Criminology* (pp. 620–656). London: Oxford University Press.

Burgess, E.W. (1925). The growth of the city: An introduction to a research project. In R.E. Park, E.W. Burgess and R.D. McKenzie (eds) *The City* (pp. 47–62). Chicago: University of Chicago Press.

Chainey, S.P. (2004). Using geographic information to support the police response to community cohesion. Proceedings of the 2004 Association for Geographic Information Conference, London, 12–14 October.

Christopher, S. (2004). A practitioner's perspective of UK strategic intelligence. In J.H. Ratcliffe (ed.) *Strategic Thinking in Criminal Intelligence*. Sydney: Federation Press.

Clarke, R.V. and Eck, J. (2003). *Becoming a Problem Solving Crime Analyst*. London: Jill Dando Institute.

Clayton, R.R., Cattarello, A.M. and Johnstone, B.M. (1996). The effectiveness of Drug Abuse Resistance Education (Project DARE): Five-year follow-up results. *Preventive Medicine*, 25, 307–318.

Cohen, L.E. and Marcus, F. (1979). Social change and crime rate trends: A Routine Activity Approach. *American Sociological Review*, 44, 588–608.

CPC (1994). *Understanding Community Policing: A Framework for Action*. Washington, DC: Community Policing Consortium.

Eck, J.E. and Spelman, W. (1987). *Problem Solving: Problem-Oriented Policing in Newport News*. Washington, DC: Police Executive Research Forum.

Groff, E.R. and LaVigne, N. (2001). Mapping an opportunity surface of residential burglary. *Journal of Research in Crime and Delinquency*, 38, 257–278.

Heldon, C.E. (2004). Exploratory analysis tools. In J.H. Ratcliffe (ed.) *Strategic Thinking in Criminal Intelligence* (pp. 99–118). Sydney: Federation Press.

HMIC (1997). *Policing With Intelligence*. London: Her Majesty's Inspectorate of Constabulary.

LaGrange, T.C. (1999). The impact of neighbourhoods, schools, and malls on the spatial distribution of property damage. *Journal of Research in Crime and Delinquency*, 36, 393–422.

Leigh, A., Read, T. and Tilley, N. (1996). Problem-oriented policing. *Police Research Group: Crime Detection and Prevention Series*, Paper 75, 62.

Leigh, A., Read, T. and Tilley, N. (1998). Brit Pop II: Problem-orientated policing in practice. *Police Research Group: Police Research Series*, Paper 93, 60.

Maguire, M. (2000). Policing by risks and targets: Some dimensions and implications of intelligence-led crime control. *Policing and Society*, 9, 315–336.

Maguire, M. and John, T. (1995). Intelligence, surveillance and informants: Integrated approaches. *Police Research Group: Crime Detection and Prevention Series*, Paper 64: 58.

Martin, D. (1997). From enumeration districts to output areas: Experiments in the automated creation of a census output geography. *Population Trends*, 88, 36–42.

Martin, D. (1998). Optimizing census geography: The separation of collection and output geographies. *International Journal of Geographical Information Science*, 12, 673–685.

NCIS (2000). *The National Intelligence Model*. London: National Criminal Intelligence Service.

Nelson, A.L., Bromley, R.D.F. and Thomas, C.J. (2001). Identifying micro-spatial and temporal patterns of violent crime and disorder in the British city centre. *Applied Geography*, 21, 249–274.

Openshaw, S. (1984). The modifiable areal unit problem. *Concepts and Techniques in Modern Geography*, 38, 41.

Ouimet, M. (2000). Aggregation bias in ecological research: How social disorganization and criminal opportunities shape the spatial distribution of juvenile delinquency in Montreal. *Canadian Journal of Criminology*, 42, 135–156.

Pawson, R. and Tilley, N. (1997). *Realistic evaluation*. London: Sage.

Ratcliffe, J.H. (2002a). Damned if you don't, damned if you do: Crime mapping and its implications in the real world. *Policing and Society*, 12(3), 211–225.

Ratcliffe, J.H. (2002b). Intelligence-led policing and the problems of turning rhetoric into practice. *Policing and Society*, 12, 53–66.

Ratcliffe, J.H. (2004). *Strategic Thinking in Criminal Intelligence.* Sydney: Federation Press.

Rice, K.J. and Smith William, R. (2002). Socioecological models of automotive theft: Integrating routine activity and social disorganization approaches. *Journal of Research in Crime and Delinquency*, 39, 304–336.

Rosenbaum, D.P. and Hanson, G.S. (1998). Assessing the effects of school-based drug education: A six-year multilevel analysis of project D.A.R.E. *Journal of Research in Crime and Delinquency*, 35, 381–412.

Rosenbaum, D.P., Flewelling, R.L., Bailey, S.L., Ringwalt, C.L. and Wilkinson, D.L. (1994). Cops in the classroom: A longitudinal evaluation of drug abuse resistance education (DARE). *Journal of Research in Crime and Delinquency*, 31, 3–31.

Sampson, R.J., Stephen W. Raudenbush and Felton, E. (1997). Neighborhoods and violent crime: A multilevel study of collective efficacy. *Science* 277, 918–924.

Sampson, R.J., Morenoff, J.D. and Thomas, G. Rowley. (2002). Assessing neighborhood effects: Social processes and new directions in research. *Annual Review of Sociology*, 28, 443–478.

Scott, M.S. (2000). *Problem-Oriented Policing: Reflections on the First 20 Years* (p. 214). Washington, DC: COPS Office.

Shaw, C.R. and McKay, H.D. (1942). *Juvenile Delinquency and Urban Areas.* Chicago: Chicago University Press.

Sheptycki, J. and Ratcliffe, J.H. (2004). Setting the strategic agenda. In J.H. Ratcliffe (ed.) *Strategic Thinking in Criminal Intelligence* (pp. 194–216). Sydney: Federation Press.

Sherman, L.W., Denise, G., Doris, M., John, E., Peter, R. and Shawn Bushway. (1998). *Preventing Crime: What works, What doesn't, What's Promising.* Washington, DC: National Institute of Justice.

Sigler, R.T. and Talley, G.B. (1995). Drug abuse resistance education program effectiveness. *American Journal of Police*, 14, 111–121.

Simpson, R. and Michael S. Scott. (1999). *Tackling Crime and Other Public-Safety Problems: Case Studies in Problem Solving.* Washington, DC: US Department of Justice: Office of Community Oriented Policing Services.

Skogan, W. (1990). *Disorder and Decline.* New York: Free Press.

Tilley, N. (2003). Community policing, problem-oriented policing and intelligence-led policing. In T. Newburn (ed.) *Handbook of Policing* (pp. 311–339). Cullompton, Devon: Willan.

Walker, S. and Charles M. Katz. (2001). *The Police in America: An Introduction.* Boston: McGraw-Hill.

12
Crime Map Cartography

Learning Objectives

One of the significant developments with modern GIS is the opportunity to explore a myriad of ways to create maps. One minor disadvantage of this increased ability to experiment with mapping is that few GIS provide any rules for map composition, and the growth of GIS into non-geographical disciplines means that few users have ever received any cartographic training. This is unfortunate, because good map design is essential if decision-makers, front-line officers and policy makers are to understand and interpret what they are seeing, and in turn make informed decisions within the crime and criminal justice system.

In this chapter we explain that there are cartographic standards that a good map will adhere to, but within these cartographic conventions the map should also be tailored to the client so that it will address the decision-making needs of the agency. This chapter provides an overview of basic cartographic notions, how the eye perceives colour, ways that both areal and point data can be effectively mapped and finally the different ways that the modern digital age can enhance efforts to create innovative maps.

12.1 Introduction – the purpose of the map

What is the purpose of a map within the criminal justice system? What is the overarching aim that the crime mapper or researcher is trying to

GIS and Crime Mapping   Spencer Chainey and Jerry Ratcliffe
© 2005 John Wiley & Sons, Ltd

address with a cartographic product? This is a key question that any researcher or analyst should ask of themselves before embarking on a mapping project. One significant aim is often to influence decision-making, so that crime reduction and prevention tactics are targeted in the right manner to the right place. As Chapter 4 showed, crime reduction and prevention policies are often place dependent, and an understanding of environmental criminology helps to interpret the spatial patterns of offending and victimisation. The map aids decision-making, and decision-making is sometimes not just a matter of knowing where the crime can be found, but what related features are nearby. Crime mapping is therefore not just about mapping crime – it can also be a process of mapping offenders, opportunities and the local environment. For example, if a crime analyst is mapping assaults and criminal damage that tend to occur at bar closing time, it would be sensible to also map the local public houses and bars. If burglaries were occurring in the late afternoon and local school children were suspected, the location of local schools might be worth mapping.

There are generally two criteria for deciding if a map of crime is to be complemented with additional information, such as school locations or the boundaries of police districts. First, these features can be added in order that an interested audience can orientate themselves to the crime features. An example of this would be a map of local burglaries that will be shown to a Neighbourhood Watch meeting. The inclusion of streets or other local identifiers will help the audience understand the location of offences. Secondly, a map could include police boundaries or other features if there is some theoretical basis to suspect that the additional information is related to the crime distribution (Eck, 1997). As John Eck points out, inclusion of some additional feature suggests that the mapper believes that there is some connection between the crime distribution and the feature. For example, a map of robberies that included the sites of local railway stations either implicitly or explicitly explores the possibility that the robberies are related in space to the location of railway stations.

While the temptation might be to include every possible criminogenic feature on a map (just in case), the analyst also has to contend with the limitations of human perception. Ability to interpret maps is quite high but congested and busy maps can easily overload the map reader. As well as crime events and crime-related features, readers also require spatial clues so that they are able to correlate places on the map with locations in the real world. These spatial reference points can be simply a label showing the name of a town, or can include rivers or the location of police stations. So the task of the crime mapper is to combine crime information, features of criminological interest and spatial reference details while balancing all of this with the limits of the map reader. Each decision to add another layer of detail can either increase the reader's ability to understand the crime picture or can tip the balance and risk losing the audience.

These days crime analysts are not limited to paper. Maps can be generated for a whole range of different mediums, as images for the Internet, PowerPoint presentations and interactive environments. Paper maps have the advantage that they are easy to create, cheap to produce with modern colour printers, and many of the default settings within GIS are set up for paper map production. There are some minor disadvantages. The layout of the page does constrain the range of layout options available, and some colour printers generate pages with considerable colour variation to that displayed on a computer monitor. Bubble jet printers are particularly susceptible to colour imbalance.

The Internet similarly provides a range of possibilities for the crime mapper. Not being required to reproduce maps on paper releases the cartographer from the traditional white background and opens up a range of cartographic colour possibilities which would otherwise require a considerable printing budget. Freedom from such constraints does not negate responsibility with regard to cartographic design. While the medium has changed, the map reader has not. Issues of confidentiality are also important to resolve before mapping on the Internet.

PowerPoint (and other presentation software packages) can be used to convey intelligence information with considerable impact and, like the Internet, releases the crime mapper from the need to have a white background. Indeed, given that dark backgrounds and light text are better for PowerPoint-type presentations, white-backed maps are not ideal. While non-paper maps provide interesting new opportunities, they also present challenges with regard to choices of colour and classifications.

12.2 Design considerations

As respected cartographer Mark Monmonier notes, 'graphics software no more guarantees good maps than word-processing software assures good writing' (1993, p. 12). In choosing to construct a cartographic output, a decision has already been made that a map is the most appropriate medium to impart some information. The next step is to consider two important factors that will influence the final product:

1. What is the nature of the audience?
2. What is the primary crime information to convey?

The audience will influence the text added to the map (will they understand jargon?), the level of spatial referencing provided (are the audience familiar with the area, or will they need spatial reference points?) and any security and confidentiality issues (is the map for 'in-house' use and able to depict sensitive information, or is it for a public website?). Identifying the nature of the audience will dictate the whole map composition. Few people in the criminal justice profession are trained cartographers, and crime mappers should strive to make the map as readable as possible. The most important feature will be the actual mapped area, but it is also important to strike a balance when composing the map so that there is still enough space to include the legend and any text. The next section explores how good design principles can be used to compose a cartographic product aesthetically, and summarises the main elements of a map.

12.2.1 Good layout and the visual hierarchy

A good layout is essential if a map is going to successfully convey the intended message. It is not a task of simply setting the study area in the middle of the page and hitting the print button: thought is required to determine the best place to site not just the study area but also the legend, title, directional guide and scale bar. Whether on the printed page or as a computer image file, the shape of the study area will be a significant factor in deciding if the output will be in *portrait* format (height greater than width) or *landscape* format (width greater than height). According to Dent (1999) there are a number of different elements to consider when composing a map. These include balance, focus of attention and internal organisation.

Balance requires the map composer to construct the elements of the map such that the whole image is balanced around the 'optical centre', an imaginary point slightly higher than the geometric centre of the visual field – the area within the borders of the map (Figure 12.1). Of course, some areas are always going to be difficult to map. As Harries (1999) suggests, an option is to separate long or awkwardly shaped areas into different pieces and use an inset map to show how they fit together. It is also possible to use other map elements, such as legend and text, to balance the mapped area and regain visual symmetry. These can help to reduce the amount of 'open space'. As Dent notes, 'complex designs require careful planning to use all spaces efficiently while retaining a visually harmonious balance' (1999, p. 245).

The *focus of attention* is the area to which map readers will be most easily drawn. The focus of attention at, or around, the optical

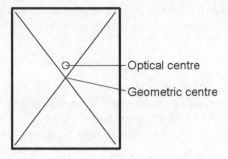

Figure 12.1 Optical and geometric centres. Source: Dent (1999, p. 243). Reproduced by permission of the McGraw-Hill Companies

centre can also be enhanced by using visual parameters such as line width, symbol size and colour, to further focus the attention. These additional features will work best when the focus is close to the optical centre, and crime mappers should be cautious when the primary information they wish to convey to the reader is significantly displaced from this central site.

Internal organisation relates to the way in which different parts of the map are aligned and structured on to the page. Map elements should be added to the page in a logical manner that displays thought and preparation, rather than placed haphazardly on to the page (Harries, 1999, p. 60). This internal organisation will translate to a better level of understanding in the map reader.

The whole map aims to convey the important facts as well as to impart other information that aids interpretation of the key facts. The crime mapper must plan to incorporate all of the necessary elements onto the map in such a way that they achieve a visual balance, and conform to what is called a 'visual hierarchy' (Dent, 1999) or a 'scale of concepts' (Monmonier, 1993). The visual hierarchy is the graphic solution that solves the problem of what gets priority on a map. By considering the visual hierarchy (or scale of concepts), the crime mapper should be able to determine what map components are given priority and prominence over others.

In nearly every case, the thematic symbols that convey the data will get the most priority and the greatest level of distinction. Other features are added to the map in hierarchical levels beneath this layer. Dent (1999) provides a visual hierarchy table as a guideline, adapted here as Table 12.1. The existence of an item in the table does not dictate that it must appear in the final map, or the eventual order of items (for example, a police marine unit may require a map that gives greater prominence to water features),

Table 12.1 Suggested organisation of mapped elements in the Visual Hierarchy (adapted from Dent, 1999, p. 252). Reproduced by permission of the McGraw-Hill Companies

Object	Visual level
Thematic symbols of crime data	1
Title, legend material, symbols and labelling	1
Base map – land areas, including administrative and jurisdictional boundaries, significant urban and physical features	2
Important explanatory materials – map sources, credits, security classification	2–3
Base map – water features such as lakes and rivers	3
Other base-map elements – labels, grids, scales, North arrow	4

but the table does provide a suggestion for the management of map layers, with Visual level 1 being the most prominent feature level.

This visual hierarchy is not a definitive rulebook: it is a suggested guideline to aid cartographers. There will be times when a more unusual map requires an approach that varies from the visual hierarchy in Table 12.1. However, as a general guide it goes some way to providing welcome guidance for map creation. The next section describes some ways in which the individual components of the visual hierarchy can be used to best effect.

12.2.2 Cartographic elements

The following section is provided for crime mappers who may not have taken a formal cartographic course. It may also serve as a helpful revision for readers that have some experience with map design.

12.2.2.1 Directional guide

Although GIS display maps with North to the top of a computer screen by default, there is no guarantee that a final product follows this convention. Many areas would be better illustrated cartographically if they had a different orientation, though the move from hand-drawn maps to GIS has undoubtedly made this harder. Directional guides, otherwise known as North arrows, are not necessary for small-scale maps that depict clearly recognised areas such as the UK, contiguous USA or Australia, as the shape alone is sufficient for the reader to determine direction. Large-scale maps should always have a North arrow or compass, especially as it may not be easy to determine the direction from the image. The arrow should be unobtrusive and not too large, as it is not intended to be a dominant

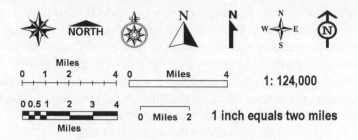

Figure 12.2 A range of different directional guides (North arrows) and scale bars

feature of the map (it has a visual level of 4 in the visual hierarchy). The top part of Figure 12.2 shows a range of North arrows.

It should be borne in mind that complicated symbols do not translate well to PowerPoint presentations. This is because some projectors are unable to correctly project the intricacy of a complex symbol and this can make it difficult for viewers at the back of a presentation room to discern the symbol. Complicated North arrows are also not suitable for maps that will be photocopied many times as they will lose clarity with each reproduction. In both of these cases a simple, clear North arrow may be preferable to a stylish and intricate compass design.

12.2.2.2 Map scale

Many people are confused by map scales, yet they are an important feature of maps because often the most significant point to convey is the position of one crime event relative to the position of another location. The distances between objects on the map in relation to distances in real space must therefore be conveyed to the map reader. Crime mappers often get confused by the interpretation of map scale because they think that small scale relates to small detail and small areal coverage. This is not the case. Large-scale maps tend to have more detail on them (e.g. a street map of a housing estate), whereas small-scale maps tend to have less local detail on them but have a larger geographic coverage (e.g. a map of the world). Scales can be represented in a number of ways.

The first is to use a *natural scale* such as the ratio of 1:633 600 (the equivalent of one inch representing 10 miles) or a *representative fraction* (1/633 600). This means that for any one unit on the map, 633 600 units are represented on the ground. This scale could also be represented verbally (using familiar units), an approach that many find easier to interpret. For example, a verbal scale equivalent of 1:633 600 or 1/633 300 is 'one inch equals ten miles'. While a verbal scale can work for paper maps, it does cause problems if your map is projected on to PowerPoint or through the Internet. With PowerPoint it is difficult to control the final size of the image when projected on the screen, and impossible to control the resolution or monitor size of a remote Internet user. Bear in mind that if a verbal scale is used, and then viewed with a portable projector (such as an overhead or computer data projector), then moving the projector closer to, or further from, the screen will change the scale. Accuracy (and therefore truth) is nearly impossible to achieve under these circumstances.

A solution is to use a *scale bar*, also known as a *graphic scale*. These visual representations of relative distance are simple to create in modern GIS and for most readers are easy to interpret. They can be as simple as a line with a distance measure, or can have alternate check boxes with tick marks and display different distance units. If a map may be viewed by an international audience, then it is worth considering inclusion of a scale for kilometres and miles. 'Kms' or 'kms' is an acceptable abbreviation for trans-Atlantic maps where the spelling of kilometres differs. Figure 12.2 illustrates these different approaches for representing scale.

12.2.2.3 Borders

Whether to envelop your map in a frame or not depends on the study area being mapped and the output format. External borders around the whole map are optional, but do help to constrain eye movement within the image area. If features such as coastlines and other significant boundaries are shown, it is helpful to include a border and to ensure that the feature extends to the frame. In this way a frame can help to avoid 'confusing space' where it is not clear what lies beyond the boundary of the study area. Figure 12.3 shows different maps of the statistical sub-divisions of Sydney, Australia. It is unclear from the first image on the left what lies

Figure 12.3 Three maps of the statistical sub-divisions of Sydney, Australia

around the city. The mapped area appears as an island. The centre map includes the coastline which is continued to the frame and this helps to anchor the city within the map. The third map uses shading to indicate water. This clarifies that there is no land to the East and helps to reduce white space.

Maps that are saved as jpeg or gif images for display on the Internet can have a border added at the time of map production, or have a border added to the image file using the border instruction within the Internet html code for the image. In the following html code, the final display would depict the image1.gif map in a frame 200 pixels wide by 50 pixels high, with a two-pixel wide border:

<img src = 'image1.gif' width = '200' height = '50' border = '2'>

Borders are usually derived from polygon files of the largest area to be mapped. As can be seen from the coastline in Figure 12.3, some borders can be very complicated and intricate. With the coastline around Sydney, each individual twist and turn increases the amount of data that have to be stored by the GIS and this can rapidly increase the size of image files and maps: a consideration when data may be e-mailed to other users. This level of intricate detail is rarely needed for most crime mapping applications and it can be worth spending some time to source more generalised background files and borders.

12.2.2.4 Text

Although the primary aim of a map is to convey graphic information, there will still be a necessity for some text. A *title* (and *sub-title*) is necessary to convey the meaning of the map. Though not essential if a caption explains the map's purpose, it is still easy for an electronically published map to become divorced from its caption. Adding a title directly to the map image is usually the best way to assure that the title and graphic never become separated. Because most languages are written from left to right and top to bottom, placing the title to the top of the map is generally expected. Placing a title at the bottom right of a map is to be avoided unless the shape of the mapped area leaves no choice. If the study area is shaped so that this really is the only option, then it may be necessary to increase the font size to give the title greater prominence.

In crime mapping there can be a tendency to employ jargon and abbreviations in titles. These should be avoided where possible so that the map can reach a wide audience. The title should also be succinct, and it is best to avoid letting the title run on for a number of lines and become a paragraph in itself. The title should aim to inform and generate an impact. Further clarification of the specific data employed could run as a sub-title below (and in a smaller font) to the main title. For example, the title 'Crime drops across the city' would get attention, while a sub-title could explain the actual data used to show the map. An example sub-title could be 'Philadelphia change in crime level by census tract, 2003 to 2004'. An alternative to a sub-title is a descriptive heading above the map key.

The name of the cartographer and the source of the data should appear on the map, but not in a prominent place. Source details are not essential if the crime data are derived locally, for example by the police department creating the map. It is also not necessary to cite the source of the municipality boundaries or census boundaries that are used as base maps, but cartographers should cite the source of the crime data that are mapped if it originates external to the agency. It is also not a hard and fast rule that the crime mapper adds their name to the map, though it is considered good cartographic practice. A simple line of text placed discretely near a bottom corner of the map is sufficient. With crime mapping it is helpful to include some measure of the reliability of the map, and for certain types of maps it may be necessary to include a *security classification*. For agencies where this is a necessary feature, the security classification should be displayed fairly prominently.

Some *location text* may be necessary to orientate the reader. Decisions regarding the quantity of text to place on the map should be a judgement call, based on a balance between helping the reader locate key areas and swamping them with too much information that will detract from the primary information. It might be helpful to name a few significant streets or towns, but it will not be necessary to name everywhere. For example, on an American urban map it would be reasonable to label 12th Street and then three roads later, 15th Street. It would not be necessary to label the two intermediate roads. Labelling roads and suburbs is less important for non-local audiences if they are unlikely to ever visit the study area.

12.2.2.5 Fonts

All of the text should be considered part of a textual hierarchy, which is itself part of the visual hierarchy. Titles are more important than sub-titles, and should therefore be more prominent. The data sources and name of the cartographer are less important than the rest of the map text. On a map of towns, some urban centres will be larger and therefore the text more significant than others, and so on. This hierarchy of text should be reflected in the choice of font. The modern cartographer is faced with a huge range of font possibilities. While there are a variety of symbol fonts, most alphanumeric fonts can be broadly classified as *serif* or *san serif*.

Serif fonts have small embellishments to parts of the text character. These small additions are predominantly decorative, but they also make large blocks of text easier to read. Garamond and Times New Roman are both examples of serif fonts. San serif fonts are more basic in design and lack the serif embellishments. They are not as easy to read in volume but are more effective when employed as headings, titles or captions. Common example of san serif fonts include Helvetica and Arial. It is worth bearing in mind that criminal justice tends to be a fairly conservative discipline. A wacky, near-illegible font may not be appropriate for a map that will be viewed by a conservative head of an agency, and informal or casual fonts will not be appropriate for serious topics such as maps of heroin overdose deaths. Too many different fonts appear unprofessional and confusing, so it is better to vary font size rather than experiment with numerous fonts on one map.

12.2.2.6 Thematic symbols

The most significant features on a map are the thematic symbols that are placed on the map to represent either features and locations (qualitative), or a measure of criminality or other quantitative data. Point symbols can not only mark specific locations, but also show the value of attributes at these places. There is a wealth of different ways that point symbols can be used, and few rules to guide the cartographer. Qualitative symbols usually have different shapes and/or colour and can be used to depict different features.

Pictorial symbols that unambiguously represent features can save the reader from constantly referring to the map key (Monmonier, 1993) though a universal and definitive set of symbols to depict specific crime events does not yet exist. The Federal Emergency Management Agency (FEMA), part of the US Department of Homeland Security, is currently developing a set of symbols relevant to emergency incident managers. See www.fgdc.gov/HSWG for more details. This work is ongoing, though users should be cautioned that there are already hundreds of different images in the symbol library. If even a tenth of them were depicted on a map, the similar colour and shape of the symbols would probably have a map reader rapidly reaching for the key.

Given the range of possible criminal acts, and the subtleties of different types of offending (for example, assaults with a weapon, without, weapon used, not used, with a knife, gun and so on), an agreed typology of symbols within the crime mapping community is unlikely. However, readers of mainstream roadmaps will be familiar with commonly-used symbols for less criminogenic features such as restaurants, hotels and public toilets.

Quantitative symbols tend to vary in length, area or volume, or occasionally hue or value. Size tends to be the predominant way of depicting differences in attribute value. Most crime mappers will be familiar with the use of graduated circles as a way of representing different quantities of data, but different volumes of point objects or line sizes can also be employed. For example, LeBeau used varying road widths to depict the number of assaults in a street (LeBeau, 2000). The thematic mapping of point data is dealt with in a later section of this chapter.

12.2.2.7 Legend (or key)

Map legends (also known as keys) require careful consideration. This is not only to ensure that they impart the intended information, but also because there is a considerable conformity with the default legends within many GIS packages that stifles innovation and experimentation.

Automated keys tend to commit a heinous crime when generating a legend for quantitative data: the ambiguous class. This is unfortunately a common problem with GIS. Imagine that a map depicts the rate of assault victimisation per 10 000 population, with four classes:

1. 40–69
2. 27–40
3. 15–27
4. 0–15

While this legend seems clear, with a little consideration problems quickly arise. First, it would be more informative to differentiate the areas with a zero rate of victimisation, as perhaps this suggests an area with little or no population, or would be useful to signal those areas of interest that do not suffer from crime. Secondly, if an area possesses an assault victimisation rate of 40 per 10 000 population, would it be included in the first or second class? A more informative legend would be as follows:

1. 40 to 69
2. 27 to less than 40
3. 15 to less than 27
4. 1 to less than 15
5. 0

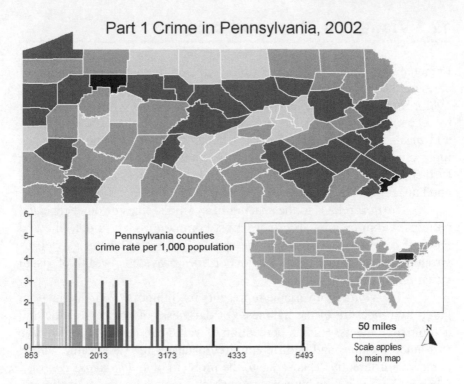

Figure 12.4 Part 1 (serious) crime rate per 1000 population in the State of Pennsylvania for 2002. The histogram doubles as both the legend and as an indication of the distribution of values

With modern GIS this usually requires refining the legend, but the extra effort will be rewarded with a clearer and more informative map that is more accurate. More imaginative keys can aid interpretation of quantitative data. Figure 12.4 shows how the legend can not only convey the attribute values for a certain class, but also indicate visually the frequency of areas within each class.

12.2.2.8 Base maps

Base maps, such as road layouts, rivers, administrative regions and vegetation maps, should be used sparingly. With modern GIS it is all too easy to complicate a map with multiple datasets that can be integrated at the click of a mouse button. Base maps are low in the 'visual hierarchy', and as such should be considered as not essential to the final image. Base maps have value in providing visual reference clues to enable users to better interpret the map, but they also have a tendency to dominate the image and distract the reader from the primary information.

12.3 Visual variables and colour

French cartographer Jacques Bertin recognised that human perception of cartographic symbols was determined by six main variables (Bertin, 1983). Human eyes are able to distinguish between one symbol and another by recognising variations in these variables. The first three are *shape*, *size* and *orientation* (Figure 12.5). The shape, size and orientation are the main clues people use to determine that items are different. The three further visual variables in addition to these main three are texture, value and hue (Figure 12.5).

Texture relates to the spacing of symbols, lines or other repeated patterns within an area. These are most commonly seen in dot, line and hatched patterns. Variations in the spacing of each element as well as variations in the thickness of the dots or lines provides a wealth of visual variation.

The *value* of a variable represents its lightness. *Value* is the term given to a measure of the lightness (or darkness) of both chromatic and achromatic objects. Human perception of value can be influenced by the surrounding values and in different circumstances the same value will be perceived differently. This is due to the problem of *simultaneous contrast*: when looking at a map humans are unable to isolate one point and ignore the surrounding areas. When the surrounding areas spill over to influence perception of an object, the effect is known as *lateral inhibition*. An object surrounded by objects with darker values will tend to appear brighter and

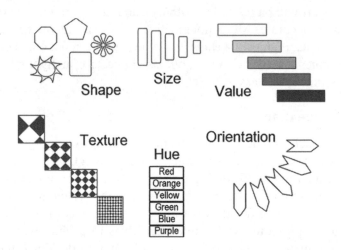

Figure 12.5 Bertin's six visual variables

slightly larger, while an object surrounded by light values will appear darker and smaller.

Colour is the term people use when they actually mean *Hue*. Hues range through our visible spectrum roughly from blue to red. Our perception of the hue is also a factor of its lightness and saturation – the amount of the colour present. A hue with low saturation will tend to appear as a washed-out colour tending towards grey.

To these six retinal variables we can add two more spatial variables: position and animation. *Position* is a variable that crime mappers have limited ability to utilise. Map readers get disoriented if crime mappers move the position of crime events too much; however, the second variable is more flexible. *Animation* is where the position, colour, shape or size of a feature changes as the animation progresses over time. Animations are most commonly used to indicate the passage of time, and any change in a map object is suggestive of a change in the variable over a set period of time. Animations are not the only way to show changes in crimes over time, and Chapter 8 describes a number of other mapping techniques for temporal patterns.

12.3.1 Shades and patterns

Thematic mapping requires some variation in the visual variables for different types of area. There are a number of the visual variables that can be adjusted, though some are better than others. Of the six variables in Figure 12.5, texture, value and hue are most commonly used in crime mapping to distinguish an area's unit value.

If a decision is made to avoid relying on colour, this does not mean that the cartographer is short of options. GIS provide a wealth of shading options that the crime mapper can draw upon to increase map comprehension. Dot patterns have variable size and dispersion and are a useful way to design a number of classes. It is usual to increase the texture (increase the number or size of the dots) as the crime variable increases. Dots are a good choice because, unlike lines, they do not follow a particular direction. They tend to fit into any polygon fairly well (unless it is very thin and long), and the pattern is easily discernable to the human eye.

Line patterns fill a polygon with lines that are parallel and flow in one direction. While the cartographer sometimes has some control over the line direction, line patterns can confuse the map reader when they run close to the boundaries of area units. They also tend to lead the eye in their predominant direction and make the eye wander. This can conflict with the

Arrests for violent
crime per 100,000 people
US states, 2001

≡ 251 to 387 (12)
☒ 169 to 251 (12)
⫽ 117 to 169 (11)
≣‖‖ 32 to 117 (14)
▨ No data

Figure 12.6 The similarity of pattern density makes it difficult to discern a meaningful distribution

perception of where administrative boundaries lie. The best way to confuse a map reader is to vary the line orientation and employ different line directions for different classes. A map that has lines running in different directions can be extremely difficult to understand. Figure 12.6 demonstrates how difficult interpretation of data can be when these simple visual rules are ignored.

Shading allows the cartographer to fill a polygon with different values (as shown in the top right of Figure 12.5). Grey shading is necessary if colour is not employed. Perception of grey shading differs and people tend to be able to better distinguish between classes at the dark end of the scale rather than the light end. One possible negative aspect of grey-scale shading is that poor or repeated photocopying of a map can quickly degrade the quality of the shading and render the class boundaries indistinguishable.

Many GIS provide the crime mapper with a plethora of different thematic options, but a number of these have limited applicability; medium- to low-quality printers will not be able to do justice to the more complicated patterns. When different colours are used, cartographers should be cautious of the impression the map might give if photocopied or faxed. Remember that it is always good practice to photocopy a colour map to examine the readability of the map in greyscale. Mixing different types of line patterns, with either different line orientation or lines and dots, is generally not a good idea and the resultant map can appear to be haphazard and unprofessional.

Human perceptual ability is different across the full range of light to dark, with perception of differences between areal units being better at the darker end than the light end. Therefore the crime mapper that wishes to have a map with light to mid-grey classes may have to compromise with fewer classes. By convention, it is traditional to make the areas with the highest amount of crime the darkest regions. There are generally two options for true value-only maps (without any hue). A white to black map uses white and black as the extreme ends of the class range, employing variations in grey to map the intermediate classes. A greyscale range does not have to extend to the white/black extremes and can run from light grey to dark grey. This may be an option if the cartographer wishes either to avoid the impression that areas shaded in white are devoid of any crime, or to use white shading to indicate that data are missing from these areas. Secondly, by avoiding the use of black for shading, it is possible to retain black for the borders of each area. Black shading makes some black borders invisible.

12.3.2 Colour, wavelength and saturation

Colour increases the range of cartographic possibilities for crime mappers, and indeed it is difficult to imagine maps of criminal patterns that do not have some element of colour, so accustomed are we to the accessibility of colour computer monitors and printers. You may remember, from earlier in this chapter, that when we talk of colour we are actually referring to the hue of an object. Colour is one of Bertin's visual variables (hue), but the practicalities of map production are such that it is best not to rely on colour alone to achieve mapping objectives. A good rule of thumb is to 'Get it right in black and white' before considering the advantages of adding colour.

We rely on our colour perception to interpret light that is reflected from a page or emitted by a computer monitor. This light is focused by the lens in our eye such that it forms a sharp image on the retina, the rear wall of the eye. On the retina are about 100 to 120 million rod cells, used to detect value but not colour (*achromatic*), and around 6 million cone cells (*chromatic*). While the cone cells can react to colour, they are less sensitive than the rods and require a greater amount of light to function effectively, which is why it is harder to detect the colour of objects at night. Individual cone cells have peak sensitivities to different wavelengths,

369

roughly in the red, green and blue bands, in the approximate proportions of 40:20:1.

Although differences in the number of cones that are effective at different colours are not usually an issue for us, the variation in wavelength does introduce a feature that is useful for deciding on a colour for presentations and map backgrounds. Our eyes are naturally tuned to focus on green, but to focus on the red end of the spectrum requires the lens in our eye to make a slight adjustment (to become more convex), as if the red object were slightly closer than in reality. Conversely, light from a blue object will focus before the retinal surface and will require the lens to adjust slightly the other way (concave), giving the impression that the blue object is slight more distant. This is useful to consider when putting together a PowerPoint slideshow, or if it is decided to have a coloured background to a map. Colours at or near the blue end of the spectrum appear to fade into the background, while colour choices to the red end appear to come towards us slightly. Reds and oranges (with long wavelengths) are therefore inappropriate choices for background colours, as they tend to dominate the foreground and do not fade into the background as well as purples or blues (with their shorter wavelengths). Blues are also more preferable as background colours as people have a reduced sensitivity to colours at the blue shift of the spectrum due to the relative paucity of 'blue' cone cells.

There can be a tendency among crime mappers to go overboard with the range of colours, especially for thematic maps. Although the human eye is able to distinguish millions of different colours, the problem for the crime mapper is to limit the number of colours so that the reader can sufficiently distinguish between different colours to accurately interpret features or areal differences in a map. Fortunately, there is a simple rule of thumb regarding the number of map classes in a thematic map: no less than four, no more than six (Dent, 1999; Harries, 1999). With this rule in mind, the crime mapper should constrain the determination of thematic classes to up to six different colour combinations.

Saturation (also more technically known as *chroma*) is an indication of the intensity in a colour. The range of saturation is from 0%, essentially grey, to 100%, where the object possesses the full colour possible. Achromatic colours have a saturation of 0%. Visually, the scope from 0 to 100% appears to us to range from mid-grey tones right through to the full radiance of the colour without a hint of grey. Saturation is one variable that can be adjusted to indicate the change in a quantitative variable (column c in Figure 12.7).

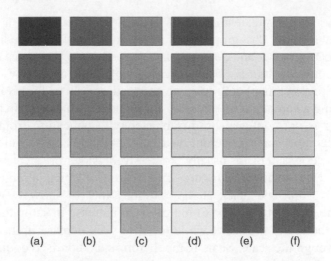

Figure 12.7 Six columns demonstrating different colour plans for thematic mapping (see Plate 4)

12.3.3 Human perception of colour

A map will rarely use just one colour, though a greyscale array can be perfectly acceptable for a range of classes. As Bertin notes, 'since visual order derives from value perception, a monochrome series running from white to black is itself sufficient for the representation of an ordered component' (Bertin, 1983, p. 90). A range of colours is more commonly used to differentiate base map features and thematic classifications. Choice of hue and saturation should be dictated not only by the degree of contrast between areas (though this is a good basis for colour selection), but also by the way that human beings perceive individual colours. As said in the previous section, colours with longer wavelengths (reds, oranges, yellows) appear to advance towards the reader, while shorter wavelength colours (blue, purple, green) recede. The longer wavelength colours can be termed the *warm* colours, while the shorter wavelength colours are termed the *cool* colours.

People tend to associate the warm colours with excitement, stimulation or aggression. This fact, combined with the advancing nature of the warm colours, make these hues ideal for significant symbols and classifications that the crime mapper wishes to give the greatest prominence. In contrast, the cool colours are associated with serenity and security. These receding colours make better backgrounds, graphics and symbols of lesser importance. The best backgrounds are either very light or very dark as this allows for the greatest contrast between background and a range of foreground features. Backgrounds that have a mid-tone or intermediate value limit the cartographer and do not make for good maps. The most effective combinations of colour are between colours of a high and a low value.

Groups of maps collected into crime atlases can be further enhanced by retaining a commonality of colour throughout each image. In this way, the reader is not confused by different colour application from map to map. If thematic shading is employed, it is helpful to always run the scale in one direction, and that is usually progressively darker the greater the level of crime. This is shown in Figure 12.7 where (a) depicts a monochrome scale with white and black at the extremes and (b) shows a greyscale range. The range of column (c) indicates an increase in saturation of red closer to the top of the scale, while (d) shows an increase in the strength of hue through the scale. Scale (d) is therefore a colourised equivalent of (b). If all of these scales were used to shade a choropleth crime map, then the darker classes would represent higher crime values.

In Figure 12.7 ranges (a) through to (d) represent sequential schemes, where there is a gradual increase in shade or colour as the value of the variable increases. The increase is in one direction. Colour scheme (e) also shows a sequential scheme; however, here the saturation is the same for each class. The only difference is in the choice of hue. The problem with such strong colours in each class is that any final map will have a strong thematic colour element, and will make the addition of other features difficult to see.

The scale shown at (f) shows a *diverging* scheme (Slocum, 1999) where a central colour represents a low value, and the hue increases away from the centre. This type of diverging scale could be used to effectively indi-cate an increase or decrease in a crime problem. Central values in this type of class definition suggest little change, but significant increases or decreases can be clearly demonstrated with the use of strong hues at the extremes of the scale.

12.3.4 Practical colour considerations

Law enforcement and criminal justice are usually organised along strict geographical hierarchies (see Chapter 2) and it is rare that all of the functions of an organisation are centrally located. This means that good maps are often passed around an organisation and from building to building. For example, a crime mapper working for a police department may be required to submit a map to a central intelligence office on a regular basis.

Often the means of communication is by fax or photocopied image. When this happens it is possible for most, if not all, of the benefit that was gained from a colour map to be lost. As stated at the beginning of this section, a crime mapper should always 'get it right in black and white' as it is impossible to guess who may fax or photocopy the map. It is good practice to photocopy a colour map in order to examine the readability of the map in greyscale. In this way the analyst can see how the receiver at the other end of a fax machine would view the image. Another way to do this is to copy the map to a word processor program and then format the picture to a greyscale image.

12.4 Thematic maps of areal data

Crime data are usually represented as either specific points indicating the location of an offence (a crime event) or the home address of an offender, or as areal data representing the amount or level of crime in a certain area. Often the areal measure of a criminal justice statistic is created by counting the number of events that have occurred within an area. For example, Figure 12.8 shows a variety of different maps of the same data: the number of arrests for burglary in 2001 per 100 000 people by US state. These data are available from the website of the US Federal Bureau of Investigation (www.fbi.gov), but are themselves sourced from individual arrests across the country. Each individual arrest could be mapped as a point location, but a map of hundreds of thousands of arrests would swamp a display and be relatively meaningless. Mapping areal data, derived from point data, is therefore often a worthwhile venture.

'Choropleth mapping' is the term given to the use of variation in shading or colour to represent values in areal units. The data are usually reduced to a number of classes in a process called *classification*. Classification is necessary to reduce large volumes of data to a level suitable to make general statements about the distribution of the crime phenomena being mapped and to aid interpretation.

Choropleth maps are useful when a crime mapper wants to display a crime distribution comprising of data measured by well-defined areal units such as police beats, county boundaries and state border lines. However, it should be borne in mind that the process of shading or colouring an area will give the impression that the offence is distributed across the whole area evenly. This was demonstrated in Figure 6.4 in Chapter 6, and is also shown in Figure 12.8 where the shading for each state is uniform,

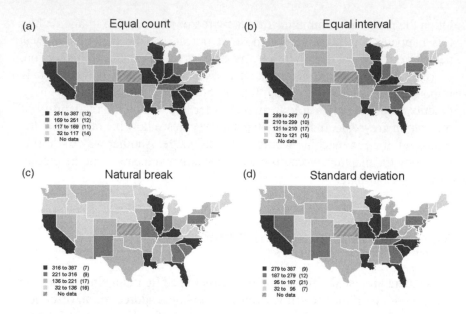

(a) Equal count

251 to 387 (12)
169 to 251 (12)
117 to 169 (11)
32 to 117 (14)
No data

(b) Equal interval

299 to 387 (7)
210 to 299 (10)
121 to 210 (17)
32 to 121 (15)
No data

(c) Natural break

316 to 387 (7)
221 to 316 (9)
136 to 221 (17)
32 to 136 (16)
No data

(d) Standard deviation

279 to 387 (9)
187 to 279 (12)
95 to 187 (21)
32 to 95 (7)
No data

Figure 12.8 Four maps of the same data: Arrests for violent crime per 100 000 people in US states, 2001. The four images depict different classification schemes using MapInfo automated methods. Source: FBI Uniform Crime Reports

but the reality is more likely that arrests will be concentrated in the large urban areas where population density is highest.

Care needs to be taken when using choropleth maps to map crime totals, such as numbers of arrests or monthly burglary totals. This is because larger areas will tend to dominate the map and this tends to distort the viewer's perception of the crime distribution by giving larger areas prominence, altering the impression of the crime distribution. It is better to use a proportional sized symbol located at the areal unit centroid, or better to use a ratio or rate of crime. Common ratios include a rate per 1000 residents (applicable for maps of domestic violence) or a rate per 1000 housing units (as could be used for a map of residential burglary). This is why Figure 12.8 shows a rate per 100 000 rather than the actual frequency of arrests. Choropleth maps of raw counts should be avoided.

The ratio is a simple calculation based on the number of crimes (or *frequency*), which becomes the numerator, divided by the figure used to calculate the rate, also known as the *denominator*. The choice of an appropriate denominator should be determined by the crime type. For example, if residential burglaries were the crime in question, then domestic housing units would be a good denominator. If assault or robbery were the crime type, the resident population might make a more appropriate measure. The result is usually multiplied by some factor to make the final ratio more easily understood. For example, if a neighbourhood had 3000 housing units, and in one month there were 21 residential burglaries, the burglary rate per 1000 housing units would be:

$$\frac{\text{burglaries}}{\text{housing units}} \times 1000$$

which in this example is:

$$\left(\frac{21}{3000}\right) \times 1000$$

which equals 7 burglaries per 1000 houses.

12.4.1 Choice of denominator

A few words are necessary in regard to the choice of denominator. While the number of housing units is an appropriate denominator for residential burglary, and these data can be sourced from a census, other denominators are not so easily obtained. For example, the population at risk of street robbery is the number of people in an area or on the street at the time of the offence. The problem is that census data are not appropriate for calculating this value (Chainey and Desyllas, 2004). The population of a city centre is very different at 4 pm compared to 4 am. Similarly, the population at risk for vehicle crime is the number of cars, but this again changes throughout the day in relation to the exposure of vehicles at different places and at different times of the day (Clarke and Mayhew, 1998; Clarke and Goldstein, 2003). Calculating these types of denominators can be tricky but new techniques are emerging that accurately model these denominator variables. For example, Chainey and Desyllas have demonstrated the use of pedestrian modelling for generating accurate and precise on-street pedestrian counts to aid the analysis of street crime risk patterns (Chainey and Desyllas, 2004).

12.4.2 Classification of areal thematic maps

The first question to address is how many classes should be mapped. Increasing the class size increases the amount of variation that can be displayed, but it also increases the difficulty that map viewers have in distinguishing between different classes. Reducing the classes improves comprehension but reduces the level of detail that can be displayed. While people can distinguish up to 10 different classes in a shaded map (Dent,

1999, p. 163), in reality an appropriate number is much less than this. As mentioned in section 12.3.2, a suggested level is no less than four classes, and no more than six.

GIS have the ability to automatically create thematic maps of areal data with little effort from the user. Many GIS can create choropleth maps and allow the user to select a range of different classification methods. The most commonly available ones are discussed here and illustrated in Figure 12.8.

12.4.2.1 Equal ranges (or equal intervals)

With an equal range classification, the data are mapped so that the range from the top to the bottom of each class is the same. This is performed by calculating the difference between the maximum value and the minimum value in the data (the *range*), and dividing the steps from one class to another into equal stages. This method has the advantage of simple legends that are easily understood by non-cartographers, although they can produce unbalanced maps if the data are not distributed evenly throughout the data range.

12.4.2.2 Equal count

This classification method arranges the class divisions so that approximately the same number of observations are in each class. If this option is selected then each class will shade roughly the same number of areal units. This method can cause some interpretive problems if some areas are significantly larger (or smaller) than others.

12.4.2.3 Natural break

The natural break method looks through the data histogram and searches for the troughs or gaps. These are used as break points in the classification system. This method can be useful if the data occur in natural 'bands' of values. The method, also often referred to as Jenks Optimisation, is usually calculated by a computer program, runs through every permutation of class boundaries and calculates a *goodness of variance fit*. This is a measure of the suitability of the classification to map the data with the aim of minimising differences between the individual observations in a class and the mean of the observations within the class. The class boundaries with the highest goodness of variance fit (maximum value 1.0) are selected.

12.4.2.4 Standard deviation

With the standard deviation approach, class boundaries are compiled once the programe calculates the mean and standard deviation of the data values. The problem with this method is that crime data rarely follow a normal distribution, a requirement for using this method correctly. In other words, if a histogram or frequency distribution of the values to be mapped does not follow a normal distribution, then mean and standard deviation values will not reflect the true distribution of the data accurately. The legends also create difficulties for non-technical audiences who rarely understand a standard deviation class structure. Unless the data conform to distributional requirements, this method is best avoided for crime maps.

12.4.2.5 The best method?

Figure 12.8 showed four different choropleth maps of the same data. Although portraying the same data, the differences between the classification schemes can have an influence on the interpretation of the data. Although the general impression of the maps appears on quick examination to be the same, subtle differences appear on closer examination. For example, the most North-Westerly state (Washington) is bordered by states in lower classes (Oregon and Idaho) in all of the maps except the standard deviation classification (map D). The different class boundaries are shown in Figure 12.9.

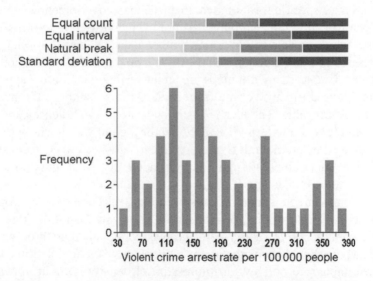

Figure 12.9 Histogram for areal values shown in Figure 12.8. Class boundaries for the four maps are shown above the histogram as horizontal bars

Choosing the best method is sometimes a process of experimentation, combined with an appreciation for the audience. The equal ranges and equal interval methods do not cope well with unusual distributions but are easy to understand, and the equal interval legend makes sense to the non-technical audience. The equal count and natural break legend labels accurately represent the data but the legends are not as orderly as the equal interval legend. The natural break method is the best for coping with unusual data distributions, as can occur with crime data. There is no one right answer and the look and feel of the map from the audience perspective will be a contributory factor to a final decision. The advantage of modern GIS is that it is easy to change mapping schemes to visualise different classifications.

Many GIS also have the functionality to set a 'custom' classification, allowing the crime mapper to arrange classes into a sequence that more logically fits with the message that the map aims to portray. For example, this can be useful when wanting to show how crime in an area compares to national average crime rate. The national average rate can be used as the middle class, with other classes showing levels below or above this national average.

12.4.3 Dot maps

Dot maps are a specific form of areal thematic mapping. Sometimes referred to as 'one-to-many' dot mapping, it is not to be confused with one-to-one dot maps which place a single dot or symbol at each event location. Crime mappers will be more familiar with the usual term for one-to-one mapping as 'pin maps'. One-to-many dot maps are an interesting quantitative approach that can be used to show crime frequencies in areas without the need to calculate a ratio value. The idea is easily understood by audiences, the process is simple with modern GIS and it can be a good way to show overall density of a crime problem. It is also possible to fairly accurately reconstitute the original data values, though counting numerous small dots on a map would be tedious.

Some caution is required, however. First, selection of an appropriate dot value can be difficult. The aim should be to establish a dot rate such that the administrative region with the least value still has three or four dots. Mappers should check that the most populous area, in dot terms, looks crowded enough to convey an impression of density without appearing swamped. Once dots coalesce too much they appear like a large blob and accurate interpretation of the data becomes impossible. On occasions like this, varying the dot size can help. Some minor variations in dot value can also help with map interpretation if the cartographer chooses an appropriately rounded number. For example, a map where one dot represents 20 offences will be easier to interpret than one where one dot represents 18.9 offences. This will often require the user to abandon the default choice of the GIS.

Secondly, from a crime mapping perspective, the most significant problem with dot maps lies in their spatial inaccuracy. A dot that represents 300 burglaries is inevitably a spatial proxy – it is unlikely that all 300 offences occurred where the dot is placed. This is a significant problem in crime mapping because crime is not randomly distributed (see Chapter 4), yet a random distribution of dots (such that there is an even spread of dots in an area) is unfortunately the method that GIS use to generate dot maps. A final consideration should be that for accuracy – all dots should be visible. This limits the amount of information in the lower visual hierarchical levels that can be displayed, and one common problem with dot maps is the lack of sufficient base map information. Figure 12.10 has two dots maps showing the pattern of theft from cars in suburbs of the Australian Capital

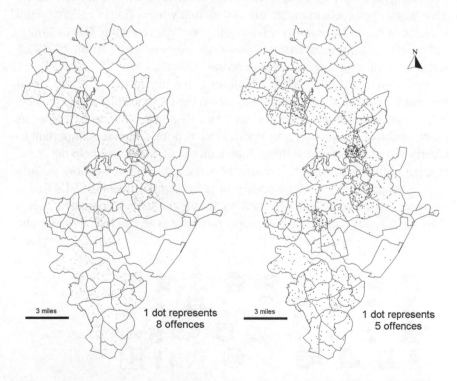

Figure 12.10 Two dot maps showing theft from cars in the Australian Capital Territory, 1998–2002

Territory (Canberra) from 1998 to 2002. In the first map, the dot size is too small and this has allowed the administrative boundaries of the suburbs to dominate the map. The choice of dot value is also inappropriate (too high) and determination of crime hotspots (by administrative area) is difficult. In the second map, a more appropriate choice of dot size and value better shows the crime hotspots in particular suburbs.

It is worth considering that if the number of dots becomes too large to reasonably display on a dot map, then a kernel density estimation surface may be an appropriate substitute (see Chapter 6).

12.5 Thematic maps of point data

Inevitably there will come a time when point data have to be mapped. This can be because the incidence of a crime type is infrequent enough that individual points can be reasonably shown on a map. This could be the case with a map aiding a serial homicide investigation. At other times, even when choropleth maps are employed, many users find occasional point locations helpful in order to orientate themselves to a map. For example, police officers invariably find it useful to indicate the location of police stations on maps for operational purposes.

Point data as significant features come under level 2 of the visual hierarchy, so must be clear enough to determine their position and relevance, but not overwhelm any level 1 visual data. If crime data are portrayed as points and local features are also portrayed as points, then it is important to clearly distinguish between them. Judicious use of colour can help the crime mapper in this task, though variation in shape and size can be more useful.

Figure 12.11 shows a variety of point symbols using ESRI's Crime analysis font. Point symbols such as these are useful characters when showing a small number of point locations; however, the meaning of some

Figure 12.11 Examples of different point symbols available in the ESRI Crime analysis font

symbols is more intuitive than others. A reasonable approach is to choose only symbols that have a clear meaning, and even then only use a few select characters. If the reader is overwhelmed by different point symbols then they will be continually referring back to the legend to understand the meaning of each symbol, an activity that gets tiring very quickly. By reducing the number of significant features to three or four, and by reducing the number of different crime types displayed, a map will improve in readability. With point data, less is most definitely more.

The Philadelphia Police Department have a sensible approach when mapping point data that only show one crime type at a time. Any incidents that involved the use of a firearm have a black cross through the symbol, though the symbol remains the same. The reader is not therefore required to interpret a new symbol and this allows the map reader to easily appreciate an additional feature of the crime (the use of a firearm or not).

The main point to consider when mapping point data is that the crime locations should have priority, as they do in the visual hierarchy. Mapping more than three or four crime types on one map should be avoided, as should overloading the map reader with a large number of additional features. Additional features that are relevant (such as the location of bars or taverns on a map of late-night assaults) should be indicated with a symbol that is unambiguous and cannot be confused with a crime site.

12.6 Getting away from paper: The digital age

Traditional mapping has been paper-based for hundreds of years, but the exponential growth in computing power has enabled cartography to explore different digital mediums, including online publishing with the Internet and in presentation environments with products such as Microsoft PowerPoint. These outlets provide an expanded range of opportunities for crime analysts in terms of expression with colour and animation. As with paper maps, the main consideration should be to decide what is the primary information that the cartographer wishes to convey to the map reader. In this, the non-paper map is no different from the paper product.

12.6.1 Internet crime maps

The digital domain of the Internet enables a full colour map to be disseminated across international boundaries almost instantly. Local communities can be supported with crime mapping available from interactive websites, and police intranet services can disseminate intelligence in the form of colour maps within a secure internal framework. The Internet and the tools that can be used to access it (such as the World Wide Web and e-mail) are rapidly becoming an essential weapon in the arsenal of the crime analyst.

There are two predominant file formats for Internet map images. GIF files are an Internet standard. GIF (Graphics Interchange Format) is a simple type of file which supports up to 256 colours and a range of various resolutions. It is usually used for smaller images such as buttons and logos because the limitation on the number of colours prohibits its use for photographs. However, it is possible to use the GIF format for maps.

An acronym for Joint Photographic Experts Group, JPEG files are a computing innovation. JPEG algorithms are able to reduce complex image files so that they can be up to a tenth of their original file size without adversely affecting picture quality. The format was designed as a way to more easily transfer large photograph files across the Internet. The JPEG format (actually a range of file formats that are constantly under development) is ideal for mapping purposes due to its worldwide compatibility. The only consideration is that with high levels of JPEG compression there is some loss of detail and occasionally a reduction in clarity of image.

For most Internet mapping applications, the most significant consideration is the pixel resolution of the map. The pixel (short for 'picture element') resolution is the number of individual pixels that make up the image. The pixel is the smallest piece of an image that can be individually coloured. In a map image, the pixels are so close together that they appear to be connected as one complete image. Larger numbers of pixels generally translate to a bigger and clearer image; however, the price for this is a much larger file size. For example, a map with a resolution of 400 pixels by 400 pixels will have 160 000 pixels, while an image that is half the height and half the width will only have 40 000 pixels.

12.6.2 Interactive mapping

One recent innovation on the Internet is interactive mapping. Interactive applications accept input and decisions from the user, in essence allowing the user to customise the image. This means that an individual can ask a crime map website to only show a certain type of crime, or to limit the temporal range of offences, as long as these data options are available in the underlying database and that these choices have been programed by the designer.

The US National Institute of Justice's MAPS website (www.ojp. usdoj.gov/nij/maps/) has web links to a number of police services that provide interactive mapping of crime and sex offender locations. One example is the mapping tool provided by the Automated Regional Justice Information System (ARJIS), a collaboration of over 30 local, state and federal agencies in the San Diego region (Wartell and McEwen, 2001). ARJIS provide a publicly accessible crime mapping tool known as the ARJIS Interactive Mapping Application. The tool allows a user with access to the Internet to view maps covering a wide range of police business such as traffic accidents, crime incidents and arrest and citation locations. The ARJIS mapping tool can be seen at www.arjis.org.

Companies that provide GIS software are rapidly moving towards automating and simplifying the process of providing interactive crime mapping tools; however, at this stage it still requires implementation by users with a fair degree of computer knowledge and programing ability. If the effort is deemed worthwhile, interactive mapping can reduce the workload on staff who often spend considerable time producing static maps for users, and can inspire individuals to explore crime patterns in a more flexible environment. If the final result is more time for crime analysts to concentrate on other crime analysis tasks, then this may be a good thing; however, crime analysis is a profession requiring training and skills, and interactive mapping should not be used to devolve all of the functions of an analyst to untrained users.

A new format that is seeing increasing use on the Internet for displaying maps, allows for some interaction, but does not require specialised Internet GIS software, is Scaleable Vector Graphics (SVG). SVG applications can allow users to interact with features on a map, offering impressive results. For an example of this type of application visit:

- http://www.statistics.gov.uk/census2001/censusmaps/index_new.html
- http://www.statistics.gov.uk/populationestimates/svg_pyramid/ indexpyramid.html (a non-crime example, but an excellent demonstration of SVG use)
- http://www.met.police.uk/crimefigures/svg_info.htm (the SVG version of London's Metropolitan Police's crime statistics page).

12.6.3 Crime mapping for presentations

Presentations that use digital projectors require particular types of maps in order to be most effective. Microsoft PowerPoint™ is the industry standard tool in the crime mapping field. PowerPoint has no significant built-in mapping facility, and so it is necessary to export maps from GIS as image files prior to insertion into a PowerPoint presentation.

The earlier discussion of colour suggested that dark blues and colours with similar hues are an appropriate background for presentation slides, as they tend to fade into the background. If so employed, this increases the range of mapping possibilities. Without the constraints of a paper background, cartographers can explore different map surrounds.

An example can be found in Figure 12.12. The map shows census tracts of the City of Philadelphia, coloured according to the distribution of the home addresses of people who were mailed a letter to attend jury service but never replied to the mail out. The distribution is expressed as a location

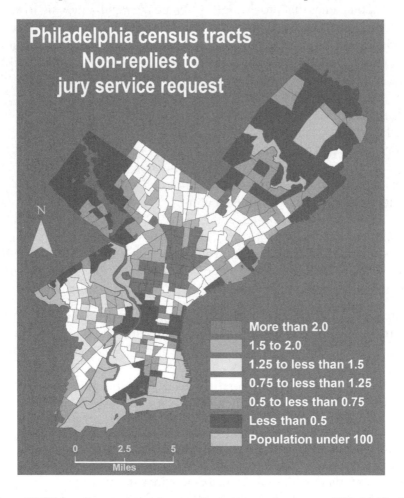

Figure 12.12 Location quotients of non-replies to jury service requests, Philadelphia, 2002 (see Plate 5)

quotient, a technique that shows each area's value in relation to the overall mean. Red shading indicates a value higher than the city mean, while blue shading indicates a value lower than the city mean. See Ratcliffe (2004) for further explanation and another example. The lack of necessity to have a white background provides two opportunities. First, as much colour as is necessary can be employed without saturating a paper page with ink. This allows a colour background to be employed without significant cost. Secondly, the ability to have a non-white background allows white to be used to indicate the middle class in a diverging scale. In this case, white is used to show areas where the ratio of people who did not respond to a mail-out is roughly the same as the overall city ratio (in other words, these areas have a location quotient near one).

Presentation maps (like Figure 12.12) should avoid using small font sizes, and try and use san serif fonts where possible. This will help audience members near the back of a room read the map. Subtle changes in hues for classes should be avoided, as the digital projector may not be able to cope with slight hue differences. Cartographers should always remember that the image seen on a computer monitor is likely to be of a higher quality in both resolution and colour than is possible to display through a projector.

12.7 Summary

If an analyst spends weeks considering and analysing a crime problem, all of that hard work counts for nothing if they are unable to convey the information to other people who can go out and solve the crime problem. This can occur if a map is the primary source of information dissemination and the map fails to adhere to some simple standards. Every map should:

- have a map scale (a scale bar is easiest to interpret);
- show direction with a compass or North arrow;
- contain a legend that is clear and unambiguous;
- include a title, the source of the crime data and the cartographer's name;
- consider including the security classification of the map;
- be organised around the visual hierarchy; and
- be conscious of the harmony of the colours.

Within the range of possible output options, from traditional paper maps right up to the use of the Internet or presentation hardware, these basic guidelines are still valid. However the map of crime is produced, if the reader cannot understand it then the time that goes into its creation is time wasted. Conversely, a quality map can make all of the difference and can convince decision-makers in the criminal justice system to target resources to the right area, police more effectively and reduce crime.

Further reading

Monmonier, M. and De Blij, H.J. (1996). *How To Lie With Maps*, Second edition. University of Chicago: Chicago Press.

A unique view of mapping that focuses on both the deliberate and inadvertent use of maps to mislead the reader. This unusual perspective aptly demonstrates how small changes to a map can dramatically change perception.

Dent, B.D. (1999). *Cartography: Thematic Map Design*. Boston: WCB, McGraw-Hill.

A classic text on cartography. Covers most aspects of cartography in a clear and readable style.

MacEachren, A.M. (1995). *How Maps Work: Representation, Visualization and Design*. New York: Guilford Press.

This book explores the integration of cognitive and semiotic approaches to provide an understanding of maps and their applications for geographical visualisation, the communication of information, and the construction of knowledge. By exploring the question of spatial representation of different kinds, at multiple levels and from various approaches, the book provides a key insight into how maps work, which can then be applied for improved map design.

References

Bertin, J. (1983). *Semiology of Graphics: Diagrams, Networks, Maps*. Madison, WI: University of Wisconsin Press, translated by W.J. Berg.

Chainey, S.P. and Desyllas, J. (2004). Measuring, identifying and analysing street crime risk. Presentation at the 2004 UK National Crime Mapping Conference. London: University of London. http://www.jdi.ucl.ac.uk/news_events/conferences/index.php.

Clarke, R.V. and Goldstein, H. (2003). Thefts from cars in centre-city parking facilities: A case study in implementing problem-oriented policing. In J. Knutson (ed.) *Problem-oriented Policing: From Innovation to Mainstream*. *Crime Prevention Studies*, Volume 15. New York: Monsey.

Clarke, R.V. and Mayhew, P. (1998). Preventing crime in parking lots: What we know and what we need to know. In M. Felson and R.B. Peiser (eds) *Reducing Crime Through Real Estate Development and Management.* Washington, DC: Urban Land Institute.

Dent, B.D. (1999). *Cartography: Thematic Map Design.* Boston: WCB, McGraw-Hill.

Eck, J. (1997). What do those dots mean? Mapping theories with data. In D. Weisburd and T. McEwen (eds) *Crime Mapping and Crime Prevention* (pp. 379–406). Monsey, New York: Criminal Justice Press.

Harries, K. (1999). *Mapping Crime: Principles and Practice.* Washington, DC: US Department of Justice.

LeBeau, J. (2000). *Demonstrating the Analytical Utility of GIS for Police Operations: A Final Report* (p. 92). Carbondale, IL: NIJ.

Monmonier, M. (1993). *Mapping it Out.* Chicago: University of Chicago Press.

Ratcliffe, J.H. (2004). Pre-jury racial bias in Philadelphia's Courts?: A study using location quotients and force-field analysis. Paper presented to the 7th Annual International NIJ Crime Mapping Research Conference, Boston, March 2004. http://www.ojp.usdoj.gov/nij/maps/boston2004/papers/Ratcliffe.pdf.

Slocum, T.A. (1999). *Thematic Cartography and Visualization.* Upper Saddle River, NJ: Prentice Hall.

Wartell, J. and McEwen, J.T. (2001). *Privacy in the Information Age: A Guide for Sharing Crime Maps and Spatial Data.* Washington, DC: Institute for Law and Justice.

13
The Management and Organisation of Crime Mapping Services

Learning Objectives

This chapter identifies the ways in which crime mapping can be managed and organised in policing and crime reduction services.

The chapter covers three main areas:

1. Implementing crime mapping, including: a description of how crime mapping can be integrated into policing and crime reduction, the crime mapping and GIS implementation process, and process maintenance and implementation evaluation.

2. The role of analysis and the ingredients that go towards the effective use of analysis outputs.

3. Organising the production of crime mapping outputs, how they can be integrated into information-driven processes, supported and steered to help ensure continual effective contribution.

This chapter does not aim to prescribe a format that all readers should follow, recognising the differences and unique qualities between different implementation, organisational and management requirements. It does though aim to describe certain principles and experiences, as well as offer guidance in supporting the management and organisation of crime mapping in policing and crime reduction.

GIS and Crime Mapping   Spencer Chainey and Jerry Ratcliffe
© 2005 John Wiley & Sons, Ltd

13.1 Introduction

Surveys of the use of crime mapping for law enforcement and crime prevention recognise that although crime mapping is increasing in use, its implementation is quite difficult and can be a frustrating process (NIJ, 1995, 1999). These difficulties have often included the technical challenges and the associated effort required in implementing GIS. For example, a survey conducted by the United States' Police Foundation (2000a) noted:

- The learning curve required to adopt crime mapping technology was often underestimated.
- Substantial planning was required to implement crime mapping.
- The range of GIS products often confused those who were procuring systems, particularly when it was difficult to meaningfully differentiate between their technological details, and if any differences were relevant to user requirements.
- There was a need for better functionality to automate the production of routine tasks, such as descriptive areal statistics.
- The importance of technical assistance and relevant training.

However, technical issues, such as the ones noted above, do not necessarily act as the biggest challenge when it comes to producing crime mapping outputs and integrating them into the routine of policing and services for crime reduction. In a post-mortem of a failed GIS, Openshaw *et al.* (1990) noted a number of organisational and management issues that related to a failed implementation of a crime mapping GIS, including:

- failure to clearly specify the system's aims, objectives and requirements;
- organisational resistance to change; and
- unrealistic organisational expectations of what would be delivered.

Furthermore, in a survey of community safety officers at the UK's National Community Safety Network Conference, Chainey (2004) noted that the biggest barrier to using crime mapping within a Crime and Disorder Reduction Partnerships (CDRP) related to an organisational culture opposed to sharing information. Addressing these issues in a partnership context is described in Chapter 7.

Other areas of this book have already addressed many of the technical challenges that affect crime mapping such as: geocoding crime data, how data used in crime mapping can be made fit for purpose and how geographic data can be combined and cross-analysed. This chapter explores the typical organisational and management arrangements that determine how crime mapping is used to support policing and crime reduction services. The chapter will also examine how an agency can move forward with the many suggestions and applications we have discussed in other parts of this book. The chapter begins by offering several guiding points for implementing crime mapping into an agency. This is then extended by reviewing the role of analysis in policing and crime reduction and the challenges that need to be overcome if crime mapping products are to be used. The chapter then concludes by discussing how crime mapping can be organised within an agency and how its organisational development can be supported. The chapter offers many tips for implementing crime mapping for the first time, but also acts as a review and improvement guide to those who already actively employ crime mapping but who face persistent organisational issues over its effective use.

13.2 Implementing crime mapping

13.2.1 Integrating crime mapping into policing and crime reduction

The production cycle for capturing and using information for policing and services for crime reduction involves five stages which directly relate to the intelligence cycle (Ratcliffe, 2004, p. 6) – a process well known to police intelligence staff. In brackets we have included how these stages impact on the crime mapper.

1. Direction (where the crime mapper receives and discusses a task with management)
2. Collation (where the crime mapper undertakes a process of data acquisition)
3. Analysis (where the analytical task is completed)
4. Dissemination (where the final map or analysis product is created and used to influence a crime reduction or prevention action)
5. Feedback and review (where the actions that were undertaken after the product dissemination are evaluated to see if they had the desired impact).

As identified in many parts of this book, crime mapping can play a role in all stages, such as: aiding the collection of information, identifying patterns, exploring problem areas, exposing causal factors, prioritising action (and the type of action to implement), and evaluating the impact of any targeted action.

Integrating crime mapping into policing and crime reduction involves an appreciation of all the components of a GIS (i.e. hardware, software, data and people – as discussed in Chapter 3). Because of the hardware and software requirements, the responsibility for integrating GIS often falls to an agency's Information Technology (IT) team, which on occasion can mean that the vital and more important 'people' component (and how they use spatial data) is overlooked. This 'people' component is more important than hardware and software requirements because the technical components are flexible and easier to change, whereas if the needs of the users are not fully appreciated then the GIS solution adopted could be poorly implemented and fail to meet the needs of the community that it should serve. The integration of GIS into police (and other agency) processes is often mistakenly too focused on the technology that is adopted, rather than the practical use, management and organisation of information that it generates. Testing, purchasing and implementing GIS in the 52 UK police forces is often tasked to each forces' IT department (Ratcliffe, 1999) which, while at first may appear to be the natural home for GIS, has begun to expose several difficulties stemming from the failure to not involve users in its integration. As one police officer said, 'Our IT Branch do not really understand the user requirement in relation to crime mapping and it shows' (quoted in Ratcliffe, 1999, p. 319). This is a comment typical of the views voiced by many police crime analysts.

The implementation of a GIS that is centrally controlled from an agency's IT department runs the risk of being too technology-led, where, for example, focus can be on developing advanced functions that look impressive but are rarely used in practical analysis. This compares to a user-led approach that is coordinated by the key stakeholders of the application of GIS – people that appreciate what is required from the GIS, and who are best placed to recommend and prioritise new functions and new tools required for crime mapping, and who consider the technical abilities of users. Ratcliffe (1999) noted that the majority of successful integrations of crime mapping into UK police forces were ones where the system was under the control of analysts or the Crime Analysis/Intelligence Unit. These users were the driving force for GIS use, initiating product improvements, and innovating crime mapping analysis and outputs. The role of IT departments in these cases was often a supporting one, helping with the maintenance of the software and hardware components.

Lessons in crime mapping and GIS implementation can also be learned from the reviews of integrating GIS in other applications such as local government, public utilities and business (Campbell and Masser, 1992; Obermeyer and Pinto, 1994; Ventura, 1995; Clark, 1996, 1998;

Nedovic-Budic and Godschalk, 1996; Robey and Sahay, 1996; Grimshaw, 2000). In particular, Campbell (1994), although reviewing local government's application of GIS, identified four factors that are relevant for helping to improve the chances of the success of GIS integration into policing and crime reduction services:

1. Keep it simple and relevant.
2. Be aware of the limits of available resources.
3. Have the organisational ability to accept and adopt changes.
4. Involve the users in the implementation of the GIS.

Many suggest avoiding a centralised control of integration (Ventura, 1995; Robey and Sahay, 1996; Heywood *et al.*, 1998); however, in practice some central organisation is required to assist in systems maintenance and IT expert support. This central involvement should, however, be user-led rather than dominated by technical detail.

13.2.2 The GIS and crime mapping implementation process

Implementing a GIS for policing or crime reduction services is a process similar to many other applications of GIS. In essence, it involves nine stages:

1. Needs assessment
2. A review of crime mapping products to adopt
3. Assignment of responsibilities
4. Strategic goals and design plan
5. Staffing requirements
6. Data requirements
7. Geocoding requirements
8. System specification and costings
9. Implementation review and action plan.

Each of these stages is described below.

13.2.2.1 Needs assessment

The needs assessment should aim to meet two key purposes. It should be designed to answer the question, 'Why do we need to implement a GIS and what benefit will it bring to how we work?' Once established, the needs assessment can act as a baseline against which the implementation of the GIS can be evaluated. The needs assessment should include a review of the ways in which crime mapping is currently used, successes and failures in its use, who the users would be within the organisation, an assessment of the community they will serve and a review of resources that may help support an implementation.

13.2.2.2 A review of crime mapping products to adopt

The second stage aims to decide the crime mapping products (or applications) that will be generated from an implementation. This need not include a technical specification of each product, but should at least describe in general terms the task each product will aim to do, its content, how it would be used and who the customer for this product would be.

13.2.2.3 Assignment of responsibilities

Who takes responsibility for moving the implementation process forward should be decided as early as possible. This may not necessarily be the first task, as others may have originally initiated the implementation process, but it should certainly be considered an early priority. At this stage it is vital to include the users and ideally get them to take on the role of moving forward the coordination of implementation. This implementation team should include senior management support in order to help make any necessary quick decisions (e.g. budgeting decisions), ensure their buy-in from the beginning to support crime mapping development and allow for management input to the coordination of the crime mapping products that will be designed. Crime mappers should also be aware that it is harder for management to close a unit or remove funding from an area that they have had a personal involvement in implementing.

13.2.2.4 Strategic goals and design plan

Setting strategic goals on what a crime mapping system (or the products that will be generated) should achieve helps to establish and record the aspirations and benefits that will come from an implementation. The strategic goals should be SMART:

- Specific
- Measurable
- Achievable
- Relevant
- Time-based.

Each strategic goal should be set with targets. The design plan should then describe the strategy for the implementation process, set a work programme and describe milestones that can be monitored during the course of implementation. The design plan's timeline should extend beyond the implementation period to help support the continual development of crime mapping and establish a date for system evaluation. Gantt or PERT charts both can help in this process (see Walsh, 2004).

13.2.2.5 Staffing requirements

Staffing requirements should be reviewed to ensure that either the aspirations of the implementation can be met with existing staff or that new staff can be recruited. This stage will also include a review of training requirements and other forms of continual professional development for crime mapping users.

13.2.2.6 Data requirements

Core data requirements in terms of their type, sources, content, quality, geographic precision and cost form an important part of the GIS implementation process because they can often govern a significant part of any allocated budget, and their availability often governs the potential development of certain crime mapping products. These data requirements should include a review of base mapping information, geographic referencing files (e.g. for geocoding) and datasets that are required for the delivery of crime mapping products. At this stage, and as referenced in Chapters 3 and 7, caution is given to simply compiling a shopping list of data requirements. Instead, data requirements should be focused on the core needs of the products that will be generated, rather than getting into a situation where the implementation and development of crime mapping becomes paralysed by the efforts to collect data, when much of these data may be unnecessary or of the wrong type. It is also worth considering two lists, one that lists essential data without which the crime mapping system cannot function, and a second wish list of data that are desirable to increase the functionality of the unit, but are not essential. The first list then becomes the system priority.

13.2.2.7 Geocoding requirements

The technical details for geocoding and the importance of fitness for purpose have been discussed in Chapter 3. Any crime mapping implementation should be appreciative of the geocoding requirements for crime and other data that are to be used, and also consider the accuracy and spatial precision that can be achieved for these data. Considering the geocoding requirements at the implementation stage will help determine if certain crime mapping products are feasible and if additional investment is required to purchase geocoding tools.

13.2.2.8 System specification and costings

After the aspirations of the GIS and crime mapping implementation have been decided, and the conditions for the generation of products have been assessed, the specification for the system can begin to be designed. System specification should include those for hardware and software, but also the specification for staffing (including any necessary training) and the support structures to aid post-implementation development. This stage should also include an accurate estimation of costs associated with these specifications.

13.2.2.9 Implementation review and action plan

The final stage of the implementation process involves two parts. The first part should include a comparison of the previous stages against any budgetary constraints, as well as a review of the feasibility and specification of future aspects of the implementation in the light of the process up until this point. Once any reviews have been completed, the second part then sets out the action plan, describing:

- the organisational structure for how crime mapping will be used, the duties of users and those responsible for system implementation, and the support structures for aiding development;
- the crime mapping products that will be developed, who they will be developed for, when they will be developed and how their impact will be reviewed;
- the initial data to be acquired;
- the technology (hardware and software) that will be adopted, and its phased timings; and
- the systems training and continual professional development programme.

This action plan acts as the *living* version of the design plan, and should include the strategic goals and targets that were set out in the strategic document, as well as identifying the milestones for the implementation process and future review points to assess the impact that crime mapping is having on the agency.

An important component that should also be considered in the action plan is how the system should be promoted. Failure to recognise the benefits of GIS and crime mapping across the breadth of an agency often stems from a lack of awareness of what it can do. Many of those that have implemented crime mapping into policing and services for crime reduction have found that embarking on a marketing campaign within their agency is a useful way to promote and raise awareness of the new tools that have been developed, and how they are having an impact. Vincent (2000) used e-mail and department newsletters as an effective means of promoting the implementation of new, simple-to-use crime mapping tools to all officers across the Hayward Police Department, California. Promoting the use of crime mapping by demonstrating successful case studies of its use can also be useful to win general support.

13.2.3 Maintenance

The implementation and development of crime mapping into any agency should be viewed as an iterative process, allowing for changes to be made to the system to reflect users' needs and changes in demand. Indeed, Hughes commented that 'the most important lesson learned is that the development and implementation of an agency-wide GIS application is...[a] process of change and refinement' (Hughes, 2000, p. 4). The hardware and the software of the system will naturally require upgrading as new technology emerges, and the users may also require training in any new technology that is adopted. The maintenance and review of how crime mapping is being used may also identify certain user tasks that are repetitious and routine, and that could be automated, allowing users to apply their skills on more rewarding and innovative tasks. The maintenance process should also recognise that not all the data used in the system are perfect. Data that are used may be poor in quality or content. The maintenance process should promote strategic information improvements where it recognises that subtle changes to how data are recorded may significantly improve data quality for analysis. This maintenance process should be flexible enough to allow new data to be added, either to complement existing data or to fill a gap where data are non-existent.

13.2.4 Evaluation

Evaluating the success of an implementation is a stage that is often overlooked by many who implement crime mapping. The nine stages that describe the implementation process act as a useful benchmark on which to consider if the implementation has had the impact that was hoped, and identify any successes and failures that could feed into the ongoing maintenance and development of crime mapping in the agency. The Police Foundation's (2000b) guide to evaluating crime mapping implementation describes two parts to evaluating an implementation: part one evaluates the process of the implementation and part two evaluates its impact.

13.2.4.1 Process evaluation

The process evaluation measures the administration and resources used for crime mapping. These measures relate to quantifiable aspects of the volumes that have been used (or produced), their quality, their effectiveness and their costs. Process measures include:

- The personnel used – the number of people that are users of crime mapping, the number of people that support the generation of crime mapping products (both expressed either as the number of people or as hours dedicated to crime mapping), skill levels of users and training received.
- Technology that was implemented – the hardware that has been installed and new software licences purchased, and if any software development has been included in the implementation.
- Data sources – the number and types of data used, the variables in each data source, the length of time it takes to access data, its fitness for purpose, and timeliness in delivery and use.
- Products – the quality and quantity of crime mapping products, the types of analyses generated, the time it takes to produce products, and their content (Police Foundation, 2000b).

13.2.4.2 Impact evaluation

Capturing the impact that crime mapping has on an agency is difficult to record systematically. Its impact may require a survey of users that captures how mapping use has aided the delivery of their work, such as the products they produce, the number of products they now produce, their content and the length of time it takes to generate products. Any survey of crime mapping users could also include questions for the audience that products were designed to serve – asking if the information is of value, if it is delivered in a timely format and if it has made a difference in what it was designed to support. The survey could also be complemented with case studies that describe how the crime mapping product has made a difference in comparison to previous practices, and if possible, quantify the savings in costs, the time difference taken to capture or generate information, or the data that is now available that previously was difficult to access or interpret.

The strategic goals and design plan from the implementation process should have also captured SMART targets. These targets offer a means to systematically measure the impact of the implementation, and quantify if the implementation has been a success or a failure. Its impact should also be measured in terms of the effect it has had on reducing crime. Many evaluations of system implementation fail to include this measure. While crime mapping may not be the sole reason that crime problems have reduced, if a system's design was to support the generation of clearer and more detailed information in order to tackle a particular crime problem, it may be possible to demonstrate the value of the crime mapping system in crime reduction terms.

In addition to process and impact evaluations, an evaluation could also compare the differences between similar agencies, and a satisfaction survey that questions if the implementation met the needs and expectations of stakeholders, users and the audience that it was designed to serve (Police Foundation, 2000b).

Evaluation offers a means to support the refinement of a GIS implementation, particularly if it helps to identify areas for improvement or a user demand for specific applications, products and areas of development. Often, it can be easy get carried away on a concept that sounds good and looks as though it can be delivered. Without evaluating a mapping system, recommendations could be made for further development when little reliable evidence suggests the mapping system's original use is having the impact that many wish and perceive it to achieve. The evaluation needs to contain substance and need not purely rely on the occasional short-term achievement that may fail to indicate if success is long term.

13.3 Understanding the role of crime analysis

Many products developed in crime mapping feature as an integral subset of crime analysis. In this section we consider this broader role of crime analysis to help draw parallels that can be used to strengthen the application of crime mapping in an organisation.

Crime analysis involves a set of systematic processes that aim to identify patterns and correlations between crime data and other relevant information sources (be they *hard* datasets, statistics or *softer* information in the form of intelligence) for the purpose of supporting decision-making that informs the design, allocation and priorities of police activity and crime prevention responses (Gottlieb *et al.*, 1994; Gill, 2000; Cope, 2004). Its benefits also extend to supporting the maximum use of the limited resources available for tackling crime, providing an objective means for identifying and understanding crime problems, initiating proactive actions in detecting and preventing crime, and taking advantage of the volumes of information that are collected by the police and other agencies (Osborne and Wernicke, 2003). As a process, crime analysis has developed alongside the paradigm of intelligence-led policing, has foundations in problem-oriented policing (Goldstein, 1979) and supports the increasing requirement for measuring outcomes and the integration of research into operational strategies (Laycock, 2001). Crime analysis is also a useful medium within which to cross-reference multi-agency datasets for partnership responses to crime problems. The developments in information technologies used in policing and the requirement to monitor police performance have also contributed to its growth (Fletcher, 2000; Gill, 2000; Manning, 2000, 2001; Cope, 2004). Crime analysis endeavours to provide the 'right information...to the right people at the right time' (Fletcher, 2000, p. 114).

The role of the crime analyst is to be an expert on crime, the facts person, a source of advice and a scientist who applies rigour to generate verifiable and reliable intelligence. They need to know what works in policing and crime reduction, understand the theory that underpins crime patterns, be informed, promote problem-solving, have the opportunity to hone their research skills, be able to communicate effectively and be in a position to ensure their opinions are heard and used (Clarke and Eck, 2003). It also requires the crime analyst to be dedicated, because performing this all-encompassing role is not always that straightforward.

13.3.1 Making use of crime mapping and analysis

No matter how good the crime analysis products, there exist significant difficulties in getting the products and recommendations that come from analysis to be used and actioned proactively. Issues with the poor management of analysis use, a police patrolling culture that questions the legitimacy of being told by desk-bound staff what is happening on the streets, a hierarchy and culture that may take little notice of non-police staff views of the criminal environment, organisation fragmentation, a reactionary rather than proactive stance on policing and tackling crime, and a failure to support innovation, amount to problems that prevent the effective use of crime analysis. Cope (2004, p. 197) captures these sentiments from a crime analyst,

> We make suggestions, we make suggestions strongly, if we believe them to be important. But...the police organise their resources how they see fit... there's nothing we can do about it. Overall, I would suggest that very few of our recommendations are actioned...and that is very frustrating.

Making use of crime mapping and crime analysis within a police or other crime reduction agency requires overcoming a number of obstacles. This section suggests ways in which these institutional, organisational and management barriers can be overcome.

13.3.1.1 Educate the audience

A vital component in designing crime analysis products is to identify the audience that it will serve. By identifying this audience it will be easier to capture their needs in the final product. The audience also needs to appreciate, and if need be educated, that crime analysis is not about creating products that merely describe and summarise the nature of current persistent problems, but instead should include forecasting, predicting and evaluating future crime issues. In other words, analysts should not simply provide management with tables of statistical analyses, colourful charts and pretty maps but a real understanding of criminal activity and a range of evidence-based options to reduce it. The crime mapper or crime analyst can become very frustrated if their job merely involves producing descriptive statistics for the weekly management report, and fails to offer the freedom to carry out research that would significantly enhance the production of crime intelligence content (Cope, 2004).

Criticisms over the quality of analysis in terms of its failure to offer an operational police officer anything they did not already know are occasionally warranted, but may also stem from the police officer's limited knowledge of analysis roles and functions, and little understanding of the associated information technology and its capability. There exists a paucity of training for police officers on analysis which in turn affects police officers' ability to ask meaningful, proactive questions of analysts. As a result, they have little appreciation for proactive analytical products. Policing is becoming more complex and with this there is a need to train police officers and recruit new staff with the skills for tackling the multifaceted and complex criminal environment (Foster, 1998; Flynn, 2000; Cope, 2004). The development of the UK's NIM (NCIS, 2000), described in Chapters 2 and 9, is partly based on the recognition that the criminal environment is becoming increasingly complex and for it to be effectively interpreted the management and use of information needs to be better organised.

It is also important to recognise that police expectations of crime analysis are influenced by a culture in which information leading to an arrest and conviction is of most value (Cope, 2004) and enforced by a (often incorrect) perception that an arrest will typically render the problem as solved. Nick Ross, a BBC journalist and Chairman of the Jill Dando Trust, compared this approach with the analogy that 'when aircraft crashed we contended ourselves with finding someone to blame rather than changing procedures or amending designs' (Clarke and Eck, 2003, p. 1). Crime analysis should chase after the problem, and not necessary just chase after a single offender, because it is the analysis of the problem that will generate the real understanding of criminal activity and how to direct the solution to the problem. Educating the audience about this more effective role is vital if crime analysis is to have the freedom to develop, have impact and innovate policing and crime reduction. If a police officer's (often limited) experience is solely relied upon and there is failure to record or systematically document officer knowledge, the result can frequently involve falsely rounding up the usual suspects (Gill, 2000) and undermine any analysis process that attempts to consider a broader range of facts, is based on good evidence and is intelligence-led.

> The whole intelligence led process can be corrupted by the banter that goes on and the self-fulfilling prophecy.... someone can become a [target] because everybody talks about them, and then we start targeting them and because we target them they become something that they are not...you could be going in the wrong direction because we believe something rather than know something. (Cope, 2004, pp. 199–200)

13.3.1.2 Consistency in the definition of the role of the analyst

The lack of any consistency in the definition of a crime analyst's role can often lead to confusion in their duties. Because analysts typically have basic IT skills they can often be tasked with acting as the key providers of management and administrative data from the agencies' information systems, be required to respond to ad hoc requests because they have access to certain software or data (e.g. responding to requests to create a spreadsheet for someone) or in some cases act as a source of IT technical support. The requirements of a crime analyst may vary due to the size of the agency, requiring those working in smaller agencies to multi-task. This can be reasonable if proportional time is also given for analysis; however, such requests should be challenged if they restrict the production of crime analysis. Analysts can become easily frustrated if all they ever seem to do is produce random pieces of information in response to requests – requests that are rarely in support of the main crime reduction aims of the department. Providing clarity and structure to the definition of an analyst's role and offering clear guidance in their tasks are important to help ensure consistency in their role.

The New Zealand Police has undertaken to provide consistent job descriptions for all its crime and intelligence analysts. The value of clarity in the definition of the role of the analyst is also highlighted in the Manchester case study later in this chapter. However, many of those that manage analysts often fail to recognise the role of analysts, are not experienced to provide appropriate guidance to an analyst, and in some cases have even required the analyst to write their own job description.

13.3.1.3 Educating the analyst

It is important for an analyst to have the opportunity and freedom to learn new techniques, theoretical concepts and develop communication channels with their colleagues. Eck (1998) noted that the lack of theory incorporated incorporated into crime analysis products, such as failing to describe why hotspots were persistent in certain areas (rather than just describing the fact that a hotspot existed in an area), meant that analysis products often lacked substance and tended to be merely descriptive. Additionally, analysts typically do not have a policing or applied crime reduction background. It is vital for analysts to understand policing approaches and practical opportunities for reducing crime so that any products or recommendations they develop are created in the context of how they can impact on the criminal environment. Analysts should develop their products and recommendations in consultation with police officers (or their other audiences). This helps to bring legitimacy to their analysis products. Many police officers are sceptical about analysis and find it uncomfortable to accept recommendations from non-police personnel (Cope, 2004), particularly when recommendations that they receive fail to appreciate the practicalities of policing or targeted crime reduction initiatives. Cultural barriers do exist between crime analysts and other police officers. An analyst should be encouraged to develop communication channels with their operational colleagues to help remove these barriers and legitimise the products they develop. For example, many police forces have discovered the simple task of allowing analysts to join police patrols in the field or giving them freedom to organise time with police officers to visit hotspots, visit a crime scene or shadow them, helps to not only educate the analyst but proactively improves communication channels.

13.3.1.4 Access to and sharing of information

It is important that an analyst has as much access as possible to all the information sources that are required to study a crime problem. Understanding a crime problem requires the analysis to become specific about the patterns of crime and explore the small details, as it is often these small details that provide the key to tackling the crime problem. Analysts have to be able to review the full picture of the crime problem, regardless of how sensitive the information may be.

The primary need in this area is often to encourage officers not to withhold information on what is happening on the streets. Police officers that patrol the streets are an important source of primary information (Manning and Hawkins, 1989), but power associated with holding information and the reluctance to release it restricts the interpretation of this information and the ability to integrate it with other knowledge sources. The role of the analyst as an information interpreter can potentially challenge this officer power and can lead to power struggles (Chan, 2001). 'Information can restrict the discretion and autonomy of street level officers, while at the same time enhancing the status of the information technology specialist' (Chan, 2001, p. 146).

13.3.1.5 Data quality

All analysis products need good quality data as these data are a main factor in the quality of information and intelligence that can be generated. Poor quality data undermines analysis. If a police force's information and intelligence systems are poor in quality and detail then this inevitably limits the insight that analysis can bring. Those that are sceptical about crime analysis are often also the same people who know, or at least have a perception, that data entered into their intelligence and information systems are poor. In their eyes, this knowledge merely increases their scepticism in regard to the value of intelligence.

Many police agencies, and others working with crime data, perform data cleaning processes after data entry to help improve their quality. Yet if operational officers are not aware that these cleaning tasks occur, then they may continue to question the viability of crime analysis products. 'Nobody trusts the analysts' stuff because they get their information from the [computer systems] and officers know they put crap on the system' was the comment from a criminal intelligence database supervisor, recorded by Cope (2004, p. 193). Data entry requires careful management (the Dumfries and Galloway Police Constabulary case study in Chapter 3 provides an excellent example of this type of data input management). It is important to raise awareness in those who enter data as to the extent to which these data are relied upon, and the importance in capturing and being consistent in how crime details are entered. Often this is a relatively simple matter of reinforcing how data should be entered in a certain format or by using standards or templates for entering such details as the make and model of a motor vehicle, rather than allowing free text entry.

13.3.1.6 Managing and being organised with analysis

Crime analysis needs to be managed and organised to help recognise where it fits in the day-to-day operational delivery of policing and crime reduction services as well as the agency's strategic direction. Analysis needs to be viewed as an integral part to an intelligence-led process, so that its products are not overlooked or ignored or just act as wallpaper. The intelligence cycle (direction, collation, analysis, dissemination, feedback and review, described in section 13.2.1 of this chapter) helps to identify the key areas where analysis has a role and the outputs that need to be generated before moving to the next stage. A process such as this helps to organise the role of analysis and manage the outputs that are to be delivered to ensure that other stages are not held up. Approaching analysis in this way means that it is easier to identify requests that fit under the function of analysis and those that do not. The need for this type of structure and direction in crime analysis is important because the hierarchy within policing or other agencies can be an intimidating environment to work in. Requests may come from many directions and because the person that asks for information from an analyst looks, sounds or is important, it may result in analysts taking on inappropriate requests.

The organisation of analysis and its use needs to be proactive and supportive of the intelligence process. By approaching the management and organisation of analysis functions in this way helps to proportionately weigh responsibilities for meeting other ad hoc requests and discourages analysis being used solely as an after-thought to try to justify any actions that have been decided (without any pre-analysis into a problem).

13.3.1.7 Feedback

A vital part in the production of analysis products is gathering feedback from the audience that used the product. In section 13.3.1.3 we discussed the need for consultation with the audience or users of analysis products. Feedback should be gathered on whether the analysis was used, how it was employed, in what way the information was useful (e.g. did it reveal something different that was not known), was its content and tone pitched correctly (e.g. was the content level sufficient and timely) and whether the analysis helped to achieve some success (Osborne and Wernicke (2003) capture a number of stories that describe successful uses of crime analysis).

An analyst should not become too defensive if constructive criticism is offered on their work. Achieving the expectations of all can be difficult,

405

and the presentation of information does require practice. Evaluating the use and effectiveness of crime analysis products will help to improve and legitimise their content. One possible option is to temporarily remove any time-consuming output from production and see if anyone complains that it is no longer available! This can work to identify active users of a product as well as identify those maps that are no longer, or rarely, used by decision-makers.

Case study: Crime mapping and analysis in the Glendale Police Department, Arizona

Material supplied by Bryan Hill, Crime Analyst, Glendale Police

The Glendale Police Department police a population of 232 000 people with 374 sworn officers and 133 non-sworn staff. The Glendale Crime Analysis Unit, staffed with three analysts, is responsible for several different types of activities associated with the police department. These include:

1. Resource allocation support – Assist with personnel projections, patrol staffing, and other planning and forecasting projects when required.
2. Uniform Crime Reporting Coding – Making sure the police reports that are automatically coded by our Records Management System (RMS) are coded correctly.
3. Committees – Participate on several committees within the department such as the Technology Committee, RMS revisions, GIS Steering Committee, and the Burglary Prevention Group.
4. Maps and Analysis Products – Create maps and analysis products for administrative, tactical, and strategic purposes. These include:

   - Routine burglary reports
   - Routine top 100 subjects reports
   - Routine administrative reports for City Council, Command Staff, other city staff, and citizens
   - Strategic analysis for auto theft, robbery, and other crime types
   - Tactical analysis on crime series.

5. Databases – Create, maintain, and repair databases throughout the department.
6. GIS – Act as a committee member and consultant to the City GIS department to support the development of a more robust GIS environment within the City of Glendale.

7. Grant Writing – Write grants for funding for homeland security, regional data sharing, and communications interoperability.
8. Regional Data Sharing – Work as the chairperson for a West Valley Regional Data Sharing Initiative group creating a regional data sharing infrastructure.
9. Graphics – Create a multitude of graphics needed for crime analysis products and other department needs.
10. Help desk – Work as an alternate 'help desk' for police employees that need technical assistance with computers or software.
11. AACA – Participate where possible with the Arizona Association of Crime Analysts group to provide training and share information.
12. IACA – Participate in projects, and attend International Association of Crime Analysts conferences and assist other analysts across the country with questions in a wide-range of subject matters when requested.
13. Programing – Develop applications in-house where needed for ArcView 3x and ArcGIS 8x.

A common request is to help police units identify 'focus areas' that need to be addressed either because the area is experiencing an increase in crime, or it is already a high crime neighbourhood. The crime analysts will run initial statistics, create maps of the areas, and provide several analysis products to assist our police colleagues in identifying, tracking, and reporting on successes or failures within the identified areas. This information is often presented in our Community Activity Review (CAR) meetings that loosely resemble the New York City CompStat model. These monthly meetings are run by the police department; however, all city departments are invited to attend and participate, as well as neighbouring police departments.

The validity and usefulness of mapping in a police department by most analysts is probably not questioned; however, in some cases its actual practical application is sometimes lost. Over the years, I have learned a great deal about mapping and analysis products of all sorts related to crime analysis. Taking this theory and knowledge and breaking it down so that police administrators, detectives, and patrol officers can see the usefulness is another matter.

Tactical mapping and analysis is what the majority of us really want to do and find the most exciting. There are many things we can do that mostly border on common sense, by looking at an offender's series of crimes and figuring out what they may do next. An analyst's common sense and practical experience can be augmented by GIS methods and processes to help us predict the next location a criminal will go to commit a new crime (Figure 13.1).

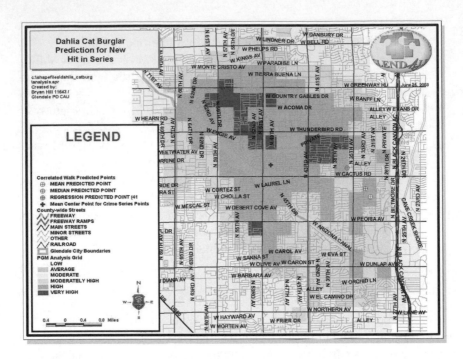

Figure 13.1 A tactical prediction map generated to aid a crime series investigation in Glendale, Arizona

Human nature is human nature, and although tactical predictions are valid and do work a lot of the time, they do not work every time and most detectives and patrol officers are very adept at reminding the analyst when they do not. I have also found that trying to explain to a detective how I go about making a tactical predictive map is usually a mistake, recognising that it is often just enough to tell them what you think is happening and offer your advice. The hardest part is actually making the investigator a part of the analytical effort and challenge their preconceptions about certain offending behaviour. This just takes time. A good analyst will not become frustrated with a detective's lack of interest in the tactical predictions, and will find a way to assist them with their investigations, without stepping on the control of *their* cases. Something that I have found is that when several levels of supervision are aware of and validate the analytical products and methods the crime analyst uses, the investigators and patrol officers are more interested in using them. I would suggest that if you are an analyst working toward making tactical predictions and other forms of crime mapping analysis as a daily way of life in your department, you should make a great effort to help commanders, lieutenants, and sergeants responsible to investigative units be better trained and able to understand the limits, practicalities and benefits of using mapping and analytical products. It is also a good idea to work at helping your chief understand that these supervisors and officers may need to attend a conference on crime analysis and crime mapping now and again, so that they more fully commit to, and invest in, GIS. In Glendale, GIS has certainly proven to be a useful tool in problem-solving for our police department.

Although a significant part of the tone of this section has been on crime analysis, its relevance directly influences and is parallel to the use and development of crime mapping. The relationship is a symbiotic one: crime mapping is now integral and essential to, and has significantly enhanced, crime analysis; and analysis is the umbrella under which crime analysis flourishes. As crime analysis and crime mapping continue to grow in strength, it is important that people who produce analytical products also take care to ensure that the interpretation of the criminal environment is accurate and continues to improve, it is used successfully to influence its core audience and this information is actively used to support crime reduction. Analysis and crime mapping have several challenges to overcome in police and other crime reduction agencies where the culture may be one that is not used to relying on evaluated, evidence-based approaches to tackling crime. In addition, misunderstandings between police and analysts can create a depressing self-fulfilling prophecy. If analysis is limited by data quality and only produces descriptive information, and police officers are unable to ask the right questions of analysis and mistrust its worth because the analysis could only offer descriptive summaries, police officers can become reluctant to share information for use in analysis (Cope, 2004). Crime analysis adds value to the original source form by synthesising information about crime to identify patterns and trend correlations between crime data. Recognising the legitimacy of good crime analysis will require organisational differences to be worked out.

13.4 Organising the production of crime mapping products

Part Three of this book has described a wide range of ways in which crime mapping is used and other parts of this chapter have helped to define the role of crime mapping in the wider context of analysis. We have also described the ingredients that need to be in place for effective mapping. Crime mapping outputs can suit operational, tactical, investigative and strategic requirements, which are often symptomatic of short-, medium- and long-term requirements. Requests for crime mapping products may also be ad hoc or come from external or partnership agencies. There may also be the requirement to organise outputs that are disseminated to the public, such as managing the information updates to an Internet server where online crime maps can be accessed. In following the tone of this chapter, organising the use of crime mapping can be viewed in parallel to the broader application area: crime analysis. It is from this wider field of crime analysis that lessons can be learned and processes followed that can help crime mapping develop and become more effectively used.

13.4.1 Organising the structures around analysis

The 3i Model (Ratcliffe, 2004), introduced in Chapter 9, offers a useful mechanism for organising the role of analysis to support policing and crime reduction (Figure 13.2). In this model the criminal environment is assumed as a permanent feature, though the boundaries are fluid and dynamic, requiring continual analysis and observation. This environment needs to be understood for any policing (or other agency) action to be effective. The first stage requires this criminal environment to be interpreted and relies on a range of information sources being available. The arrow in the diagram goes from the analysis unit to the criminal environment, signifying the need for active information gathering. It is not sufficient for an analyst to sit back and wait for information to flow to them. They must actively seek

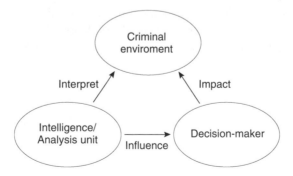

Figure 13.2 The 3i Model. The model contains three structures (criminal environment, intelligence/analysis unit and decision-makers) and three processes (interpret, influence and impact)

it out. Analysis of this information creates a pool of intelligence. The second stage requires the intelligence to influence the decision-makers. These decision-makers are often operational and leadership ranks, although analysts tend to have more success influencing those that brief patrols and those that decide on the tactical strategies for policing and crime reduction (Ratcliffe, 2005). To be a true decision-maker, an individual must possess the skills to understand crime, know how to intervene and reduce crime, and have a positive impact on the criminal environment.

Viewing the structures and processes that are involved in policing and crime reduction in this manner helps to identify where analysis has its role and how it contributes to the other parts of an organisation. Analysis should help in interpreting the criminal environment and its outputs, and should form a major part of the intelligence that is used to influence the actions of the decision-maker who then brings about a positive impact on the criminal environment.

13.4.2 Managing and organising crime mapping

The role of performing and managing the delivery of crime mapping and crime analysis tasks often falls to the 'Crime Analysis Unit' or 'Intelligence Unit'. In this section we identify the key roles of this unit, and also suggest a role for supporting groups who can help take forward crime mapping and analysis.

13.4.2.1 The Crime Analysis Unit

The Crime Analysis (or Intelligence) Unit is where information is interpreted and turned into value-added products that support decision-making for targeting, designing, resourcing and prioritising police activity and crime prevention responses. The Crime Analysis Unit should be appropriately staffed and possess the skills required to produce the analysis products that are required. The Unit should also be in a position of easy access to timely information sources, have tools that support any post-processing required for these data (e.g. data cleaning and geocoding), have IT systems in place that are responsive to analysts' requirements, be physically established in a place in the organisation where their products can be easily accessed and used at the decision-making stage, be appropriately steered, promoted and supported in the tasks that they perform and be supported in their personal professional development.

411

13.4.2.2 The Crime Analysis Management Group

If the Crime Analysis Unit is where the day-to-day production of tasks are completed, this can often lead to a situation where the longer-term outlook and development of the Unit is overlooked. The role of a Crime Analysis Management Group is to steer the development of analyst training, analysis tools and information improvements to ensure a continual process of development. The Group provides the user recommendations for the development of new functions and new tools required for analysis (e.g. the development of functionality required for their GIS), identifies core training needs and skills requirements to support analyst development (including conference and seminar opportunities for analysts) and identifies strategic information improvements required in order to improve information-driven intelligence delivery and problem-solving (e.g. the need to improve the recording of address and location information in a crime report). This Management Group would also identify and prioritise non-police data that are of use for crime analysis (e.g. census data, correction data on parolees) and would activate the steps for accessing this information for police or other agency use (e.g., a crime and disorder reduction partnership). Membership of this Management Group should include a representative selection of crime analysts, a senior member from the agency's IT department and chaired by a senior member from the agency such as the Intelligence Manager.

A group of this type can bring significant benefits to the development of analysis by helping to add professionalism to analysis and crime mapping development, promoting awareness of the products it produces, developing new and improved contacts with key data sources and continually adding to the legitimacy of analytical outputs.

Case study: Project Spectrom – a new operational policing model for West Midlands Police, England

Material supplied by Andy Brumwell and Steve Rose, West Midlands Police

In December 2003, West Midlands Police (WMP) established a small team under the operational banner of Project Spectrom. Its purpose was to introduce a new policing model across the force, based on 3 geographical typologies that would be identified, around which patrol strategies and resources would be based. These typologies were:

1. *Priority Areas* – areas experiencing relatively high crime and disorder over a sustained period.
2. *Hotspots* – areas experiencing short term high levels of crime and disorder which, if not reduced, would impinge on local and force performance and lead the area into longer term problems.
3. *Patrol Areas* – other areas experiencing relatively low levels of crime and disorder, but still requiring a police presence in order to provide reassurance to the community and prevent the area becoming a hotspot.

The shift away from policing geographically fixed beats to more fluid and changing geographies has required a change in the mind set of both management and operational officers. Moreover, it has required change in the role and competencies of the crime analyst, who previously performed little crime mapping other than plotting the location of crimes that occurred over the past few days or producing an attractive hotspot map which, in reality, while giving the air of sophistication, can be produced at the press of a few buttons.

The first requirement was to identify the three geographic typologies across the force. This required a fair degree of spatial analysis, with the results used to assist individual Operational Command Units (OCUs) to understand their own areas and allocate resources accordingly. This initial analysis acted as a starting point from which OCUs could undertake further analysis to refine areas and patrol strategies including providing a list of short and long-term tactical options relevant to each typology. What emerged from this initial analysis was a useful insight into analysts' skills and the management of information across WMP. These included:

- Analysts possessed only limited skills in crime mapping, spatial analysis and presentation techniques.
- There was a general lack of understanding by both managers and crime analysts over what is possible utilising readily available desktop mapping software.
- A reluctance or difficulty in learning yet another software system.
- A lack of understanding over what other data existed and could be used in addition to standard police data.
- Data quality problems associated with police and non-police data.
- The need for rapid access to data, information and intelligence with standard functionality to answer specific questions relating to the importance of *place* or *location*. This included crime locations, home locations of offenders and profiling of offenders and criminal networks.
- The need for more flexible GIS software (as opposed to the current corporate GIS), to enable other questions to be investigated and other data to

be included in analysis e.g. partnership data from Fire Service, Ambulance Service, and the Local Authority.

- Difficulties associated with the introduction of GIS software which met the WMP's requirements but was not the GIS software supported centrally by the IT department under the police's National Strategy for Information Systems (NSPIS).

The Project Spectrom team addressed these issues at a number of different levels:

- The FLINTS (Force Linked Intelligence System) team were involved at an early stage to provide additional mapping and spatial analysis functionality to the already successful FLINTS software. An additional two day training course in the use of the new mapping functionality of FLINTS was made available for all analysts.
- MapInfo desktop GIS was introduced for all local OCU and force analysts, including a one day course covering an introduction to the theories and principles of crime mapping and a three day MapInfo course specific to crime analysis.
- The creation of a central unit to support local OCUs and the force in the use of crime mapping techniques and collection of partnership data.
- The inclusion of crime mapping awareness training for all new sergeants and inspectors seeking promotion.

The introduction of crime mapping into an organisation, even one which already had some elements of mapping available to it, was certainly not without its problems. Different needs of different customers within a police force can lead to the use of a number of disparate systems all supported by different interested parties. This situation, whilst not ideal, can be tolerated in the short term, as long as everybody is getting what they require in order to work effectively and the message of what can be achieved using GIS continues to spread. From a centralised force perspective, the ultimate goal of one GIS satisfying all needs is still a number of years away. What Project Spectrom has achieved is more important – that is to improve the quality of the analytical products routinely produced by analysts, improve the understanding of management as to what is possible utilising GIS and thereby allocate resources to where they are best utilised.

The use and management of crime mapping and analysis in the West Midland's largest city, Birmingham, has also extended to the creation of a unit to support Birmingham's Community Safety Partnership's requirements. The Birmingham Partnership's Information and Intelligence

Team manage the collation and analysis of information and intelligence in support of the Partnership's objectives. Although still in its infancy, the team has already made an impact with the use of crime mapping techniques as a basis for many of their analytical products. The composition of the Information and Intelligence Team is also vital, with a manager and three analysts showing the investment the partnership is making in analysis and the proper use of information: one member of the team is a dedicated fire analyst employed by the West Midlands Fire Service, and another member is a Drugs Action Team Analyst employed by WMP but situated within the local health care trust. The other two members are located within the Partnership Hub and employed by the West Midlands Police. The mix of agency representation aids the collection and flow of information between the partner agencies allowing crime mapping and GIS to enhance the data sharing possibilities and the generation of intelligence products.

Case study: The importance of management to support crime analysis

Material supplied by Superintendent Dave Flitcroft, Greater Manchester Police

Greater Manchester's Crime and Disorder Reduction Partnerships (CDRPs) have developed the UK's National Intelligence Model into a model for partnership working. The Greater Manchester Against Crime (GMAC) Partnership Business Model provides a standard method for organising and managing a work programme that is focused on addressing the key partnership priorities. The model is used both at CDRP level (of which there are 10 across Greater Manchester) and the Greater Manchester conurbation level. The GMAC Partnership Business Model is underpinned by 14 Strategic Analytical Coordinators trained and equipped to a common standard.

Particular importance is placed on a commissioning approach for the development and delivery of analytical products. The commissioning approach helps to ensure the focus for analytical requests is based around the Partnerships' core business functions (Figure 13.3). Partnership Business Groups are the body that commission analytical products with the aim to meet operational and strategic outcomes. Support is also offered to analysts and members of the partnerships from a panel of experts, enabling a depth and diversity of skills, knowledge and research to be tapped across GMAC.

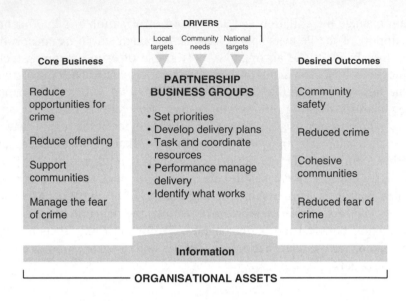

Figure 13.3 The GMAC Partnership Business Model

The commissioning of work requests through Partnership Business Groups has several purposes:

- It helps to ensure that focus is maintained on partnership priorities.
- It ensures that careful and deliberate thought is given to identifying the questions that require answering from analysis.
- It provides direction. The analyst is clear on what information is required.
- It identifies which analytical resource is most appropriate to answer the question, or part of the question.
- Commissioning helps to manage the workload of analysts.

From an analyst's viewpoint, commissioning also enables an analyst to identify and collect relevant data and information, identify relevant support from the panel of experts, identify the limitations of data and adopt alternative methods of collating information.

Organising a work programme that focuses on core business functions helps to ensure that the questions asked by the commissioning group are relevant. For example, these could include:

- *Reducing opportunities for crime* – this requires questions to be tailored that help understand where, when and how crimes are occurring to whom and why. Once this is understood the knowledge can be applied to vulnerable people and places to reduce the likelihood of crime.

416

- *Reducing offending* – this requires an understanding of who is committing offences and tackling these people in the most effective way. It is also important to use knowledge of when, where, why and how offenders act to reduce the opportunities for crime.
- *Supporting communities* – this requires an understanding of the context of a community in which crime occurs in order to protect communities against the fragmentation and division caused by crime, disorder and tension. A key issue in addressing community cohesion is to identify and address issues of disproportionate criminality, victimisation and tension.
- *Managing the fear of crime* – this requires an understanding of communities, their fears and concerns, and recognising that certain members of a community have different fears and perspectives on crime and disorder.

The concept of a panel of experts is not to identify a fixed group of people to support the analyst, but to ensure that the right people with the relevant knowledge and skills are involved as consultants during the development of the information. During the commissioning process it may be appropriate to identify the right people to be involved. The strategic analyst will recommend members to this panel whilst drawing up the aim, purpose and scope of the product.

It is critical that decision-makers have confidence in the information presented to them, the information's provenance, and that they understand this information and how it can be used to help deliver the desired outcomes. It would be wrong to place the responsibility for delivering recommendations that impact on these outcomes solely on the shoulders of analysts, no matter how skilled they are.

All documents produced by analysts aim to present key findings or judgments, make sound and evidence-based recommendations and identify knowledge gaps. All three aspects are equally important and complementary. For example, the identification of knowledge gaps drives activity for further research. For this reason it is important that the questions asked by the commissioning group are not restricted to the data available. The GMAC Partnerships also recognise that analysts are a valuable and scarce resource. The GMAC approach is to support them with active and responsible management, while helping them in the organisation of their analysis duties.

13.4.2.3 The Analysts' Forum

Analysts can often find it difficult to meet other analysts where they can learn from each other, find out about new techniques and upcoming events of interest. An Analysts' Forum offers the opportunity for analysts to meet, develop their contacts, share information on good practice and new analytical techniques, exchange problem-solving ideas and hone their presentation skills. In the United States the International Association of Crime Analysts offers an excellent forum for analysts to exchange these ideas. In other parts of the world less formal arrangements have been established. For example, in London (UK), analysts from the 33 borough CDRPs meet every three to four months to hear about new developments. Other forums include those that operate online, such as the NIJ-administered CRIMEMAP list server.

An Analysts' Forum could operate across a large police force, across a region or at the national level. It could also be a point of contact where feedback is received from the Crime Analysis Management Group, including updates on training development, data, tools and technology, conferences, latest research, as well as a forum for general questions and answers. A forum of this type may need financial support for it to be administered and may also require the funding body to promote its importance to the analysts' managers to ensure that analysts are given the freedom to participate. Forums such as these also offer the added opportunity for supporting continual professional development of analysts, as well as maintaining and building the knowledge of analysis developments and problem-solving approaches to policing and crime reduction in the geographic area they serve.

13.5 Summary

The application of crime mapping and the outputs it produces are generally only as good as the information contained within the information systems of the contributing agencies. Even if this information is of good quality, for crime mapping applications to be effective requires there to be clarity in its direction of purpose, and for its use to be supported and promoted. In this final chapter we have described the process for implementing GIS and the organisational challenges that can influence the success of crime mapping. Recognising the structures and processes that are involved in bringing about impact on the criminal environment will help to organise, coordinate and manage the effective use of crime mapping.

Further reading

Leipnik, M.R. and Albert, D.P. (2003). *GIS and Law Enforcement: Implementation Issues and Case Studies*. London: Taylor & Francis.

Police Foundation (2000). Guidelines to implement and evaluate crime analysis and mapping in law enforcement agencies. Washington, DC: United States Department of Justice.

In addition to the material provided in this chapter, these two useful and specific sources offer comprehensive descriptions of the GIS and crime mapping implementation process for policing and crime reduction agencies.

Osborne, D.A. and Wernicke, S.C. (2003). *Introduction to Crime Analysis: Basic Resources for Criminal Justice Practice*. New York: Haworth Press.

This book offers detail on the components to consider when creating a Crime Analysis Unit, advice for the new crime analyst, and education and training resources for crime analysts.

References

Campbell, H. (1994). How effective are GIS in practice? A case study of British local government. *International Journal of Geographical Information Systems*, 8(3), 309–325.

Campbell, H. and Masser, I. (1992). GIS in local government: Some findings from Great Britain. *International Journal of Geographical Information Systems*, 6(6), 529–546.

Chan, J. (2001). The technology game: How information technology is transforming police practice. *Criminal Justice*, 1/2, 139–159.

Clark, M. (1996). Professional integrity and social role of hydro-GIS. Proceedings from HydroGIS 96: Application of Geographic Information Systems in Hydrology and Water Resources Management, April 1996, IAHS, Vienna.

Clark, M. (1998). GIS: Democracy or delusion? *Environment and Planning A*, 30(2), 303–316.

Clarke, R.V. and Eck, J. (2003). *Become a Problem Solving Crime Analyst*. London: Jill Dando Institute of Crime Science. www.jdi.ucl.ac.uk.

Chainey, S.P. (2004). Crime mapping use in Crime and Disorder Reduction Partnerships in the UK: Results from an interactive survey conducted at the National Community Safety Network Conference, Cardiff, Wales 2004. www. jdi.ucl.ac.uk.

Cope, N. (2004). Intelligence led policing or policing led intelligence?: Integrating volume crime analysis into policing. *British Journal of Criminology*, 44(2), 188–203.

Eck, J. (1998). What do these dots mean? Mapping theories with data. In D. Wiesburd and T. McEwen (eds) *Crime Mapping and Crime Prevention, Crime Prevention Studies*, Volume XIII. Devon: Willan Publishing.

Fletcher, R. (2000). An intelligent use of intelligence: Developing locally responsive information systems in the post-Macpherson era. In A. Marlow and B. Loveday (eds) *After Macpherson: Policing After the Stephen Lawrence Inquiry*. Dorset: Russell House Publishing.

Flynn, P. (2000). *Education in Policing and for the Millennium and Beyond.* Cropwood Paper, Institute of Criminology, University of Cambridge.

Foster, J. (1998). Memorandum on police training. Submitted to The Home Affairs Select Committee.

Gill, P. (2000). *Rounding up the Usual Suspects? Developments in Contemporary Law Enforcement Intelligence.* Aldershot: Ashgate.

Goldstein, H. (1979). Improving policing: A problem-oriented approach. *Crime and Delinquency,* 25, 236–258.

Gottlieb, S.L., Arenberg, S. and Singh, R. (1994). *Crime Analysis: From First Report to Final Arrest.* Montclair, California: Alpha Publishing.

Grimshaw, D.J. (2000). *Bringing Geographical Information Systems into Business.* Chichester: John Wiley & Sons.

Heywood, I., Cornelius, S. and Carver, S. (1998). *An Introduction to Geographical Information Systems.* Essex: Longman.

Hughes, K. (2000). *Implementing a GIS Application: Lessons Learned in a Law Enforcement Environment.* In Crime Mapping News, 2:1. The Police Foundation, United States Department of Justice, Washington, DC.

Laycock, G. (2001). Research for police. Who needs it? *Trends and Issues in Crime and Criminal Justice,* No. 211, Australian Institute of Criminology.

Manning, P. (2000). Policing new social spaces. In J. Sheptycki (ed.) *Issues in Transitional Policing.* London: Routledge.

Manning, P. (2001). Technology's ways: Information technology, crime analysis and the rationalizing of policing. *Criminal Justice,* 1, 83–104.

Manning, P. and Hawkins, K. (1989). Police decision making. In M. Weatheritt, (ed.) *Police Research: Some Future Prospects.* Aldershot: Avebury.

National Criminal Intelligence Service (2000). *The National Intelligence Model.* London: NCIS.

National Institute of Justice (1995). *Use of Computerized Mapping in Crime Control and Prevention Programs.* United States Department of Justice, Washington, DC.

National Institute of Justice (1999). *The Use of Computerized Crime Mapping by Law Enforcement.* United States Department of Justice, Washington, DC.

Nedovic-Budic, Z. and Godschalk, D.R. (1996). Human factors in adoption of geographic information systems: A local government case study. *Public Administration Review,* 56(6), 554–567.

Obermeyer, N.J. and Pinto, J.K. (1994). *Managing Geographical Information Systems.* New York: Guildford Press.

Openshaw, S., Cross, A., Charlton, M. and Brunsdon, C. (1990). Lessons learnt from a post mortem of a failed GIS. 2nd AGI Conference, October 1990, Brighton.

Osborne, D.A. and Wernicke, S.C. (2003). *Introduction to Crime Analysis: Basic Resources for Criminal Justice Practice.* New York: Haworth Press.

Police Foundation (2000a). *Integrating Community Policing and Computer Mapping: Assessing Issues and Needs Among COPS Office Grantees.* United States Department of Justice, Washington, DC.

Police Foundation (2000b). *Guidelines to Implement and Evaluate Crime Analysis and Mapping in Law Enforcement Agencies*. United States Department of Justice, Washington, DC.

Ratcliffe, J.H. (1999). Implementing and integrating crime mapping into a police intelligence environment. *International Journal of Police Science and Management*, 2(4), 313–323.

Ratcliffe, J.H. (2004). *Strategic Thinking in Criminal Intelligence*. Sydney: Federation Press.

Ratcliffe, J.H. (2005). The effectiveness of police intelligence management: A New Zealand case study. *Police Practice and Research*, 6(4).

Robey, D. and Sahay, S. (1996). Transforming work through information technology: A comparative case study of geographic information systems in county government. *Information Systems Research*, 7(1), 93–110.

Ventura, S.J. (1995). The use of geographic information systems in local government. *Public Administration Review*, 55(5), 461–467.

Vincent, K. (2000). Implementing crime mapping: Hayward Police Department. In *Crime Mapping News*, 2:1. The Police Foundation, United States Department of Justice, Washington, DC.

Walsh, P.F. (2004). Project management. In J.H. Ratcliffe (ed.) *Strategic Thinking in Criminal Intelligence*, First edition pp. 163–176. Sydney: Federation Press.

Index

Address cleaning   55, 56, 59, 64
ADDRESS-POINT   46, 47, 49
Aggregate criminal spatial
    behaviour   86
Alley-gating   28
Analysis, *see* Geographic profiling
    aoristic analysis   251
    case study, Glendale, Arizona   406
    cost benefit   30
    of crime, *see* Crime analyst
    criticisms   401
    definition   399
    Hierarchical Linear Modelling
        344–5
    management of   405, 411
    operational, *see* Operational analysis
    problem-solving   27
    SARA   71, 291, 331
    spatial, *see* Spatial analysis
    see also Crime analyst; Spatial
        analysis; Time
Anchor points   301
Anti-Social Behaviour Act   50
Areal interpolation   216
    areal units   340, 343
    see also Modifiable Areal Unit
        Problem (MAUP)
Australia, coordinate system   46

Boundaries
    cartography   360
    Great Britain   22
British Crime Survey   28, 65

British National Grid   45
British Youth Lifestyles Survey   295
Burglary
    distraction   27
    prevention   28
    residential   291
    spatial patterns   122, 335
    temporal patterns   87, 293

Cartographic school, *see* Criminology
Cartography
    and aims of crime mapping   354
    choropleth mapping   233, 244, 373
    font styles   362
    human perception of   354, 369, 371
    internet   355, 361, 382
    legends   364, 376–8
    map layout   356
    map scales   359
    north arrows   358
    patterns   367
    point maps   380
    for PowerPoint   355, 359–60,
        383–5
    selecting features   354
    shading   368
    symbol options   363
    text options   361
    visual hierarchy   357
    see also Colour
Census
    UK geography   341
    US geography   342

GIS and Crime Mapping   Spencer Chainey and Jerry Ratcliffe
© 2005 John Wiley & Sons, Ltd

Index

CHEERS 73
Chicago School, *see* Criminology
Choropleth maps 233, 244, 373
Closed Circuit Television 68, 88
Colour
  associations 371
  hue 367
  human perception of 371
  rods and cones 369
  rule of thumb 370
  schemes 371–2
  simultaneous contrast 366
  wavelengths 370
Community safety, definition of 17
CompStat 26, 259–71, 407
Coordinate systems 45–6
CRAVED 73, 90, 295
Crime
  attractors 106
  audits 26
  crime reduction 19
  diffusion of benefits 19, 28
  displacement 19, 28, 130, 224
  environmental 50
  fear of 19
  four dimensions of 79
  generators 106
  graffiti 50
  hotspots 146, 147–8, 263, 278
  journey to 100, 103, 296–7, 301
  juvenile delinquency 83
  offender behaviour 102
  perception of 187
  rate 173–6, 190, 213, 364,
    374–5, 378
  reasons for not reporting 66
  reporting rate 66
  robbery 27, 122
  street crime risk, *see* Crime, rate
  triangle 89
  unreported 65
  vehicle crime 122, 266
  see also Burglary
Crime analyst
  forum 417
  and policing 403
  role of 400, 402
Crime and Disorder Act 17, 185

Crime and Disorder Reduction
  Partnerships (CDRP)
  benefits of mapping 186–7
  data 189–95
  data sharing 200–3, 332
  data sharing guidelines 203
  management 415–18
  organisational culture 390
  purpose 185, 330
  web solution 211
Crime awareness newsletters 31
Crime mapping
  administrative 25
  as evidence 309
  as image files 242
  definition 2–4
  forecasting, *see* Predictive
  in Glendale, Arizona 406
  graduated symbol maps 149
  history of 2–3
  implementation evaluation 397–9
  implementation issues 390, 413
  implementation process 393–7
  integration with analysis 409
  in Lincoln, Nebraska 7, 125, 268
  point maps 146, 380
  predictive 177, 408
  Research Centre, Crime Mapping,
    see MAPS (Mapping and
    Analysis for Public Safety)
    program
  tactical applications 288–9, 407
  thematic mapping 150, 153, 233, 367,
    371, 378–80
  see also CompStat; CrimeStat; Kernel
    density
Crime Mapping Research Centre, *see*
  MAPS (Mapping and Analysis for
  Public Safety) program
Crime Pattern Theory 116
  awareness spaces 97
  cognitive maps 97
  crime attractors 106
  crime generators 106
  definition 96
  nodes, pathways and edges 105, 117
  templates 99, 101
  see also Distance decay effect

Crime prevention
  and crime reduction   19
  definition   15
  primary prevention   16
  secondary prevention   16
  tertiary prevention   17
  see also Crime Prevention Through
    Environmental Design
Crime Prevention Through Environmental
  Design   85, 96
Crime reduction
  alley-gating   28
  crime prevention and   19
  policing for   325
  see also Crime and Disorder Reduction
    Partnerships (CDRP)
CrimeStat   124, 128–9, 131–2, 157, 175
Criminology
  cartographic school   81
  and cartography   354
  Chicago School   1, 82, 101, 107, 335, 344
  circle theory   102, 308
  collective efficacy   336
  commuter hypothesis   102, 308
  crime chemistry   88
  environmental   80–1
  GIS School   85
  Guerry, Andre-Michel   81
  marauder hypothesis   102, 308
  Quetelet, Adolphe   81
  rational choice   91–2, 304
  routine activity theory   87, 304, 334
  social capital   337
  social disorganisation   335
  strain theory   80
  zonal (concentric) model   82
  zone in transition   83
  see also Crime pattern theory

Data
  combining   213
  privacy   206, 209
  quality   404
  requirements, *see* Geographical
    Information System(s) (GIS)
  sharing   403–4
  structures, *see* Geographical Information
    System(s) (GIS)

Defensible Space   85, 94
Department of Homeland Security   363
Distance
  as time   298
  decay effect   104, 296, 305
  Euclidean   298
  Manhattan   298
  street route   298
  see also Crime, journey to
Distance decay effect   104, 296, 305
Downgrading   84
Dragnet   308
Drug Abuse Resistance Education
  (DARE)   326

Ecological correlation   153, 218
Ecological fallacy   101, 152
Environmental backcloth   105
Environmental criminology   80–1
Epidemiology   3

Federal Emergency Management
  Agency   363
First Law of Geography, Tobler's   117

Gang members   8
Gentrification   84
Geocoding
  accuracy   60–3, 395
  address cleaning   56, 64
  ADDRESS-POINT   46
  address reference files   56–7
  completeness   61
  consistency   61
  gazetteer   47, 58, 173
  precision   60–1
  problems   52–5, 64
  reliability   60
  scrubbing, *see* Address cleaning
  TIGER   48, 63
Geographic profiling
  buffered distance decay   306
  criminal geographic targeting   306
  Dragnet software   308
  jeopardy surface   307
  least effort principle   100
  Operation Lynx   305
  origins   27, 103, 303

Geographic profiling (*Continued*)
  quantitative and qualitative
    components 303
  Rigel software 306–9
Geographical information science (GISc),
    origins of 2
Geographical Information System(s)
    (GIS) 3
  base maps 38
  coordinate systems 44
  data requirements 395
  data structures 41–3
  definition 38
  hardware 40
  implementation 392
  layers 38–9
  projections 44
  in public housing 93
  querying 69
  raster 43
  school, *see* Criminology
  software 40, 392
  training 41, 392
Global Positioning System (GPS) 2, 25
  handheld 50–1

Heterogeneity 338–9
Hierarchical Linear Modelling 344–5
Home Office Research, Development and
    Statistics Directorate 20
Hotspot matrix 277, 281
Hue, *see* Colour
Hypothesis setting 72

In-filling 84
Intelligence
  cycle 391, 405
  intelligence-led policing 271, 327
  and the 3i model 267, 410
  see also National Criminal Intelligence
    Service; National Intelligence
    Model
Internet mapping 9–10, 31, 195,
    211, 382–3
  case study, Cornwall, England 211
  case study, Sussex, England 195
  ethical issues 326
  Scaleable Vector Graphics 383

Intranet mapping
  case study, Cornwall, England 211
  case study, Incident Management And
    General Enquiry system
    (IMAGE) 58
  case study, Lincoln, Nebraska 7–10
  case study, Sussex, England 195

Jenks Optimisation 376

Kernel density
  bandwidths 157
  case study, Camden, London 162–3
  cell size 159
  dual 176
  methods 155–6, 229
  problems 161
  statistical significance 169
  underlying population 172, 175

Lateral inhibition 366
Latitude 44
Law enforcement
  definition of 18
  see also Police; Policing
Least effort principle 100
Local Indicators of Spatial Association
    (LISA)
  Bonferonni test 168
  Getis and Ord Gi 164
  Gi* example 168
  overview 164
  Rook's Case 166
Longitude 44

MAPS (Mapping and Analysis for
    Public Safety) program 3,
    121, 383
Mental Map 305
Merseyside 29
Modifiable Areal Unit Problem (MAUP)
    146, 151, 154, 173, 340
Modus operandi 9

National Criminal Intelligence
    Service 22
  National Threat Assessment 22
National Institute of Justice 20

National Intelligence Model
  and hotspot matrix   283
  levels   21
  National Threat Assessment   22
  products   271
  see also Policing
Nearest Neighbour Index   126
Neighbourhood change   84
Neighbourhood studies   342
Neighbourhood Watch   326
Neural networks   179
North London Strategic Alliance   50

Offender
  aggregate criminal spatial behaviour   86
  anchor points   301
  behaviour   102
  profiling   26, 273, 290
  risks   295
  self-selection   312
  sex offenders   8, 102
  typology   305
Operation Lynx   305
Operation Safe Streets   345
Operational analysis   25, 26
Ordnance Survey   46
  see also ADDRESS-POINT

Parolees   8
Pedestrian counts   174
Philadelphia   24, 137, 139, 231,
    259–60, 264, 345
Pin maps   8, 148
Police
  American   18
  basic command unit   21
  British police services   18
  City of London   22, 82
  command and control unit   25
  and community safety   19
  crime control   18
  definition of   18
  Dumfries and Galloway   58
  Greater Manchester   415
  Metropolitan   22, 82, 174
  Pennsylvania   24
  Philadelphia   24, 137, 139, 231,
    259–60, 264, 345

Police Foundation   390
  private   18
  West Midlands   412
  see also Law enforcement; Policing
Policing   20
  community policing   325, 328
  federal   20, 373
  intelligence-led   271, 327
  municipal   20
  partnerships   184
  problem oriented   27, 328
  problem-solving   27
  Project Spectrum   412–14
  SARA process   71, 291, 331
  state   20
  3i model   267, 410
  see also Law enforcement; Police
PowerPoint   355, 359–60, 383–5
Predictive   177–9
Privacy, data   206, 209
Problem-oriented policing, *see* Policing,
    problem-oriented
Profiling
  offender   291
  offending   291–2
  see also Geographic profiling

Redistricting   9
Rehabilitation   84
Repeat victimisation   273, 313
Rigel   306, 308
Rookeries   81, 82, 106
Routine activity
  handlers   88
  place manager   89
  theory   87, 116, 304, 334

SARA, *see* Analysis
Scalable Vector Graphics   383
Self-containment index   302
Sex offenders   8, 102
Simultaneous contrast   366
Situational crime prevention   15, 92
SMART   394
Spatial analysis
  buffers   29–30
  centrography   101, 119, 245
  coefficient of Areal correspondence   247

Spatial analysis (*Continued*)
distance measurement   298
edge effects   129
Geographical Analysis Machine
   175, 177
K-order distances   159
mean centre   120
Moran's *I*   130, 132, 157
nearest neighbour analysis   126–30, 157
regression, geographically
   weighted   138
regression, ordinary least squares   134
spatial error models   136
spatial joins   70
spatial lag models   135
spatial scan statistic   175
STAC   161
standard deviational ellipse   102,
   123–6, 247
standard distance   102, 122, 246
weighted displacement quotients   247–50
see also Ecological fallacy; Kernel
   density; Local Indicators of
   Spatial Association (LISA);
   Modifiable Areal Unit
   Problem (MAUP)
Spatial distributions   126
Spatial hierarchy   19–21
Spatial patterns, types of   126
Spatial processes
autocorrelation   118, 165
dependency   117, 126
heterogeneity   117
see also Modifiable Areal Unit Problem
   (MAUP); Spatial analysis
STAC   161
Strategic crime control   324, 332

Temporal analysis, *see* Time
Territoriality   94–5
Theory, *see* Crime pattern theory;
   Criminology
Thiessen polygons   346
Thinning out   84
3i model   267, 410
TIGER, *see* Geocoding
Time
animation   240
aoristic analysis   251
aoristic example   253
burglary patterns   87
distance as   298
formats   68
histograms   235–40
hotspot matrix categories   279
mapping change   231–49
notation   224–7
querying   229, 231
rape patterns   102
temporal relationships   230
temporal resolution   224, 228
time line   225
time span   227, 251
vehicle crime patterns   91
Tobler's First Law of Geography   117

Urbanisation   84

Visual hierarchy   357

Weighted displacement quotients   247–50

Yorkshire Ripper   302

Zodiac Killer   303